Springer Series in
MATERIALS SCIENCE 71

Springer
Berlin
Heidelberg
New York
Hong Kong
London
Milan
Paris
Tokyo

Springer Series in
MATERIALS SCIENCE

The Springer Series in Materials Science covers the complete spectrum of materials physics, including fundamental principles, physical properties, materials theory and design. Recognizing the increasing importance of materials science in future device technologies, the book titles in this series reflect the state-of-the-art in understanding and controlling the structure and properties of all important classes of materials.

61 **Fatigue in Ferroelectric Ceramics and Related Issues**
By D.C. Lupascu

62 **Epitaxy**
Physical Principles and Technical Implementation
By M.A. Herman, W. Richter, and H. Sitter

63 **Fundamentals of Ion Irradiation of Polymers**
By D. Fink

64 **Morphology Control of Materials and Nanoparticles**
Advanced Materials Processing and Characterization
Editors: Y. Waseda and A. Muramatsu

65 **Transport Processes in Ion Irradiated Polymers**
By D. Fink

66 **Multiphased Ceramic Materials**
Processing and Potential
Editors: W.-H. Tuan and J.-K. Guo

67 **Nondestructive Materials Characterization**
With Applications to Aerospace Materials
Editors: N.G.H. Meyendorf, P.B. Nagy, and S.I. Rokhlin

68 **Diffraction Analysis of the Microstructure of Materials**
Editors: E.J. Mittemeijer and P. Scardi

69 **Chemical–Mechanical Planarization of Semiconductor Materials**
Editor: M.R. Oliver

70 **Applications of the Isotopic Effect in Solids**
By G.V. Plekhanov

71 **Dissipative Phenomena in Condensed Matter**
Some Applications
By S. Dattagupta and S. Puri

72 **Predictive Simulation of Semiconductor Processing**
Status and Challenges
Editors: J. Dabrowski and E.R. Weber

73 **SiC Power Materials and Devices**
Some Applications
Editor: Z.C. Feng

74 **Plastic Deformation in Nanocrystaline Materials**
by M. Gutkin and I. Ovid'ko

Volumes 10–60 are listed at the end of the book.

Sushanta Dattagupta
Sanjay Puri

Dissipative Phenomena in Condensed Matter

Some Applications

With 72 Figures

Springer

Professor Sushanta Dattagupta
S.N. Bose National Centre for Basic Sciences
Sector - III, Block - JD,
Salt Lake Kolkata, 700 098
India
email: sdgupta@bose.res.in

Professor Sanjay Puri
Jawaharlal Nehru University
School of Physical Sciences
New Delhi, 110067
India
email: puri@mail.jnu.ac.in

Series Editors:

Professor Robert Hull
University of Virginia
Dept. of Materials Science and Engineering
Thornton Hall
Charlottesville, VA 22903-2442, USA

Professor Jürgen Parisi
Universität Oldenburg, Fachbereich Physik
Abt. Energie- und Halbleiterforschung
Carl-von-Ossietzky-Strasse 9–11
26129 Oldenburg, Germany

Professor R. M. Osgood, Jr.
Microelectronics Science Laboratory
Department of Electrical Engineering
Columbia University
Seeley W. Mudd Building
New York, NY 10027, USA

Professor Hans Warlimont
Institut für Festkörper-
und Werkstofforschung,
Helmholtzstrasse 20
01069 Dresden, Germany

ISSN 0933-033X

ISBN 3-540-20381-8 Springer-Verlag Berlin Heidelberg New York

Cataloging-in-Publication Data
Dattagupta, S. (Sushanta), 1947– .
Dissipative phenomena in condensed matter: some applications / S. Dattagupta, S. Puri
p.cm. – (Springer series in materials science, ISSN 0933-033X; 71)
Includes bibliographical references and index.
ISBN 3-540-20381-8 (acid-free paper)
1. Energy dissipation. 2. Condensed matter. 3. Nonequalibrium statistical mechanics.
4. Correlation (Statistics) I. Puri, S. (Sanjay), 1961– . II. Title. III. Series.
QC173.458.E53D38 2004 530.4'16–dc22 2003069135

Springer-Verlag is a part of Springer Science+Business Media
springeronline.com

Typesetting by the authors
Final processing by Frank Herweg, Leutershausen
Cover concept: eStudio Calamar Steinen
Cover production: *design & production* GmbH, Heidelberg

Printed on acid-free paper 57/3141/tr 5 4 3 2 1 0

To Akshay, Nikaash, Sharmishtha, Shahana,

and the memory of Alpha

Preface

The interaction of a physical system with its environment generally results in *dissipation*. In statistical physics, there is a symbiotic relationship between spontaneous fluctuations and dissipation, which is described by the *fluctuation-dissipation theorem*. For example, a large particle immersed in a fluid experiences frictional drag and velocity fluctuations, both resulting from repeated impacts of the fluid particles. Dissipation plays a crucial role in determining the properties of both equilibrium and nonequilibrium systems.

This book discusses dissipative phenomena in a range of problems in condensed matter physics. We will consider both classical and quantum-mechanical systems in this context. The level of discussion here assumes that the reader is acquainted with the basic formalism of *stochastic* and *Markov processes*. We also assume a familiarity with the concept of the *correlation function*, and its relationship to the experimentally-measured *susceptibility* and *structure factor*. With this background, we will discuss cooperative effects in strongly-interacting systems, which exhibit physical phenomena like pattern formation, phase ordering or domain-growth kinetics, slow dynamics, non-Debye spectral response, etc. We will also discuss single-particle systems which have a strong interaction with their environment in the context of *quantum dissipation*. Stochastic models will be used extensively, as these are an integral part of the current framework of statistical physics. Furthermore, the connection with experimental results will be strongly emphasized at each stage of this exposition.

Before proceeding, it is useful to provide a broad classification of nonequilibrium systems. We consider systems which have been disturbed from equilibrium by an external force. If the system is weakly disturbed, the nonequilibrium properties can be obtained in a perturbative approximation. This is the underlying principle of *linear response theory*, which provides a framework for determining response functions in terms of equilibrium correlation functions. On the other hand, if the system is taken far from equilibrium by the application of a large perturbation, the subsequent approach of the system to equilibrium is determined by the interplay of nonlinear and dissipative effects. Typically, the evolution of such systems is characterized by complex *spatio-temporal pattern formation*, which cannot be analyzed within the equilibrium framework.

A major aim of this book is to study the evolution of far-from-equilibrium systems. Examples of these include (i) a ferromagnet which is quenched from the (disordered) paramagnetic phase to the (ordered) ferromagnetic phase, (ii) a homogeneous mixture which is quenched below the coexistence curve, (iii) disordered systems like *spin glasses* or *structural glasses*, (iv) systems exhibiting *quantum phase transitions*, etc. We will discuss dissipative phenomena in models which are formulated in the context of the above examples, but which are applicable to a variety of other condensed matter systems. In most cases, we will be interested in nonequilibrium phenomena far from *critical points* in the corresponding phase diagram. Therefore, analytical and numerical studies of *mean-field models* will be useful in understanding these systems.

The subject of dissipative phenomena is too large to cover in a single book. Hence, we focus on topics which overlap with our research interests. This book is broadly divided into three parts. Chapters 1 and 2 provide the basic framework for modeling dissipative phenomena in condensed matter physics. Chapter 1 reviews the formalism of the *Langevin equation* and the *two-state jump process*, and derives these stochastic processes from microscopic models using the *system-plus-bath* approach. In Chap. 2, we focus on *kinetic Ising models*. In this context, we discuss the Glauber spin-flip model and the Kawasaki spin-exchange model. We use the relevant *master equations* to derive coarse-grained counterparts of these microscopic models.

In Chaps. 3–6, we discuss problems of phase ordering dynamics in *classical dissipative systems*. Chapter 3 is an overview of simple aspects of ordering dynamics, and is a reference point for Chaps. 4–6. In Chap. 4, we discuss domain growth in systems with quenched disorder. Chapter 5 focuses on the kinetics of phase separation near surfaces with a preferential attraction for one of the components of the mixture. Chapters 4 and 5 illustrate how experimentally relevant effects can drastically alter the properties of phase ordering systems. Chapter 6 studies the dynamics of ordering in the complex Ginzburg–Landau equation, which is a paradigm for pattern formation in systems characterized by spiral defects.

In Chaps. 7–10, we turn our attention to *quantum dissipative systems* and discuss novel developments in this area. In Chap. 7, we focus on the *transverse Ising model* (TIM) as a natural extension of the usual Ising model. However, unlike its classical counterpart, the TIM shows explicit quantum features. We also study a disordered version of the TIM as a model for *quantum glasses*, which exhibit characteristic spin-glass properties modified by quantum fluctuations. In Chap. 7, the TIM is studied in the system-plus-bath approach, where the system is quantum-mechanical but the heat bath is assumed to be classical. In Chaps. 8 and 9, we clarify the novel features which arise when the heat bath is also taken to be a quantum system. In this context, Chap. 8 studies *dissipative tunneling* in a double-well potential, and Chap. 9 focuses upon *quantum Brownian motion*. Finally, Chapter 10 concludes this book with a general discussion of the crossover from *coherence* to *decoherence* in quantum dissipative systems.

Acknowledgments

This book is an outcome of the time spent by us at the School of Physical Sciences, Jawaharlal Nehru University, New Delhi. Many of the problems discussed here were studied jointly with our students and colleagues: Rajeev Ahluwalia, Varsha Banerjee, Bibhu Biswal, Debashish Chowdhury, Michael Cross, Subir Das, Deepak Kumar, Nita Parekh, Tabish Qureshi, Jitendra Sharma, Radhika Sharma, Jagmeet Singh and Kumela Tafa. Some of the material presented here is based on publications co-authored with these students and colleagues. We are grateful to them for these productive research collaborations.

Sporadic trips to Japan, Europe and North America have also resulted in fruitful associations with colleagues, the results of which have been included in this book. In this context, we wish to thank Barbara Aichmayer, Sorin Bastea, Kurt Binder, Robert Blinc, Alan Bray, Rashmi Desai, Hans Werner Diehl, Burkhard Dünweg, Peter Fratzl, Harry Frisch, Hermann Grabert, Hisao Hayakawa, Volker Heine, Ray Kapral, Joel Lebowitz, Peter Nielaba, Akira Onuki, Yoshi Oono, Rasa Pirc, Heiko Rieger, Ekhard Salje, Herbert Schober, Bosa Tadic and Ulrich Weiss. Apart from these friends, we have also benefited immensely from scientific discussions with many other colleagues, both in India and outside.

Chapter 5 (on *surface-directed spinodal decomposition*) is based on a long-standing research interest of SP. He is particularly grateful to Kurt Binder for an ongoing collaboration on this problem, and his kind hospitality in Mainz. Chapter 10 (on *coherence and decoherence*) is based on a recent research interest of SD. In this context, his collaboration (at the S.N. Bose Centre, Kolkata) with Gautam Gangopadhyay, Arun Jayannavar, Narendra Kumar and M. Sanjay Kumar is gratefully acknowledged. SD would like to express his gratitude to the Alexander-Humboldt-Stiftung for its support during visits to Germany, which allowed uninterrupted time for writing this book. He is also grateful to Heiner Müller-Krumbhaar and other colleagues at the Forschungszentrum, Jülich, for their hospitality.

We would like to thank Subir Das and R.K. Brojen Singh for help in drafting some of the figures in this book. Finally, we are grateful to our wives Bindu and Ranu for their patience and support during the prolonged period over which this book was written.

Kolkata *Sushanta Dattagupta*
New Delhi *Sanjay Puri*

November 2003

Contents

1 Diffusion Processes

1.1 Introduction

Diffusion is usually studied in the context of Brownian motion. However, it is a paradigm for dissipative behavior in a variety of physical systems, e.g., spin diffusion in a magnet, hydrodynamic motion in a fluid, interstitial movement in a defect lattice, atomic jumps in a solid via a vacancy mechanism, rotational diffusion of macromolecules in fluids, etc. A large number of diffusion processes and their application to spectroscopy have been discussed by Dattagupta [1].

The most familiar example of a diffusive system is a heavy tagged particle in a fluid. This particle undergoes incessant collisions with the (much lighter) particles of the fluid. These collisions are random, and result in what is referred to as the *Brownian motion* of the tagged particle [2–4]. At the microscopic level of description, the particle is in constant interaction with the constituents of the fluid, which comprise a *heat bath.* This heat bath is in equilibrium at a fixed temperature. However, for a specified initial condition, the tagged particle is *not* in equilibrium. The motion of the latter is *diffusive*, leading to a linear increase of its mean-squared displacement in time. Furthermore, the mean-squared velocity and the velocity auto-correlation function evolve from their initial values to the expected equilibrium values, determined by the temperature of the bath.

The mathematical formulation of Brownian motion consists of a stochastic differential equation (or *Langevin equation*) for the momentum of the particle [5–9]. This equation is driven by a random force, simulating the effect of collisions between the tagged particle and the particles of the heat bath. The effects of additional systematic forces can also be incorporated in the Langevin formulation [5–9].

The Langevin equation provides a description involving an equation of motion for a dynamical variable, e.g., the momentum. An equivalent approach is to formulate an evolution equation for the time-dependent probability distribution in the phase space of the system. This equation is referred to as the Fokker–Planck (FP) or the Smoluchowski equation in different contexts [10]. In this book, we will use the Langevin equation as the basis for most of our discussions of dissipative phenomena. In diverse applications, models of dissipation in condensed matter physics are obtained as physically-motivated generalizations of the basic ideas of Brownian motion.

As Brownian motion is used as a paradigm for our subsequent treatment of nonequilibrium effects, Sect. 1.2 presents a discussion of the Langevin equation. This will be done in the familiar context of a tagged particle moving under the influence of both systematic and random forces. We will summarize the relevant properties of various physical quantities, e.g., mean-squared displacement, velocity correlation, etc.

It is also physically relevant to provide a microscopic basis for the Langevin equation. Section 1.3 demonstrates that the Langevin equation arises from an approximate description of an interacting many-body system, which is described by a complicated time-independent Hamiltonian or Lagrangian [11,12]. The crucial point is that the Langevin equation is dissipative, whereas the underlying microscopic equations of motion have time-reversal symmetry. The mechanism for emergence of dissipation in a system, which has unitary evolution at the microscopic level, is a fundamental physical problem. The resolution of this problem provides important principles for modeling physical systems.

Section 1.4 briefly discusses the Fokker–Planck and master equations, and their equivalence to the Langevin equation. It should be emphasized that the Fokker–Planck equation follows directly from the theory of Markov processes [13]. As a matter of fact, most models of dissipative phenomena can be formulated in terms of a stationary Markov process, in conjunction with detailed balance. Non-Markovian effects, if they do arise, can often be viewed as a consequence of the projection of a larger Markov process on a smaller variable space – at least in the limit of a full phase-space description.

Brownian motion is described by a continuous diffusion process. There are situations, however, which require analysis of *jump diffusion processes* or *discrete* Markov processes. The most common examples are magnetic spin dynamics, and the hopping motion of a tracer atom on a lattice. As an illustrative example, we shall discuss in Sect. 1.5 the stochastic dynamics of a single spin or a two-state jump process. The simplicity of this example already elucidates many of the physical features of the more complicated case with multiple interacting spins, discussed in Chap. 2.

Having shown (in Sect. 1.3) how the stochastic Langevin equation follows from a microscopic Hamiltonian description, it is useful to undertake a similar exercise for single-spin dynamics. The resultant analysis, presented in Sect. 1.6, provides a calculable expression for the jump rate. It also constitutes a theoretical tool for treating more complex cases, e.g., derivation of a master equation for the Ising model (Chap. 2) or the quantum generalization of the Ising model (Chap. 7). It turns out that, even though the stochastic process under consideration is a classical one, our analysis requires the language of quantum mechanics. This is not unexpected, because discrete jumps at the microscopic level correspond to quantum transitions. The analysis in

Sect. 1.6 is carried out with the help of Appendix 1.A, in which we discuss a quantum master equation for the density operator. Finally, Sect. 1.7 concludes this chapter with a discussion of the subsystem density operator for stochastic spin dynamics.

1.2 Brownian Motion and the Langevin Equation

Consider a large (Brownian) particle of mass M immersed in a fluid. This particle is continuously bombarded by smaller particles constituting the fluid. As a result of its velocity $v(t)$ being in the *forward* direction, the large particle encounters more collisions from the *front* than from the *rear*. This results in a systematic retardation or frictional force proportional to the instantaneous velocity of the particle. Hence the equation of motion can be written as [14]

$$M\frac{dv}{dt} + M\gamma v(t) = \theta(t)\,, \tag{1.1}$$

where γ is the coefficient of friction, and $\theta(t)$ is a random force (or a *noise*) arising from the collisions. For simplicity of presentation, we will focus on the 1-dimensional case. The generalization to arbitrary dimensionality is straightforward.

In a more formal picture, the Brownian particle may be viewed as a *subsystem* in contact with a thermal bath which drives noisy fluctuations in the subsystem (see Sect. 1.3). We can then set up an initial condition $v(t=0) = v_0$, which corresponds to an ensemble of the subsystem, and obtain the solution of (1.1) as

$$v(t) = v_0 e^{-\gamma t} + \frac{1}{M}\int_0^t dt' e^{-\gamma(t-t')}\theta(t')\,. \tag{1.2}$$

Since $\theta(t)$ is a noise term, we expect that its average is zero, i.e.,

$$\overline{\theta(t)} = 0\,, \tag{1.3}$$

where the bar denotes an average over the bath but for a fixed subensemble, i.e., a fixed initial velocity v_0. (The complete statistical average can be obtained by performing an additional average over v_0.) Thus

$$\overline{v(t)} = v_0 e^{-\gamma t}\,, \tag{1.4}$$

defining a relaxation time γ^{-1}. Equation (1.4) is the solution of the constitutive equation:

$$\frac{d\overline{v}}{dt} + \gamma\overline{v(t)} = 0\,, \tag{1.5}$$

which breaks the time-reversal symmetry explicitly, and hence describes irreversible behavior.

To study fluctuations in $v(t)$, we first examine the mean-squared velocity:

$$\overline{v^2(t)} = v_0^2 e^{-2\gamma t} + \frac{1}{M^2} \int_0^t dt' \int_0^t dt'' e^{-\gamma(t-t')-\gamma(t-t'')} \overline{\theta(t')\theta(t'')} , \tag{1.6}$$

where the cross-terms vanish because of (1.3). The simplest approximation for $\theta(t)$ is to assume that the noise is purely random, i.e.,

$$\overline{\theta(t')\theta(t'')} = \overline{\theta(t')}\ \overline{\theta(t'')} = 0 , \quad t' \neq t'' . \tag{1.7}$$

When $t' = t''$, the quantity on the left-hand-side (LHS) reduces to the mean-squared noise which should be nonzero. Hence, we set

$$\overline{\theta(t')\theta(t'')} = \nu \delta(t' - t'') , \tag{1.8}$$

where ν is a constant which measures the strength of the noise. In that case, (1.6) simplifies to

$$\overline{v^2(t)} = v_0^2 e^{-2\gamma t} + \frac{\nu}{2\gamma M^2} \left(1 - e^{-2\gamma t}\right) . \tag{1.9}$$

Equation (1.9) fixes the constant ν if we recognize that the Brownian particle should asymptotically thermalize with the heat bath, maintained at a fixed temperature T [14,15], i.e.,

$$\lim_{t \to \infty} \overline{v^2(t)} = \frac{\nu}{2\gamma M^2} = \frac{k_B T}{M} . \tag{1.10}$$

The last equality follows from the equipartition theorem, with k_B denoting the Boltzmann constant. This yields

$$\nu = 2M\gamma k_B T , \tag{1.11}$$

and

$$\overline{\theta(t')\theta(t'')} = 2M\gamma k_B T \delta(t' - t'') . \tag{1.12}$$

Two comments can now be made regarding (1.12). Firstly, it reiterates that the source of the systematic damping term (proportional to γ) and the noise term in the Langevin equation is indeed the same, i.e., the heat bath. Secondly, γ is a parameter which characterizes dissipation, whereas the LHS of (1.12) describes fluctuations. Hence, (1.12) constitutes an example of a *fluctuation-dissipation relation* [7].

We next consider the auto-correlation of $v(t)$ from (1.2):

$$\overline{v(t)v(\tau)} = v_0^2 e^{-\gamma(t+\tau)} + \frac{1}{M^2}\int_0^t dt' \int_0^\tau dt'' e^{-\gamma(t-t')-\gamma(\tau-t'')}\overline{\theta(t')\theta(t'')} \,. \tag{1.13}$$

Without loss of generality, we consider $t > \tau$ and use (1.12) to obtain

$$\overline{v(t)v(\tau)} = v_0^2 e^{-\gamma(t+\tau)} + \frac{k_B T}{M}\left[e^{-\gamma(t-\tau)} - e^{-\gamma(t+\tau)}\right] . \tag{1.14}$$

To obtain the auto-correlation function in equilibrium, we must average further over the appropriate Maxwellian distribution in v_0. This yields

$$\langle v(t)v(\tau)\rangle_0 = \frac{k_B T}{M} e^{-\gamma(t-\tau)} , \quad t > \tau , \tag{1.15}$$

where the angular brackets denote an averaging over *both* the noise distribution and the initial velocity distribution. Furthermore, the subscript "0" refers to an averaging over the Maxwellian distribution for v_0 rather than an arbitrary distribution. A few remarks are in order at this stage.

1. The solution in (1.15) exhibits the property of stationarity, i.e., the correlation function in equilibrium depends only on the difference of the two times t and τ.
2. The correlation function shows *mixing*, i.e.,

$$\lim_{|t-\tau|\to\infty} \langle v(t)v(\tau)\rangle_0 = \langle v(t)\rangle_0 \langle v(\tau)\rangle_0 = 0 \,. \tag{1.16}$$

3. Although the noise is delta-correlated, the driven variable $v(t)$, which is also a stochastic process, is exponentially-correlated. This is a general property of a 1-dimensional Gaussian-Markov stationary process [16,17].
4. Both $\overline{v(t)}$ and $\langle v(t)v(\tau)\rangle_0$ are characterized by the same relaxation time γ^{-1}. Correspondingly, they satisfy an identical rate equation. This is consistent with the underlying theme of *linear response theory* (LRT), which describes the behavior of a statistical system for small departures from equilibrium, and is implicit in the Langevin formulation [1].

The simplest form of the Langevin equation, as given in (1.1), needs to be generalized if the Brownian particle is also under the influence of a systematic force. The resultant equation reads

$$\frac{dP}{dt} + \gamma P(t) = F(Q) + \theta(t) \,, \tag{1.17}$$

where $P(t)$ is the momentum, and $F(Q)$ is the force. We assume that the force only depends upon the coordinate Q. (For simplicity, we will work in a Cartesian coordinate system.) Equation (1.17) should be supplemented by the Newtonian definition:

$$\frac{dQ}{dt} = \frac{P(t)}{M} . \tag{1.18}$$

Taken together, (1.17) and (1.18) provide a complete phase-space description of Brownian motion. Clearly, this is not analytically tractable, except for a few special cases, e.g., a particle confined by a harmonic potential [7–9].

1.3 Langevin Equation from a System-plus-Bath Approach

Let us now provide a microscopic "derivation" for the Langevin equation [11,12]. We consider a particle of mass M moving in an arbitrary potential $U(Q)$, where Q is the coordinate. The particle is assumed to interact with N harmonic oscillators, of mass m_j and frequency ω_j $(j = 1 \to N)$, through a *linear* coupling. In our microscopic modeling, this set of oscillators will constitute a heat bath. The situation is schematically illustrated in Fig. 1.1. Since the vector nature of the momentum and coordinates is unimportant for our present discussion, we shall again assume all variables to be 1-dimensional. Thus, the Hamiltonian of the entire system can be written as

$$\mathcal{H} = \frac{P^2}{2M} + U(Q) + \sum_j \left[\frac{{p_j}^2}{2m_j} + \frac{1}{2} m_j {\omega_j}^2 \left(q_j - \frac{C_j Q}{m_j {\omega_j}^2} \right)^2 \right] , \tag{1.19}$$

where $P(t)$ is the particle momentum; $\{p_j, q_j\}_{j=1}^N$ are harmonic-oscillator coordinates; and C_j's are arbitrary coupling constants. We can use the canonical transformation

$$\tilde{q}_j = \frac{m_j {\omega_j}^2}{C_j} q_j , \tag{1.20}$$

$$\tilde{p}_j = \frac{C_j}{m_j {\omega_j}^2} p_j , \tag{1.21}$$

to rewrite the Hamiltonian as

$$\tilde{\mathcal{H}} = \frac{P^2}{2M} + U(Q) + \sum_j \left[\frac{\tilde{p}_j^2}{2\tilde{m}_j} + \frac{1}{2} \tilde{m}_j {\omega_j}^2 \left(\tilde{q}_j - Q \right)^2 \right] . \tag{1.22}$$

In (1.22), the "masses" are defined as

$$\tilde{m}_j = \frac{C_j^2}{m_j {\omega_j}^4} . \tag{1.23}$$

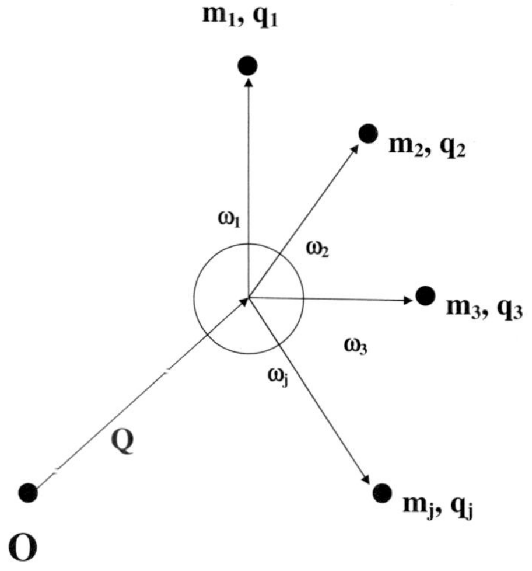

Fig. 1.1. Large particle with coordinate $\boldsymbol{Q}$, interacting with a set of N harmonic oscillators with coordinates $\{\boldsymbol{q}_j\}_{j=1}^{N}$. In general, the harmonic oscillators have different masses $\{m_j\}$ and frequencies $\{\omega_j\}$

Using Hamilton's equations [18], we immediately derive from (1.19)

$$\frac{dQ}{dt} = \frac{P}{M}, \tag{1.24}$$

$$\frac{dP}{dt} = -\frac{\partial U(Q)}{\partial Q} + \sum_j C_j \left(q_j - \frac{C_j Q}{m_j \omega_j^2} \right), \tag{1.25}$$

$$\frac{dq_j}{dt} = \frac{p_j}{m_j}, \tag{1.26}$$

$$\frac{dp_j}{dt} = -m_j \omega_j^2 \left(q_j - \frac{C_j Q}{m_j \omega_j^2} \right). \tag{1.27}$$

Using (1.27), we can rewrite (1.26) as

$$\frac{d^2 q_j}{dt^2} = -\omega_j^2 \left(q_j - \frac{C_j Q}{m_j \omega_j^2} \right), \tag{1.28}$$

which is merely the harmonic-oscillator equation with an inhomogeneous term proportional to $Q(t)$. The solution of this equation is

$$q_j(t) = q_j(0) \cos(\omega_j t) + \frac{p_j(0)}{m_j \omega_j} \sin(\omega_j t) + \frac{C_j}{m_j \omega_j} \int_0^t dt' Q(t') \sin[\omega_j (t - t')]. \tag{1.29}$$

An integration by parts enables us to express (1.29) in the following form:

$$q_j(t) - \frac{C_j Q(t)}{m_j\omega_j^2} = \left[q_j(0) - \frac{C_j Q(0)}{m_j\omega_j^2}\right]\cos(\omega_j t) + \frac{p_j(0)}{m_j\omega_j}\sin(\omega_j t)$$
$$- \frac{C_j}{m_j\omega_j^2}\int_0^t dt' \frac{P(t')}{M}\cos\left[\omega_j(t-t')\right], \qquad (1.30)$$

where we have used (1.24). Finally, substituting (1.30) in (1.25), we obtain

$$\frac{dP}{dt} = -\frac{\partial U(Q)}{\partial Q} - \int_0^t dt' \frac{P(t')}{M}\sum_j \frac{C_j^2}{m_j\omega_j^2}\cos[\omega_j(t-t')]$$
$$+\sum_j \left\{C_j\left[q_j(0) - \frac{C_j Q(0)}{m_j\omega_j^2}\right]\cos(\omega_j t) + \frac{C_j p_j(0)}{m_j\omega_j}\sin(\omega_j t)\right\}. \qquad (1.31)$$

Equation (1.31) has the form of a *generalized* Langevin equation:

$$\frac{dP}{dt} = -\frac{\partial U(Q)}{\partial Q} - \int_0^t dt' \frac{P(t')}{M}\zeta(t-t') + \theta(t). \qquad (1.32)$$

However, the "noise" $\theta(t)$ now explicitly depends on the initial coordinates and momenta of the harmonic oscillators, i.e.,

$$\theta(t) = \sum_j \left\{C_j\left[q_j(0) - \frac{C_j Q(0)}{m_j\omega_j^2}\right]\cos(\omega_j t) + \frac{C_j p_j(0)}{m_j\omega_j}\sin(\omega_j t)\right\}, \qquad (1.33)$$

and the "friction" $\zeta(t)$ appears as a *memory function*:

$$\zeta(t) = \sum_j \frac{C_j^2}{m_j\omega_j^2}\cos(\omega_j t). \qquad (1.34)$$

We now make the crucial step towards introducing *irreversibility* by considering an initial ensemble of states, in which $P(0), Q(0)$ are held fixed, but the harmonic-oscillator variables are drawn at random from a canonical distribution with probability

$$\mathcal{P}\left(\{p_j(0)\},\{q_j(0)\}\right) = \frac{e^{-\beta\mathcal{H}}}{\prod_j \int\int dp_j(0)dq_j(0)e^{-\beta\mathcal{H}}}, \qquad (1.35)$$

where $\beta = (k_B T)^{-1}$. Now, it is appropriate to refer to the harmonic-oscillator system as a *heat bath.* Introducing "bars" to denote averages over the initial phase-space distribution, we have for any arbitrary operator A:

$$\overline{A} = \prod_j \int\int dp_j(0)dq_j(0)A(\{q_j\},\{p_j\})\mathcal{P}(\{p_j(0)\},\{q_j(0)\}) \,. \quad (1.36)$$

It is then straightforward to verify that

$$\overline{\theta(t)} = 0 \,, \quad (1.37)$$

and

$$\overline{\theta(t)\theta(t')} = k_B T \zeta(t-t') \,. \quad (1.38)$$

Equation (1.32) now has all the attributes of the Langevin equation, with (1.38) as an explicit statement of the fluctuation-dissipation relation. This derivation emphasizes the fact that noise in a thermodynamic system originates from microscopic variables, which are governed by the probabilistic laws of statistical mechanics. Later, in Sect. 1.6, we will provide a similar treatment in the context of quantum spin systems.

1.3.1 Ohmic Dissipation Model

The frictional term in our derivation of (1.32) contains a memory function $\zeta(t)$, which depends upon the masses m_j, frequencies ω_j and the coupling constants C_j (see (1.34)). Thus, the Langevin equation (1.32) is non-Markovian. However, a special choice of the oscillator-frequency distribution can lead to an approximately Markovian description, yielding the usual Langevin equation with constant friction [19]. Assume that there are an infinitely large number of oscillators in the heat bath, with a continuous distribution of frequencies. Thus, sums over oscillator frequencies are replaced by integrals according to

$$\sum_j \to N \int_0^\infty g(\omega) d\omega \,, \quad (1.39)$$

where $g(\omega)$ is the density of states. Furthermore, assume that all masses and coupling constants are equal:

$$m_j = m \,, \quad C_j = \frac{C}{\sqrt{N}} \,. \quad (1.40)$$

Equation (1.34) then yields

$$\zeta(t) = \frac{C^2}{m} \int_0^\infty d\omega \frac{g(\omega)}{\omega^2} \cos(\omega t) \,. \quad (1.41)$$

Finally, the density of states $g(\omega)$ is taken to be quadratic in ω and has an upper cut-off ω_c, i.e.,

$$g(\omega) = \frac{3\omega^2}{\omega_c^3}\,, \quad \omega < \omega_c\,,$$
$$= 0\,, \quad \omega > \omega_c\,. \tag{1.42}$$

From (1.41), we obtain

$$\zeta(t) = \frac{3C^2}{m\omega_c^3}\frac{\sin(\omega_c t)}{t}\,. \tag{1.43}$$

If ω_c is sufficiently large, the system momentum $P(t)$ varies slowly over time-scales of order ω_c^{-1}. In that limit, $\int_0^{\omega_c} d\omega \cos(\omega t) \simeq \pi\delta(t)$, and

$$\zeta(t) \simeq 2\zeta_0\delta(t)\,, \tag{1.44}$$

where

$$\zeta_0 = \frac{3\pi C^2}{2m\omega_c^3}\,. \tag{1.45}$$

Under these approximations, (1.32) now simplifies to the usual Langevin equation:

$$\frac{dP}{dt} = -\frac{\partial U(Q)}{\partial Q} - \gamma P(t) + \theta(t)\,, \tag{1.46}$$

where $\gamma = \zeta_0/M$, and

$$\overline{\theta(t)} = 0\,,$$
$$\overline{\theta(t')\theta(t'')} = 2M\gamma k_B T\delta(t' - t'')\,. \tag{1.47}$$

The model described above with (a) the density of states in (1.42), and (b) the assumption concerning the time-scale ω_c^{-1} which led to (1.44), is known as the *Ohmic dissipation* model [20]. The name derives from the fact that, if the system under consideration were a harmonic oscillator,

$$U(Q) = \frac{1}{2}M\Omega^2 Q^2\,. \tag{1.48}$$

Equations (1.24) and (1.46) would be analogous to the equation of an LCR circuit driven by a noisy voltage. The analogy applies when we identify Q as the electric charge, P/M as the electric current, M as the inductance, $(M\Omega^2)^{-1}$ as the capacitance, and ζ_0 as the Ohmic resistance.

1.4 Fokker–Planck and Master Equations

The Langevin equation (1.1) is an equation of motion for a dynamical variable, e.g., velocity $v(t)$ in the case considered. On the other hand, the underlying stochastic process driven by the noise must be defined through a probability-distribution function $\mathcal{P}(v,t)$, which obeys an evolution equation. The two descriptions (viz., the Langevin equation and the equation for $\mathcal{P}(v,t)$) are equivalent. An analogous situation arises in quantum mechanics, where the Heisenberg and Schrödinger pictures yield identical information about a physical system [21,22]. As stated earlier, the evolution equation for $\mathcal{P}(v,t)$ is known as the FP equation. For the Brownian particle described by (1.1), the FP equation has the form [10]:

$$\frac{\partial}{\partial t}\mathcal{P}(v,t) = \gamma\frac{\partial}{\partial v}\left[v\mathcal{P}(v,t)\right] + \frac{\gamma k_B T}{M}\frac{\partial^2}{\partial v^2}\mathcal{P}(v,t)\,. \tag{1.49}$$

The stochastic process described by (1.49) is called the Ornstein–Uhlenbeck process [16], and has the solution

$$\mathcal{P}(v,t) = \left[\frac{2\pi}{M\beta}\left(1-e^{-2\gamma t}\right)\right]^{-1/2} \exp\left[-\frac{M\beta}{2}\frac{\left(v-v_0e^{-\gamma t}\right)^2}{1-e^{-2\gamma t}}\right], \tag{1.50}$$

corresponding to the initial condition

$$\mathcal{P}(v,0) = \delta(v-v_0)\,. \tag{1.51}$$

In the more general case of the complete phase-space dynamics governed by (1.17)–(1.18), the generalized version of (1.49) is the so-called Fokker–Planck–Smoluchowski–Kramers equation [7,8,10]:

$$\frac{\partial}{\partial t}\mathcal{P}(Q,P,t) = \left\{-\frac{P}{M}\frac{\partial}{\partial Q} + \frac{\partial}{\partial P}\left[-F(Q)+\gamma P\right] + M\gamma k_B T\frac{\partial^2}{\partial P^2}\right\}\mathcal{P}(Q,P,t)\,. \tag{1.52}$$

In (1.52), $\mathcal{P}(Q,P,t)$ is generally interpreted as a *conditional* probability, i.e., the probability that the position and momentum of the particle are Q and P at time t, *given* that they were Q_0 and P_0 at $t=0$. To emphasize this point, it is sometimes conventional to write the probability as $\mathcal{P}(Q_0,P_0,0|Q,P,t)$.

Equation (1.52) is a special case of a more general master equation, which characterizes a stationary Markov process [7]:

$$\frac{\partial}{\partial t}\mathcal{P}(x_1,t_1|x,t) = \int dx'\left[\mathcal{P}(x_1,t_1|x',t)W(x'|x) - \mathcal{P}(x_1,t_1|x,t)W(x|x')\right]\,. \tag{1.53}$$

Equation (1.53) is usually referred to as the Smoluchowski–Chapman–Kolmogorov equation, and has the following interpretation. The first term on the right-hand-side (RHS) is the *gain term* for entering the state x at time t from any arbitrary state x'. On the other hand, the second term on the RHS is the *loss term* for exiting the state x at time t to any arbitrary state x'.

The quantity $W(x'|x)$ denotes the transition probability from $x' \to x$, and depends upon the system Hamiltonian. By identifying x (which can in general be a multi-dimensional stochastic process) as the pair of 1-dimensional variables (Q, P), and assuming that only small jumps occur, (1.53) can be reduced to a partial differential equation of the form in (1.52).

It may be noted that x is a *continuous* stochastic process in the above formulation of the FP and master equations. If the spectrum of allowed values of x is *discrete* rather than continuous, we have a *jump process* and the master equation assumes the form [1]:

$$\frac{\partial}{\partial t}\mathcal{P}(x_1,t_1|x,t) = \sum_{x'}{}' \left[\mathcal{P}(x_1,t_1|x',t)W(x'|x) - \mathcal{P}(x_1,t_1|x,t)W(x|x')\right] , \tag{1.54}$$

where the prime on the summation sign indicates that the term with $x' = x$ is excluded. For the purpose of making calculations in a given physical situation, one has to model the jump probability $W(x'|x)$. We next illustrate this formulation using the simple example of a *two-state jump process.* In the literature, this is also referred to as the *telegraph process* or *dichotomic Markov process* [1].

1.5 Two-State Jump Process

Consider a single spin (with spin angular momentum $\hbar/2$), placed in a magnetic field H pointing along the z-axis. The corresponding Hamiltonian is given by

$$\mathcal{H}_s = -H\sigma_z , \tag{1.55}$$

where σ_z is the corresponding Pauli matrix. For the sake of brevity, we have absorbed the prefactor in (1.55) into the definition of H. The magnetization is $\langle\sigma_z\rangle$, where the angular brackets denote an average over the canonical ensemble. If the field is turned on at time $t = 0$, the magnetization will evolve from zero to a nonzero value. This evolution must be described in the context of nonequilibrium statistical physics. Therefore, we assume that the subsystem described by (1.55) is in contact with a heat bath which causes transitions (at random instants of time) amongst the eigenstates of σ_z, denoted by $|+\rangle$ and $|-\rangle$. Because of these heat-bath-induced fluctuations, the eigenvalue of σ_z (say, m) evolves as a two-state jump process, stochastically changing between the values $m = +1$ and -1, as sketched in Fig. 1.2.

The appropriate master equation for this process can be written from (1.54) as [1]

$$\frac{d}{dt}\mathcal{P}(n,0|m,t) = W(-m|m)\mathcal{P}(n,0|-m,t) - W(m|-m)\mathcal{P}(n,0|m,t) , \tag{1.56}$$

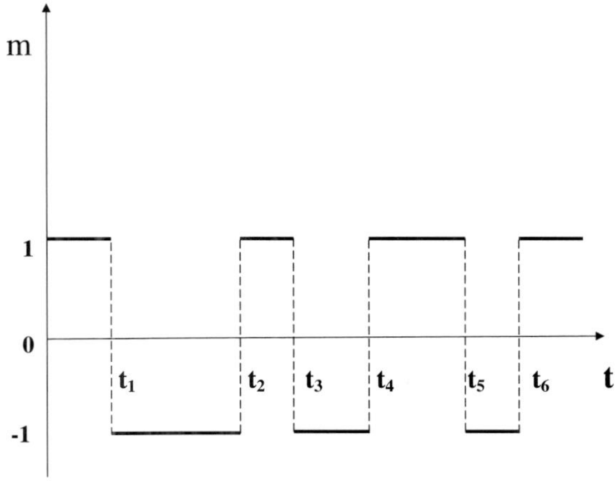

Fig. 1.2. Stochastic evolution of a spin in contact with a heat bath. The eigenvalue (m) of the Pauli matrix σ_z switches from ± 1 to ∓ 1 at randomly-distributed times t_1, t_2, t_3, etc.

where we consider a situation in which the spin is in state n at time $t = 0$. For notational convenience, we suppress the labels $(n, 0)$ and write (1.56) as

$$\frac{d}{dt}\mathcal{P}(m,t) = W(-m|m)\mathcal{P}(-m,t) - W(m|-m)\mathcal{P}(m,t)\,. \tag{1.57}$$

This has the familiar form of a rate equation if we consider, for instance, $m = 1$:

$$\frac{d}{dt}\mathcal{P}(1,t) = W(-1|1)\mathcal{P}(-1,t) - W(1|-1)\mathcal{P}(1,t)\,. \tag{1.58}$$

In equilibrium, we know that $\mathcal{P}(m,t) \to p^{\text{eq}}(m)$, the Boltzmann factor. Therefore, it follows from (1.57) that

$$\frac{W(m|-m)}{W(-m|m)} = \frac{p^{\text{eq}}(-m)}{p^{\text{eq}}(m)}\,. \tag{1.59}$$

Equation (1.59) is known as the *detailed-balance condition*, and must be satisfied by the transition probability so that the underlying stochastic process describes the asymptotic evolution to a thermal equilibrium state [7]. In the present context,

$$p^{\text{eq}}(m) = \frac{e^{\beta H m}}{2\cosh(\beta H)} = \frac{1}{2}[1 + m\tanh(\beta H)]\,. \tag{1.60}$$

This suggests that a reasonable choice for the transition probability is

$$W(-m|m) = \lambda p^{\text{eq}}(m)\,, \tag{1.61}$$

where λ is a parameter assumed to be independent of m.

It should be stressed that, though (1.59) follows from (1.61), the converse is not true. Equation (1.61) is merely an ansatz – a consistency condition which is in conformity with detailed balance. Furthermore, λ is a phenomenological parameter that has the same status as (say) the friction coefficient γ in the Langevin equation. A *calculation* of λ requires a separate model for the subsystem-heat bath coupling (see Sect. 1.7 below).

With these remarks, we now calculate the magnetization as it evolves from an arbitrary initial value. The magnetization is defined as

$$m_z(t) = \langle \sigma_z(t) \rangle = \mathcal{P}(1,t) - \mathcal{P}(-1,t) . \tag{1.62}$$

Thus, from (1.57),

$$\frac{d}{dt} m_z(t) = 2[W(-1|1)\mathcal{P}(-1,t) - W(1|-1)\mathcal{P}(1,t)] , \tag{1.63}$$

which, upon using (1.60)–(1.61), yields

$$\frac{d}{dt} m_z(t) = -\lambda \left[m_z(t) - \tanh(\beta H) \right] . \tag{1.64}$$

This is the familiar rate equation which describes the relaxation of the magnetization to its equilibrium value:

$$m_z^{\mathrm{eq}} = \tanh(\beta H) . \tag{1.65}$$

This formulation also allows us to compute the auto-correlation function in equilibrium:

$$C(t) = \langle \sigma_z(0) \sigma_z(t) \rangle_0 = \sum_{n,m=\pm 1} n p^{\mathrm{eq}}(n) m \mathcal{P}(n,0|m,t) . \tag{1.66}$$

Using (1.57), (1.60) and (1.61), we obtain

$$\frac{d}{dt} C(t) = -\lambda \left[C(t) - \langle \sigma_z \rangle_0^2 \right] , \tag{1.67}$$

where

$$\langle \sigma_z \rangle_0 \equiv m_z^{\mathrm{eq}} = \tanh(\beta H) . \tag{1.68}$$

Equation (1.67) can be easily solved as follows:

$$C(t) = \langle \sigma_z \rangle_0^2 + \left(1 - \langle \sigma_z \rangle_0^2 \right) e^{-\lambda t} . \tag{1.69}$$

Several interesting physical results may now be derived on the basis of (1.69) as follows:

1. The *correlation time* for the underlying process, defined as

$$\tau_c = \frac{\int_0^\infty dt[C(t) - C(\infty)]}{C(0) - C(\infty)}, \tag{1.70}$$

equals λ^{-1} in the present case.
2. Employing results from LRT [1], the response function is given by

$$\Psi(t) = \beta \mathrm{sech}^2(\beta H)\left(1 - e^{-\lambda t}\right). \tag{1.71}$$

3. Correspondingly, the non-transient frequency-dependent susceptibility can be written as

$$\chi(\omega) = \beta \mathrm{sech}^2(\beta H)\left(\frac{\lambda}{\lambda - i\omega}\right). \tag{1.72}$$

A response characterized by a single correlation time τ_c is usually referred to as a *Debye response* [23].

1.6 Single-Spin Kinetics from a System-plus-Bath Approach

In the previous section, the kinetics of a single spin, and the corresponding rate equations for the magnetization (1.64) and correlation function (1.67), were discussed in the context of a stochastic model. In this model, the rate constant λ appears as a phenomenological parameter which specifies the (mean) jump frequency in the two-state jump process. It is instructive to examine the same problem within a system-plus-bath approach (SBA) [24–27], as we have done in Sect. 1.3 for the Brownian particle. This not only provides a microscopic meaning for the rate λ, but also suggests a derivation of the master equation. This approach is also useful in more complicated cases involving, e.g., interacting spin systems [28], as will be evident in subsequent chapters.

The SBA we discuss here is analogous to that adopted in Sect. 1.3 for the derivation of the classical Langevin equation. We start with a Hamiltonian $\mathcal{H}$ which consists of three parts [24,27]:

$$\mathcal{H} = \mathcal{H}_s + V + \mathcal{H}_B. \tag{1.73}$$

In (1.73), $\mathcal{H}_s$ is the subsystem Hamiltonian for a single spin in a magnetic field (see (1.55)), V is the interaction Hamiltonian between the subsystem and the bath, and $\mathcal{H}_B$ is the bath Hamiltonian. As the concept of spin is intrinsic to quantum mechanics, we find it convenient to use a quantum-mechanical approach in the present context. Furthermore, because one of our goals is to derive a master equation for the probability of spin configurations, we need to introduce the density operator ρ. Thus, our approach is similar to employing the Schrödinger picture of quantum mechanics, rather than the Heisenberg picture [21,22].

Keeping in mind the underlying jump process, in which the spin is expected to undergo spin-flip interactions with the heat bath, it is natural to choose the coupling term V in (1.73) as

$$V = g\hat{b}\sigma_x \, . \tag{1.74}$$

The presence of the term σ_x, which is purely off-diagonal in the representation in which σ_z is diagonal, guarantees spin-flip processes. In (1.74), g is a coupling constant and $\hat{b}$ is an operator which acts on the Hilbert space of the bath. For our present purpose, it is not necessary to specify in detail the nature of either $\hat{b}$ or the Hamiltonian $\mathcal{H}_B$, in contrast to the case treated in Sect. 1.3. It is sufficient to assume that $\hat{b}$ causes transitions amongst the eigenstates of $\mathcal{H}_B$.

The derivation of the master equation for the density operator of the subsystem is relegated to Appendix 1.A. We should stress that the master equation formulated in (1.A.24) is more generally applicable than to the single-spin case. For further calculations in the single-spin problem, we need to compute the form of the coupling term V in the interaction picture [22]:

$$V_I(\tau) = \frac{1}{2} g\hat{b}(\tau) \left(\sigma_+ e^{-2iH\tau/\hbar} + \sigma_- e^{2iH\tau/\hbar} \right) , \tag{1.75}$$

where H is the magnetic field. Equation (1.75) is obtained by using (1.A.4), (1.55), and the properties of Pauli matrices. We have also introduced the notation

$$\hat{b}(\tau) = \exp\left(\frac{i}{\hbar} \mathcal{H}_B \tau \right) \hat{b} \exp\left(-\frac{i}{\hbar} \mathcal{H}_B \tau \right) , \tag{1.76}$$

where $\hat{b}(0) = \hat{b}$.

1.7 The Subsystem Density Operator

It is evident that the subsystem density operator $\rho_s(t)$ or $\tilde{\rho}_s(t)$ (defined in (1.A.10)–(1.A.11)) has a 2×2 matrix representation in the present case. Let us next formulate evolution equations for the four matrix elements of $\tilde{\rho}_s(t)$. These are obtained by substituting the form of $V_I(\tau)$ from (1.75)–(1.76) in the expression for the relaxation matrix $\hat{R}$ in (1.A.21). Then, (1.A.22) yields

$$\frac{\partial}{\partial t} \langle + | \tilde{\rho}_s(t) | + \rangle = - \langle + | \hat{R} \tilde{\rho}_s(t) | + \rangle =$$

$$= \frac{g^2}{\hbar^2} \int_0^\infty d\tau \mathrm{Tr}_B \left[\left(e^{-2iH\tau/\hbar} \hat{b}(\tau) \rho_B \hat{b}(0) + e^{2iH\tau/\hbar} \hat{b}(0) \rho_B \hat{b}(\tau) \right) \langle - | \tilde{\rho}_s(t) | - \rangle \right.$$
$$\left. - \left(e^{-2iH\tau/\hbar} \hat{b}(\tau) \hat{b}(0) \rho_B + e^{2iH\tau/\hbar} \rho_B \hat{b}(0) \hat{b}(\tau) \right) \langle + | \tilde{\rho}_s(t) | + \rangle \right] . \quad (1.77)$$

Employing the cyclic property of the trace operation over the bath, we have

$$\frac{\partial}{\partial t} \langle + | \tilde{\rho}_s(t) | + \rangle \quad (1.78)$$
$$= \frac{g^2}{\hbar^2} \langle - | \tilde{\rho}_s(t) | - \rangle \int_0^\infty d\tau \left[e^{-2iH\tau/\hbar} \langle \hat{b}(0) \hat{b}(\tau) \rangle + e^{2iH\tau/\hbar} \langle \hat{b}(\tau) \hat{b}(0) \rangle \right]$$
$$- \frac{g^2}{\hbar^2} \langle + | \tilde{\rho}_s(t) | + \rangle \int_0^\infty d\tau \left[e^{-2iH\tau/\hbar} \langle \hat{b}(\tau) \hat{b}(0) \rangle + e^{2iH\tau/\hbar} \langle \hat{b}(0) \hat{b}(\tau) \rangle \right] ,$$

where we have introduced bath correlation functions as, e.g.,

$$\langle \hat{b}(\tau) \hat{b}(0) \rangle = \mathrm{Tr}_B \left[\rho_B \hat{b}(\tau) \hat{b}(0) \right] . \quad (1.79)$$

We may rewrite (1.78) as

$$\frac{\partial}{\partial t} \langle + | \tilde{\rho}_s(t) | + \rangle = W(-1|1) \langle - | \tilde{\rho}_s(t) | - \rangle - W(1|-1) \langle + | \tilde{\rho}_s(t) | + \rangle , \quad (1.80)$$

where the *transition probability* is

$$W(-1|1) = \frac{g^2}{\hbar^2} \int_0^\infty d\tau \left[e^{-2iH\tau/\hbar} \langle \hat{b}(0) \hat{b}(\tau) \rangle + e^{2iH\tau/\hbar} \langle \hat{b}(\tau) \hat{b}(0) \rangle \right] , \quad (1.81)$$

and $W(1|-1)$ is obtained from (1.81) by replacing H with $-H$ in the integrand. We can similarly show that

$$\frac{\partial}{\partial t} \langle - | \tilde{\rho}_s(t) | - \rangle = W(1|-1) \langle + | \tilde{\rho}_s(t) | + \rangle - W(-1|1) \langle - | \tilde{\rho}_s(t) | - \rangle . \quad (1.82)$$

Thus the sum of the RHS of (1.80) and (1.82) is identically zero, signifying that the total probability is conserved.

Our next task is to express the RHS of (1.81) as a symmetric integral over the range $[-\infty, \infty]$. We note that

$$\langle \hat{b}(0) \hat{b}(\tau) \rangle = \mathrm{Tr}_B \left[\rho_B \hat{b}(0) e^{i\mathcal{H}_B \tau/\hbar} \hat{b}(0) e^{-i\mathcal{H}_B \tau/\hbar} \right]$$
$$= \langle \hat{b}(-\tau) \hat{b}(0) \rangle , \quad (1.83)$$

where we use the cyclic property of the trace, and the fact that $\mathcal{H}_B$ commutes with ρ_B. Hence,

$$W(-1|1) = \frac{g^2}{\hbar^2} \int_{-\infty}^{\infty} d\tau e^{2iH\tau/\hbar} \langle \hat{b}(\tau)\hat{b}(0) \rangle \,. \tag{1.84}$$

Similarly,

$$W(1|-1) = \frac{g^2}{\hbar^2} \int_{-\infty}^{\infty} d\tau \; e^{-2iH\tau/\hbar} \langle \hat{b}(\tau)\hat{b}(0) \rangle \,. \tag{1.85}$$

Furthermore, the quantities $W(-1|1)$ and $W(1|-1)$ are related by a detailed-balance factor. In order to establish this, we rewrite (1.84) as (cf. (1.A.13))

$$W(-1|1) = \frac{g^2}{\hbar^2} \int_{-\infty}^{\infty} d\tau e^{2iH\tau/\hbar} \frac{1}{Z_B} \mathrm{Tr}_B \left[e^{-\beta \mathcal{H}_B} \hat{b}(\tau)\hat{b}(0) \right] \,. \tag{1.86}$$

By changing the integration variable from $\tau \to -\tau$, and employing (1.83), we have

$$W(-1|1) = \frac{g^2}{\hbar^2} \int_{-\infty}^{\infty} d\tau e^{-2iH\tau/\hbar} \frac{1}{Z_B} \mathrm{Tr}_B \left[e^{-\beta \mathcal{H}_B} \hat{b}(0)\hat{b}(\tau) \right] \,. \tag{1.87}$$

Next, we use the cyclic property of the trace operation to obtain

$$W(-1|1) = \frac{g^2}{\hbar^2} \int_{-\infty}^{\infty} d\tau e^{-2iH\tau/\hbar} \frac{1}{Z_B} \mathrm{Tr}_B \left[\hat{b}(\tau) e^{-\beta \mathcal{H}_B} \hat{b}(0) \right] \,. \tag{1.88}$$

Finally, using (1.76),

$$W(-1|1) = \frac{g^2}{\hbar^2} \int_{-\infty}^{\infty} d\tau e^{-2iH\tau/\hbar} \langle \hat{b}(\tau - i\hbar\beta)\hat{b}(0) \rangle \,, \tag{1.89}$$

which, upon a change of integration variable, yields

$$W(-1|1) = e^{2\beta H} W(1|-1) \,. \tag{1.90}$$

Equation (1.90) is now the detailed-balance condition, and can be rewritten as

$$p^{\mathrm{eq}}(-1) W(-1|1) = p^{\mathrm{eq}}(1) W(1|-1) \,, \tag{1.91}$$

where $p^{\mathrm{eq}}(\pm 1)$ denote the equilibrium probabilities (cf. (1.60)):

$$p^{\mathrm{eq}}(\pm 1) = \frac{e^{\pm \beta H}}{2\cosh(\beta H)} \,. \tag{1.92}$$

Having fixed the detailed-balance condition, we now make the Markovian assumption that the bath is a memory-less entity, i.e., fluctuations in the bath have very short life-times. In that case, the quantity

$$W(-1|1) + W(1|-1) = \frac{g^2}{\hbar^2} \int\limits_{-\infty}^{\infty} d\tau \left(e^{2iH\tau/\hbar} + e^{-2iH\tau/\hbar} \right) \langle \hat{b}(\tau)\hat{b}(0) \rangle \tag{1.93}$$

can be further simplified as

$$W(-1|1) + W(1\ -1) \simeq \frac{2g^2}{\hbar^2} \int\limits_{-\infty}^{\infty} d\tau \langle \hat{b}(\tau)\hat{b}(0) \rangle \,. \tag{1.94}$$

To demonstrate this explicitly, we assume that the bath variables have an exponential correlation function:

$$\langle \hat{b}(\tau)\hat{b}(0) \rangle = \langle \hat{b}(0)^2 \rangle e^{-|\tau|/\tau_c} \,, \tag{1.95}$$

where the time-scale τ_c satisfies

$$\tau_c \ll \frac{\hbar}{2H} \,. \tag{1.96}$$

This leads directly to (1.94), and furthermore,

$$W(-1|1) + W(1|-1) \simeq \frac{4g^2 \langle \hat{b}(0)^2 \rangle \tau_c}{\hbar^2} \,. \tag{1.97}$$

We now identify the characteristic relaxation rate:

$$\lambda = \frac{4g^2 \langle \hat{b}(0)^2 \rangle \tau_c}{\hbar^2} \,. \tag{1.98}$$

Equations (1.91) and (1.97) then yield (cf. (1.61))

$$W(-1|1) = \lambda p^{\mathrm{eq}}(1) \,, \tag{1.99}$$

$$W(1|-1) = \lambda p^{\mathrm{eq}}(-1) \,. \tag{1.100}$$

The important point here is that λ is independent of the magnetic field H. Thus, the quantities $W(-1|1)$ and $W(1|-1)$, which are interpreted as transition probabilities, depend on the subsystem energy (characterized by H in the present case) only through the equilibrium probabilities $p^{\mathrm{eq}}(\pm 1)$.

Substituting for $W(-1|1)$ and $W(1|-1)$ from (1.99)–(1.100) into (1.80) and (1.82), we finally have

$$\frac{\partial}{\partial t}\langle+|\tilde{\rho}_s(t)|+\rangle = -\frac{\partial}{\partial t}\langle-|\tilde{\rho}_s(t)|-\rangle$$
$$= \lambda\left[p^{\mathrm{eq}}(1)\langle-|\tilde{\rho}_s(t)|-\rangle - p^{\mathrm{eq}}(-1)\langle+|\tilde{\rho}_s(t)|+\rangle\right] . \quad (1.101)$$

Since the diagonal matrix elements of the density operator have the interpretation of classical probabilities, (1.101) is analogous to a classical rate equation [26]:

$$\frac{d}{dt}\mathcal{P}(1,t) = -\frac{d}{dt}\mathcal{P}(-1,t)$$
$$= W(-1|1)\mathcal{P}(-1,t) - W(1|-1)\mathcal{P}(1,t) , \quad (1.102)$$

where

$$\mathcal{P}(\pm 1,t) = \langle\pm|\tilde{\rho}_s(t)|\pm\rangle . \quad (1.103)$$

Equation (1.102) is identical to (1.58), as expected.

On the other hand, the off-diagonal elements of the density operator arise only in the quantum-mechanical context. Using the approximations described above, the off-diagonal elements can be shown to satisfy

$$\frac{\partial}{\partial t}\langle+|\tilde{\rho}_s(t)|-\rangle = -\frac{\partial}{\partial t}\langle-|\tilde{\rho}_s(t)|+\rangle$$
$$= \frac{\lambda}{2}\left[\langle-|\tilde{\rho}_s(t)|+\rangle - \langle+|\tilde{\rho}_s(t)|-\rangle\right] . \quad (1.104)$$

Our knowledge of the four elements of the density operator can now be used to calculate the magnetization, *including* the transverse components, which was not possible within the stochastic approach of Sect. 1.5. The magnetization is defined as

$$m_\mu(t) = \mathrm{Tr}[\rho(t)\sigma_\mu] = \mathrm{Tr}_s[\rho_s(t)\sigma_\mu] , \quad \mu = x, y, z , \quad (1.105)$$

where $\rho_s(t)$ is the reduced density operator defined in (1.A.9). As far as the longitudinal component is concerned, $\rho_s(t)$ in (1.105) can be replaced by $\tilde{\rho}_s(t)$ (cf. (1.A.11)). Therefore,

$$\frac{d}{dt}m_z(t) = \frac{\partial}{\partial t}\left[\langle+|\tilde{\rho}_s(t)|+\rangle - \langle-|\tilde{\rho}_s(t)|-\rangle\right] , \quad (1.106)$$

where the RHS is given by (1.101). Substituting for $p^{\mathrm{eq}}(\pm 1)$ from (1.92), we obtain the rate equation given earlier in (1.64). On the other hand, the transverse components can be obtained by rewriting (1.105) as

$$m_\mu(t) = \mathrm{Tr}_s\left[\tilde{\rho}_s(t)e^{i\mathcal{H}_s t/\hbar}\sigma_\mu e^{-i\mathcal{H}_s t/\hbar}\right] . \quad (1.107)$$

Using the properties of Pauli matrices [22], we have

$$e^{i\mathcal{H}_s t/\hbar}\sigma_x e^{-i\mathcal{H}_s t/\hbar} = \sigma_x \cos\left(\frac{2Ht}{\hbar}\right) + \sigma_y \sin\left(\frac{2Ht}{\hbar}\right),$$
$$e^{i\mathcal{H}_s t/\hbar}\sigma_y e^{-i\mathcal{H}_s t/\hbar} = \sigma_y \cos\left(\frac{2Ht}{\hbar}\right) - \sigma_x \sin\left(\frac{2Ht}{\hbar}\right). \quad (1.108)$$

Substituting (1.108) into (1.107), we obtain

$$m_x(t) = \cos\left(\frac{2Ht}{\hbar}\right)[\langle+|\tilde{\rho}_s(t)|-\rangle + \langle-|\tilde{\rho}_s(t)|+\rangle]$$
$$+i\sin\left(\frac{2Ht}{\hbar}\right)[\langle+|\tilde{\rho}_s(t)|-\rangle - \langle-|\tilde{\rho}_s(t)|+\rangle], \quad (1.109)$$

$$m_y(t) = i\cos\left(\frac{2Ht}{\hbar}\right)[\langle+|\tilde{\rho}_s(t)|-\rangle - \langle-|\tilde{\rho}_s(t)|+\rangle]$$
$$-\sin\left(\frac{2Ht}{\hbar}\right)[\langle+|\tilde{\rho}_s(t)|-\rangle + \langle-|\tilde{\rho}_s(t)|+\rangle]. \quad (1.110)$$

Now, from (1.104), it is clear that

$$\langle+|\tilde{\rho}_s(t)|-\rangle + \langle-|\tilde{\rho}_s(t)|+\rangle = \text{const.}. \quad (1.111)$$

In general, we can start from a pure state which satisfies

$$\langle+|\tilde{\rho}_s(0)|-\rangle = \langle-|\tilde{\rho}_s(0)|+\rangle = 0, \quad (1.112)$$

so the constant in (1.111) can be taken to be zero. Thus

$$m_x(t) = i\sin\left(\frac{2Ht}{\hbar}\right)[\langle+|\tilde{\rho}_s(t)|-\rangle - \langle-|\tilde{\rho}_s(t)|+\rangle], \quad (1.113)$$

$$m_y(t) = i\cos\left(\frac{2Ht}{\hbar}\right)[\langle+|\tilde{\rho}_s(t)|-\rangle - \langle-|\tilde{\rho}_s(t)|+\rangle]. \quad (1.114)$$

Taking the time-derivatives of (1.113) and (1.114), and using (1.104), we obtain

$$\frac{d}{dt}m_x(t) = \frac{2H}{\hbar}m_y(t) - \lambda m_x(t), \quad (1.115)$$

$$\frac{d}{dt}m_y(t) = -\frac{2H}{\hbar}m_x(t) - \lambda m_y(t). \quad (1.116)$$

Note that, while the longitudinal component has *relaxational* dynamics only, the transverse components also exhibit *oscillatory* dynamics on a frequency scale $2H/\hbar$.

1.A Appendix: Quantum Master Equation Through a System-plus-Bath Approach

1.A.1 Time-Domain (Cumulant Expansion)

The discussion here follows the work of Kubo [29–31] and van Kampen [32]. Recall that the total Hamiltonian is

$$\mathcal{H} = \mathcal{H}_s + V + \mathcal{H}_B \,, \tag{1.A.1}$$

where $\mathcal{H}_s$ is the subsystem Hamiltonian, V is the interaction, and $\mathcal{H}_B$ is the heat-bath Hamiltonian. The Liouville–von Neumann equation for the overall density operator is as follows [1]:

$$i\hbar \frac{\partial}{\partial t}\rho(t) = [\mathcal{H}, \rho(t)] \,. \tag{1.A.2}$$

We introduce the *interaction-picture* operators [22]

$$\rho_I(t) = \exp\left[\frac{i}{\hbar}\left(\mathcal{H}_s + \mathcal{H}_B\right)t\right]\rho(t)\exp\left[-\frac{i}{\hbar}\left(\mathcal{H}_s + \mathcal{H}_B\right)t\right] , \tag{1.A.3}$$

$$V_I(t) = \exp\left[\frac{i}{\hbar}\left(\mathcal{H}_s + \mathcal{H}_B\right)t\right] V \exp\left[-\frac{i}{\hbar}\left(\mathcal{H}_s + \mathcal{H}_B\right)t\right] , \tag{1.A.4}$$

and (1.A.2) yields

$$i\hbar \frac{\partial}{\partial t}\rho_I(t) = [V_I(t), \rho_I(t)] \,. \tag{1.A.5}$$

In terms of the Liouville-operator notation [1], (1.A.5) can be written as

$$\frac{\partial}{\partial t}\rho_I(t) = -iL_I(t)\rho_I(t) \,, \tag{1.A.6}$$

with the formal solution

$$\rho_I(t) = \exp_T\left(-i\int_0^t L_I(t')dt'\right)\rho(0) \,. \tag{1.A.7}$$

In (1.A.7), $\exp_T$ denotes a time-ordered series where the operators are placed from left to right as their time-arguments decrease. Combining this with (1.A.3), we obtain

$$\rho(t) = \exp\left[-i(L_s + L_B)t\right]\exp_T\left(-i\int_0^t L_I(t')dt'\right)\rho(0) \,, \tag{1.A.8}$$

where L_s and L_B are the Liouville operators associated with $\mathcal{H}_s$ and $\mathcal{H}_B$, respectively.

We next introduce the reduced density operator for the subsystem alone:

$$\rho_s(t) = \mathrm{Tr}_B[\rho(t)]\,, \tag{1.A.9}$$

where "Tr_B" denotes the trace operation over the bath variables. From (1.A.8),

$$\rho_s(t) = \exp(-iL_s t)\mathrm{Tr}_B\left[\exp_T\left(-i\int_0^t L_I(t')dt'\right)\rho(0)\right]. \tag{1.A.10}$$

The term $\exp(-iL_B t)$ drops out because of the cyclic property of the trace operation over the Hilbert space of the bath. It is convenient to introduce the operator

$$\tilde{\rho}_s(t) = \exp\left(\frac{i}{\hbar}\mathcal{H}_s t\right)\rho_s(t)\exp\left(-\frac{i}{\hbar}\mathcal{H}_s t\right). \tag{1.A.11}$$

From (1.A.10), we then have

$$\tilde{\rho}_s(t) = \mathrm{Tr}_B\left[\exp_T\left(-i\int_0^t L_I(t')dt'\right)\rho(0)\right]. \tag{1.A.12}$$

Let us next examine the simplification that results if we consider the case of a *factorized* initial condition, viz., we assume that the subsystem is decoupled from the heat bath for $t < 0$. The perturbation V, which couples the subsystem to the heat bath, is switched on at $t = 0$. We are interested in the subsequent time-evolution of $\rho_s(t)$. Furthermore, during this evolution, the heat bath is assumed to remain in thermal equilibrium described by the density operator:

$$\rho_B = \frac{1}{Z_B}\exp\left(-\beta\mathcal{H}_B\right), \tag{1.A.13}$$

where $Z_B = \mathrm{Tr}_B\left(e^{-\beta\mathcal{H}_B}\right)$ is the partition function for the heat bath. Thus

$$\rho(0) = \rho_B \otimes \rho_s(0)\,, \tag{1.A.14}$$

which then yields

$$\tilde{\rho}_s(t) = \left[\exp_T\left(-i\int_0^t L_I(t')dt'\right)\right]_{\mathrm{av}}\rho_s(0)\,. \tag{1.A.15}$$

In terms of explicit matrix elements,

$$\left[\exp_T\left(-i\int_0^t L_I(t')dt'\right)\right]_{\text{av}} = \sum_{n_b,n_b'}\left(n_b n_b\middle|\exp_T\left(-i\int_0^t L_I(t')dt'\right)\middle|n_b' n_b'\right) \times\langle n_b'|\rho_B|n_b'\rangle\,, \tag{1.A.16}$$

where the states $|n_b\rangle, |n_b'\rangle$, refer to the eigenstates of the bath Hamiltonian $\mathcal{H}_B$.

Applying the cumulant-expansion theorem, and keeping terms upto second-order cumulant only, we obtain

$$\tilde{\rho}_s(t) = \exp_T\Bigg(-i\int_0^t dt'\langle L_I(t')\rangle_{\text{av}} - \int_0^t dt'\int_0^{t'} dt''\,[\langle L_I(t')L_I(t'')\rangle_{\text{av}} - \langle L_I(t')\rangle_{\text{av}}\langle L_I(t'')\rangle_{\text{av}}]\Bigg)\rho_s(0)\,. \tag{1.A.17}$$

Here the angular brackets refer to an averaging over the bath degrees of freedom, as in (1.A.16). Furthermore, the subscript T implies that the time-ordering has to be maintained in writing the second-order cumulant. Thus

$$\langle L_I(t')L_I(t'')\rangle_{\text{av}} = \sum_{n_b n_b'}(n_b n_b|L_I(t')L_I(t'')|n_b' n_b')\langle n_b'|\rho_B|n_b'\rangle\,. \tag{1.A.18}$$

The above approximation is referred to as the *Born approximation* in the context of quantum optics [33]. There are two forms of the master equation usually employed in the literature, both being based on the Born approximation used in (1.A.17). We discuss these two forms below.

Form I

We assume $\langle V_I\rangle_{\text{av}} = 0$, which eliminates the first term in the exponent of (1.A.17). In addition, we take the system to be invariant under time translation (*stationarity*) and obtain from (1.A.17)

$$\tilde{\rho}_s(t) = \exp\left(-\int_0^t d\tau(t-\tau)\langle L_I(\tau)L_I(0)\rangle_{\text{av}}\right)\rho_s(0)\,. \tag{1.A.19}$$

The correlation function in (1.A.19) contains the effect of heat-bath induced fluctuations. We assume the latter to be operative on time-scales much shorter than the time-scales of interest for the subsystem dynamics (the *Markovian limit*). In that case, we can extend the upper limit of the integral in (1.A.19) to infinity. Thus

$$\tilde{\rho}_s(t) \simeq \exp\left(-t\int_0^\infty d\tau\langle L_I(\tau)L_I(0)\rangle_{\text{av}}\right)\rho_s(0) \equiv \exp(-\hat{R}t)\rho_s(0)\,, \tag{1.A.20}$$

where $\hat{R}$ is the *relaxation matrix*:

$$\hat{R} = \int_0^\infty d\tau \langle L_I(\tau) L_I(0) \rangle_{\mathrm{av}} \,. \tag{1.A.21}$$

Equation (1.A.21) is the solution of the master equation

$$\frac{\partial}{\partial t}\tilde{\rho}_s(t) = -\hat{R}\tilde{\rho}_s(t) \,, \tag{1.A.22}$$

which is formally similar to the classical master equation for a stationary Markov process [1]:

$$\frac{\partial}{\partial t}\mathcal{P}(t) = -\hat{W}\mathcal{P}(t) \,. \tag{1.A.23}$$

In terms of the original reduced density operator for the subsystem (cf. (1.A.10)), the master equation (1.A.22) has the form

$$\frac{\partial}{\partial t}\rho_s(t) = -\frac{i}{\hbar}[\mathcal{H}_s, \rho_s(t)] - e^{-i\mathcal{H}_s t/\hbar}\hat{R}e^{i\mathcal{H}_s t/\hbar}\rho_s(t) \,, \tag{1.A.24}$$

which manifestly contains (through the first term on the RHS) the systematic dynamics generated by the subsystem Hamiltonian. Furthermore, using the definitions of $\hat{R}$ and the angular brackets, we obtain

$$\begin{aligned}\hat{R}\tilde{\rho}_s(t) &= \frac{1}{\hbar^2}\int_0^\infty d\tau \mathrm{Tr}_B \left([V_I(\tau), [V_I(0), \rho_B\tilde{\rho}_s(t)]]\right) \\ &= \frac{1}{\hbar^2}\int_0^\infty d\tau \mathrm{Tr}_B \big[V_I(\tau)V_I(0)\rho_B\tilde{\rho}_s(t) + \rho_B\tilde{\rho}_s(t)V_I(0)V_I(\tau) \\ &\quad -V_I(\tau)\rho_B\tilde{\rho}_s(t)V_I(0) - V_I(0)\rho_B\tilde{\rho}_s(t)V_I(\tau)\big] \,. \end{aligned} \tag{1.A.25}$$

Form II

We take the time derivative of (1.A.17) and obtain

$$\begin{aligned}\frac{\partial}{\partial t}\tilde{\rho}_s(t) = -\Big[& i\langle L_I(t)\rangle_{\mathrm{av}} \\ & + \int_0^t d\tau \left(\langle L_I(t)L_I(\tau)\rangle_{\mathrm{av}} - \langle L_I(t)\rangle_{\mathrm{av}}\langle L_I(\tau)\rangle_{\mathrm{av}}\right)\Big]\tilde{\rho}_s(t) \,. \end{aligned} \tag{1.A.26}$$

Each term on the RHS of (1.A.26) is calculated separately below. Thus,

$$\langle L_I(t)\rangle_{\mathrm{av}}\tilde{\rho}_s(t) = \frac{1}{\hbar}\mathrm{Tr}_B(\rho_B[V_I(t), \tilde{\rho}_s(t)]) \,, \tag{1.A.27}$$

using the definition of Liouville operators and the cyclic property of the trace operation.

Now, quite generally, and in almost all problems of physical interest, the interaction Hamiltonian is the tensor product of two operators, one operating on the Hilbert space of the subsystem of interest, while the other operates on the Hilbert space of the heat bath. Thus,

$$V = \hbar \sum_{j=1}^{n} S_j B_j \,, \tag{1.A.28}$$

where n is expected to be a finite number, S_j is a subsystem operator, and B_j is a bath operator.

Note that

$$V_I(t) = \hbar \sum_{j=1}^{n} S_j(t) B_j(t) \,, \tag{1.A.29}$$

where (cf. (1.A.4))

$$S_j(t) = \exp\left(\frac{i}{\hbar}\mathcal{H}_s t\right) S_j \exp\left(-\frac{i}{\hbar}\mathcal{H}_s t\right) , \tag{1.A.30}$$

and

$$B_j(t) = \exp\left(\frac{i}{\hbar}\mathcal{H}_B t\right) B_j \exp\left(-\frac{i}{\hbar}\mathcal{H}_B t\right) . \tag{1.A.31}$$

Substituting (1.A.29) in (1.A.27), using the cyclic property of the trace and reverting back to $\rho_s(t)$, we obtain (employing the definition in (1.A.11))

$$\langle L_I(t)\rangle_{\text{av}} \tilde{\rho}_s(t) = \exp\left(\frac{i}{\hbar}\mathcal{H}_s t\right) \left(\sum_{j=1}^{n} \langle B_j\rangle_{\text{av}} [S_j, \rho_s(t)]\right) \exp\left(-\frac{i}{\hbar}\mathcal{H}_s t\right) . \tag{1.A.32}$$

Next, in order to evaluate the second term on the RHS of (1.A.26) (i.e., the first term under the integral), it is convenient to write a particular matrix element involving the eigenstates of $\mathcal{H}_s$, as defined in the following:

$$\mathcal{H}_s|n\rangle = E_s^n|n\rangle \,, \quad 1 = \sum_n |n\rangle\langle n| \,. \tag{1.A.33}$$

Thus

$$\begin{aligned} &\langle n|\langle L_I(t) L_I(\tau)\rangle_{\text{av}} \tilde{\rho}_s(t)|m\rangle \\ &= \sum_{n'm'} (nm|\langle L_I(t) L_I(\tau)\rangle_{\text{av}}|n'm')\langle n'|\tilde{\rho}_s(t)|m'\rangle \\ &= \sum_{n_b n_b'} \sum_{n'm'} (nn_b, mn_b|L_I(t) L_I(\tau)|n'n_b', m'n_b')\langle n_b'|\rho_B|n_b'\rangle\langle n'|\tilde{\rho}_s(t)|m'\rangle \,, \end{aligned} \tag{1.A.34}$$

where we have used the properties of Liouville operators [1]. From the definition of $L_I(t)$, we further obtain

$$\text{LHS} = \frac{1}{\hbar^2} \sum_{\substack{n_b\ n_b' \\ n_b''\ n_b'''}} \sum_{\substack{n'\ m' \\ n_1\ n_2}}$$

$$\Big(\langle n n_b | V_I(t) | n_1 n_b'' \rangle \delta_{m n_2} \delta_{n_b n_b'''} - \langle n_2 n_b''' | V_I(t) | m n_b \rangle \delta_{n n_1} \delta_{n_b'' n_b} \Big)$$

$$\times \Big(\langle n_1 n_b'' | V_I(\tau) | n' n_b' \rangle \delta_{n_2 m'} \delta_{n_b''' n_b'} - \langle m' n_b' | V_I(\tau) | n_2 n_b''' \rangle \delta_{n_1 n'} \delta_{n_b'' n_b'} \Big)$$

$$\times \langle n_b' | \rho_B | n_b' \rangle \langle n' | \tilde{\rho}_s(t) | m' \rangle$$

$$= \sum_{jk} \sum_{n'm'} \Big\{ \langle B_j(t-\tau) B_k(0) \rangle_{\text{av}} \left[\langle n | S_j(t) | m' \rangle \langle m' | S_k(\tau) | n' \rangle \langle n' | \tilde{\rho}_s(t) | m \rangle \right.$$

$$\left. - \langle m' | S_j(t) | m \rangle \langle n | S_k(\tau) | n' \rangle \langle n' | \tilde{\rho}_s(t) | m' \rangle \right]$$

$$+ \langle B_k(0) B_j(t-\tau) \rangle_{\text{av}} \left[\langle n' | S_j(t) | m \rangle \langle m' | S_k(\tau) | n' \rangle \langle n | \tilde{\rho}_s(t) | m' \rangle \right.$$

$$\left. - \langle n | S_j(t) | n' \rangle \langle m' | S_k(\tau) | m \rangle \langle n' | \tilde{\rho}_s(t) | m' \rangle \right] \Big\} . \tag{1.A.35}$$

The expressions inside angular brackets involve correlation functions for bath variables. In operator form

$$\langle L_I(t) L_I(\tau) \rangle_{\text{av}} \tilde{\rho}_s(t)$$

$$= \sum_{jk} \Big\{ \langle B_j(t-\tau) B_k(0) \rangle_{\text{av}} \left[S_j(t) S_k(\tau) \tilde{\rho}_s(t) - S_k(\tau) \tilde{\rho}_s(t) S_j(t) \right]$$

$$- \langle B_k(0) B_j(t-\tau) \rangle_{\text{av}} \left[S_j(t) \tilde{\rho}_s(t) S_k(\tau) S_j(t) \right] \Big\} . \tag{1.A.36}$$

Noting that the above term occurs under an integral in (1.A.26), and that under the Markovian assumption, the bath correlations are short-lived over the time-scales of interest for the system, we may replace the argument t in $\tilde{\rho}_s(t)$ by τ. Thus

$$\int_0^t d\tau \Big(\langle L_I(t) L_I(\tau) \rangle_{\text{av}} - \langle L_I(t) \rangle_{\text{av}} \langle L_I(\tau) \rangle_{\text{av}} \Big) \tilde{\rho}_s(t)$$

$$= \sum_{jk} \Big\{ \langle B_j(t-\tau) B_k(0) \rangle_{\text{av}} \left[S_j(t) S_k(\tau) \tilde{\rho}_s(\tau) - S_k(\tau) \tilde{\rho}_s(\tau) S_j(t) \right]$$

$$- \langle B_k(0) B_j(t-\tau) \rangle_{\text{av}} \left[S_j(t) \tilde{\rho}_s(\tau) S_k(\tau) - \tilde{\rho}_s(\tau) S_k(\tau) S_j(t) \right] \Big\}$$

$$= \int_0^t d\tau e^{i\mathcal{H}_s t/\hbar} \sum_{jk} \Big\{ \langle\langle B_j(t-\tau) B_k(0) \rangle\rangle \left[S_j, e^{-i\mathcal{H}_s(t-\tau)/\hbar} S_k \rho_s(\tau) e^{i\mathcal{H}_s(t-\tau)/\hbar} \right]$$

$$- \langle\langle B_k(0) B_j(t-\tau) \rangle\rangle \left[S_j, e^{-i\mathcal{H}_s(t-\tau)/\hbar} \rho_s(\tau) S_k e^{i\mathcal{H}_s(t-\tau)/\hbar} \right] \Big\} e^{-i\mathcal{H}_s t/\hbar} , \tag{1.A.37}$$

where we have used (1.A.11) and (1.A.30), and also introduced the notation:

$$\langle\langle XY\rangle\rangle = \langle XY\rangle_{\rm av} - \langle X\rangle_{\rm av}\langle Y\rangle_{\rm av}\,. \tag{1.A.38}$$

Substituting (1.A.32) and (1.A.37) in (1.A.26) and recalling (1.A.11), the master equation can be recast in the compact form [33]:

$$\begin{aligned}\frac{\partial}{\partial t}\rho_s(t) = &-\frac{i}{\hbar}\left[\mathcal{H}_s, \rho_s(t)\right] - i\sum_{j=1}^{n}\langle B_j\rangle_{\rm av}[S_j, \rho_s(t)] - \int_0^t d\tau \sum_{j,k=1}^{n} \\ &\Big\{\langle\langle B_j(t-\tau)B_k(0)\rangle\rangle\left[S_j, e^{-i\mathcal{H}_s(t-\tau)/\hbar}S_k\rho_s(\tau)e^{i\mathcal{H}_s(t-\tau)/\hbar}\right] \\ &-\langle\langle B_k(0)B_j(t-\tau)\rangle\rangle\left[S_j, e^{-i\mathcal{H}_s(t-\tau)/\hbar}\rho_s(\tau)S_k e^{i\mathcal{H}_s(t-\tau)/\hbar}\right]\Big\}\,.\end{aligned} \tag{1.A.39}$$

1.A.2 Frequency-Domain (Resolvent Expansion)

Let us next consider an expansion in the frequency-domain [34–37]. From (1.A.2), we have

$$\frac{\partial}{\partial t}\rho(t) = -iL\rho(t)\,, \tag{1.A.40}$$

where L is the Liouville operator associated with the total Hamiltonian $\mathcal{H}$. Taking the Laplace transform of (1.A.40),

$$\tilde{\rho}(s) = (s+iL)^{-1}\rho(t=0)\,. \tag{1.A.41}$$

Correspondingly, the reduced density operator (cf. (1.A.9)) is given by

$$\tilde{\rho_s}(s) = \mathrm{Tr}_B\left[(s+iL)^{-1}\rho_B\right]\rho_s(t=0)\,, \tag{1.A.42}$$

where we have used the factorization approximation in (1.A.14). Then, (1.A.42) can be simplified via some algebraic manipulations [1] as follows:

$$\tilde{\rho_s}(s) = \frac{1}{s+iL_s}\left[1+\left(\tilde{G}(s)\right)_{\rm av}\frac{1}{s+iL_s}\right]\rho_s(t=0)\,, \tag{1.A.43}$$

where

$$\tilde{G}(s) = L_I\sum_{n=0}^{\infty}(-i)^{n+1}\left[\frac{1}{s+i(L_s+L_B)}L_I\right]^n\,, \tag{1.A.44}$$

and

$$\left(\tilde{G}(s)\right)_{\rm av} = \sum_{n_b,n_b'}\left(n_b n_b|\tilde{G}(s)|n_b' n_b'\right)\langle n_b'|\rho_B|n_b'\rangle\,. \tag{1.A.45}$$

Equation (1.A.43) can be further simplified to obtain

$$\tilde{\rho_s}(s) = \left[\frac{1}{s + iL_s + \left(\tilde{G}^c(s) \right)_{\rm av}} \right] \rho_s(t=0) \, , \tag{1.A.46}$$

where

$$\left(\tilde{G}^c(s) \right)_{\rm av} = \left(\tilde{G}(s) \right)_{\rm av} \sum_{n=0}^{\infty} (-1)^{n+1} \left[\frac{1}{s + iL_s} \left(\tilde{G}(s) \right)_{\rm av} \right]^n \, , \tag{1.A.47}$$

with the superscript c referring to *connected* terms in the diagrammatic expansion. The time-domain version of (1.A.46) is as follows:

$$\frac{\partial}{\partial t} \rho_s(t) = -iL_s \rho_s(t) - \int_0^t dt' \left[G^c(t-t') \right]_{\rm av} \rho_s(t') \, , \tag{1.A.48}$$

which is the required master equation.

Furthermore, notice that upto second-order in the interaction Hamiltonian,

$$\tilde{\rho_s}(s) \simeq \left[\frac{1}{s + iL_s + \tilde{\Sigma}(s)} \right] \rho_s(t=0) \, . \tag{1.A.49}$$

Here, $\tilde{\Sigma}(s)$ is the *self-energy*:

$$\tilde{\Sigma}(s) = \left[L_I \frac{1}{s + i(L_s + L_B)} L_I \right]_{\rm av} \, , \tag{1.A.50}$$

where L_I is the Liouville operator corresponding to the interaction Hamiltonian V, $L_I \equiv L_I(t=0)$. If we make an additional Markovian approximation, the self-energy can be approximated as

$$\tilde{\Sigma}(s) \simeq \tilde{\Sigma}(0) = \int_0^{\infty} dt \left\{ L_I \exp\left[-i(L_s + L_B)t \right] L_I \right\}_{\rm av} \, . \tag{1.A.51}$$

2 Kinetic Ising Models

2.1 Introduction

One of the major aims of this book is to describe dissipative phenomena in systems which undergo phase transitions. In this context, perhaps the simplest model of an interacting many-body system is the Ising model [38–41], which was first introduced as a model for magnetic systems. However, with suitable generalizations, it has wide applications to a range of problems in condensed matter physics.

Consider a set of N spins $\{S_i\}$, which are fixed on the sites $\{i\}$ of a lattice. These spins are coupled through a position-dependent exchange interaction J_{ij} $(= J_{ji})$. The Ising model assumes that the interaction is *axially* symmetric, i.e., only the z-components of the spins are relevant. In this chapter, we will focus on the case of spin-1/2 particles. Then, the eigenvalues of the Pauli matrix σ_{zi} corresponding to spin S_i are $m_i = \pm 1$. The spin-1/2 Ising Hamiltonian has the following form:

$$
\begin{aligned}
\mathcal{H} &= -\sum_{i>j=1}^{N} J_{ij}\sigma_{zi}\sigma_{zj} - \sum_{i=1}^{N} H_i\sigma_{zi} \\
&\equiv -\sum_{i>j=1}^{N} J_{ij}\sigma_i\sigma_j - \sum_{i=1}^{N} H_i\sigma_i \,,
\end{aligned}
\tag{2.1}
$$

where we designate $\sigma_{zi} \equiv \sigma_i$ for notational convenience. The first term on the RHS of (2.1) describes the exchange interaction between spins. The negative sign of the exchange term ensures that the energy is lowered (for $J_{ij} > 0$) when the two spins S_i and S_j are in the same state, as in a ferromagnet. Furthermore, we exclude self-interaction terms by setting $J_{ii} = 0$. The second term on the RHS corresponds to a Zeeman interaction with a site-dependent magnetic field along the z-direction. Notice that $\mathcal{H}$ is essentially classical, as it does not involve any non-commuting operators. For a particular system state $\{m_i\}$, the corresponding energy is obtained from (2.1) by replacing $\sigma_i \to m_i$, $i = 1 \to N$.

Although the Hamiltonian in (2.1) is formulated for a magnetic system, it is clear that a similar description applies for any interacting two-state system, as the two states can be mapped onto $m = +1$ or -1. A well-known example in statistical physics is the lattice-gas model of the liquid-vapor phase transition [38–41]. In this model, each lattice site is occupied by either an atom (A) or a vacancy (V). The lattice gas is described by occupation variables, with values $n_i = 0$ or 1 when the site i is occupied by V or A. An equivalent description arises if we introduce spin variables $m_i = 2n_i - 1$, which assume values $m = \pm 1$. The lattice-gas Hamiltonian, assuming pairwise interactions with strength $-E_{ij}$ between atoms at sites i and j, is

$$\begin{aligned}\mathcal{H} &= - \sum_{i>j=1}^{N} E_{ij} n_i n_j \\ &= - \sum_{i>j=1}^{N} \frac{E_{ij}}{4} \left(1 + m_i\right)\left(1 + m_j\right) .\end{aligned} \tag{2.2}$$

The "spin"-Hamiltonian in (2.2) is analogous to that in (2.1), apart from an irrelevant constant term. The corresponding parameters are identified as

$$\begin{aligned}J_{ij} &= \frac{E_{ij}}{4} , \\ H_i &= \sum_{j=1}^{N} \frac{E_{ij} + E_{ji}}{8} = \sum_{j=1}^{N} \frac{E_{ij}}{4} ,\end{aligned} \tag{2.3}$$

as $E_{ij} = E_{ji}$ for physical problems of interest.

As a third application of the Ising model, we briefly discuss the binary (AB) alloy or mixture [38–41]. This is a simple generalization of the lattice-gas problem. Again, we can describe this system in terms of occupation-number variables $n_i^\alpha = 1$ or 0, depending on whether or not a site i is occupied by species α (A or B). Clearly, $n_i^A + n_i^B = 1$ for all sites. As before, a more convenient description is obtained in terms of spin variables $m_i = 2n_i^A - 1 = 1 - 2n_i^B$. We associate an energy $-E_{ij}^{\alpha\beta}$ between species α and β, located at sites i and j, respectively. The corresponding Hamiltonian is

$$\begin{aligned}\mathcal{H} &= - \sum_{i>j=1}^{N} \left[E_{ij}^{AA} n_i^A n_j^A + E_{ij}^{BB} n_i^B n_j^B + E_{ij}^{AB} \left(n_i^A n_j^B + n_i^B n_j^A\right)\right] \\ &= - \sum_{i>j=1}^{N} \frac{E_{ij}^{AA} + E_{ij}^{BB} - 2E_{ij}^{AB}}{4} m_i m_j - \sum_{i=1}^{N} \left(\sum_{j=1}^{N} \frac{E_{ij}^{AA} - E_{ij}^{BB}}{4}\right) m_i \\ &\quad - \sum_{i>j=1}^{N} \frac{E_{ij}^{AA} + E_{ij}^{BB} + 2E_{ij}^{AB}}{4} .\end{aligned} \tag{2.4}$$

Again, the Hamiltonian in (2.4) is analogous to that in (2.1) if we identify

$$J_{ij} = \frac{E_{ij}^{AA} + E_{ij}^{BB} - 2E_{ij}^{AB}}{4} ,$$

$$H_i = \sum_{j=1}^{N} \left(\frac{E_{ij}^{AA} - E_{ij}^{BB}}{4} \right) . \tag{2.5}$$

In many physical applications, one is interested in the nearest-neighbor Ising Hamiltonian, where exchange interactions are restricted to nearest neighbors on the lattice. Furthermore, one often considers the uniform (or pure) case where the exchange interactions and magnetic field are site-independent, i.e., $J_{ij} = J$ and $H_i = H$ in (2.1). Then, the appropriate Hamiltonian is

$$\mathcal{H} = -J \sum_{\langle ij \rangle} \sigma_i \sigma_j - H \sum_{i=1}^{N} \sigma_i , \tag{2.6}$$

where the subscript $\langle ij \rangle$ denotes nearest-neighbor pairs. The Ising model and its variants are not restricted to two-state systems, and can be easily generalized to the case of multiple-state systems. Thus, three-state systems can be mapped onto a spin-1 Hamiltonian; four-state systems onto a spin-3/2 Hamiltonian, etc. In general, higher-spin models have a larger number of possible interaction terms (and parameters) in the Hamiltonian.

The above discussion has clarified the utility of Ising-like models in a wide range of problems in condensed matter physics. We next consider the issue of *kinetics*. In order to keep the analysis simple, we again restrict our discussion to the spin-1/2 model described by (2.1). The Ising model variables do not have intrinsic dynamics, as is seen by constructing the relevant Poisson bracket. In order to associate kinetics with the Ising model, we generalize the stochastic formulation developed in the context of single-spin kinetics in Sect. 1.5. The resultant stochastic Ising model is referred to as the spin-flip or Glauber model [42–44]. In the Glauber model, the spin system is assumed to be in contact with a heat bath which drives thermal fluctuations in the system. The effect of these fluctuations is to flip a randomly-chosen spin at a random instant of time, i.e., $m_i \to -m_i$. Of course, the probability of a jump depends on the configuration of all other spins, in general. As in the single-spin case of Sect. 1.5, the heat bath can be interpreted as consisting of phonons which induce spin flips via a spin-lattice coupling [28]. However, the detailed structure of the heat bath is not relevant for the present discussion.

The Glauber model is appropriate for describing dissipative phenomena associated with (say) the paramagnetic $\to$ ferromagnetic transition. On the other hand, if the Ising model describes a lattice gas or a binary (AB) alloy, the appropriate microscopic kinetics involves the diffusion of atoms, e.g., atomic jumps to vacant sites in the lattice gas, or $A \leftrightarrow B$ interchanges in the binary alloy. Thus, the heat bath causes spin-exchanges rather than spin-flips, i.e., m_i jumps from $+1 \to -1$ while $m_j (j \neq i)$ *simultaneously* jumps from $-1 \to +1$. This process mimics phonon-induced atomic jumps. The resultant model is referred to as the spin-exchange or Kawasaki model [45,46].

It should be emphasized that transition probabilities in the Glauber and Kawasaki models satisfy the detailed-balance condition in (1.59) [43]. Thus, although the two models describe different time-dependent behavior, the equilibrium state is *unique*. As $t \to \infty$, we recover properties calculable from the *equilibrium* statistical mechanics of the Ising model in an appropriate ensemble. To obtain approximate values of these equilibrium quantities, it is useful to employ the mean-field approximation (MFA) [47]. Since the MFA will have an important role in our discussion of dissipative phenomena, we outline a few relevant results in Sect. 2.2 below.

2.2 Equilibrium Properties in the Mean-Field Approximation

The equilibrium properties of the Ising model in a uniform magnetic field H (cf. (2.1)) are described in the MFA by the Bragg–Williams (BW) form of the Gibbs free energy (per site) [40,41]:

$$G(T, H, \psi) = -\frac{1}{2} J(0)\psi^2 - H\psi + k_B T \left[\frac{1+\psi}{2} \ln\left(\frac{1+\psi}{2}\right) + \frac{1-\psi}{2} \ln\left(\frac{1-\psi}{2}\right) \right], \tag{2.7}$$

where

$$J(0) = \sum_{j=1}^{N} J_{ij} \,. \tag{2.8}$$

The RHS of (2.7) is a variational function of the magnetization $\psi = \langle \sigma_i \rangle$, which is assumed to be spatially *uniform*. The first two terms constitute the enthalpy, while the third term is $-TS$, where S is the site-occupation entropy. If we Taylor-expand the entropy term in (2.7), the Gibbs free energy assumes the customary ψ^4-form:

$$G(T, H, \psi) = \frac{1}{2} \left[k_B T - J(0) \right] \psi^2 - H\psi + \frac{k_B T}{12} \psi^4 + O(\psi^6) - k_B T \ln 2 \,. \tag{2.9}$$

The order parameter ψ in (2.7) or (2.9) can describe both ferromagnetic and antiferromagnetic order, with the sign of $J(0)$ being negative in the latter case. Furthermore, in the antiferromagnetic case, ψ refers to the sub-lattice or staggered magnetization in the two-sub-lattice picture [40,41].

The equilibrium value of ψ at fixed (T, H) is obtained from (2.7) by minimizing the variational form of the Gibbs free energy, i.e.,

$$\left. \frac{\partial G}{\partial \psi} \right|_{\psi=\psi_0} = 0 \,. \tag{2.10}$$

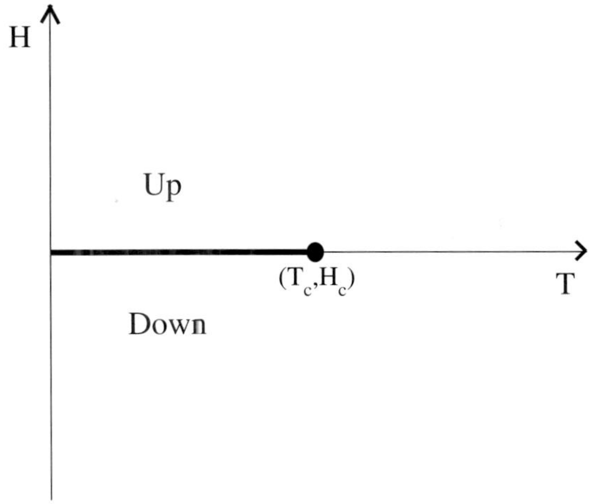

Fig. 2.1. Phase diagram of a ferromagnet in the (T, H)-plane. The point $(T_c, H_c = 0)$ is a second-order critical point, and the line $(T < T_c, H_c = 0)$ is a line of first-order phase transitions

This yields the well-known transcendental equation [47]:

$$\psi_0 = \tanh\left[\beta J(0)\left(\psi_0 + \frac{H}{J(0)}\right)\right] . \tag{2.11}$$

For $H = 0$, we identify the mean-field (MF) critical temperature

$$T_c = \frac{J(0)}{k_B} , \tag{2.12}$$

such that $\psi_0 = 0$ for $T > T_c$ and $\psi_0 \neq 0$ for $T < T_c$. The relevant phase diagram in the (T, H)-plane is shown in Fig. 2.1. For our purposes, T_c will be used as a parameter defined through the microscopic interactions J_{ij}. Furthermore, all approximations (Bragg–Williams, Weiss, etc.) which lead to an order-parameter equation as in (2.11) will be referred to as MF approximations.

Next, let us consider the case of the binary mixture (or lattice gas) with N_A $(= c_A N)$ atoms of species A, and N_B $(= c_B N)$ atoms of species B $(N = N_A + N_B)$. The appropriate order parameter in this case is the local density difference, $\psi = \langle n_i^A \rangle - \langle n_i^B \rangle$. The above analysis will have to be modified somewhat because the appropriate ensemble is characterized by a fixed *magnetization* rather than a fixed *magnetic field*. The relevant free energy to be minimized is the Helmholtz potential $F(T, \psi) = G(T, H, \psi) + H\psi$. At temperatures $T > T_c$, the system is in a *homogeneous* (or *one-phase*) state with order parameter $\psi_h = c_A - c_B$. At temperatures $T < T_c$, the equilibrium state consists of A-rich and B-rich regions characterized by order parameter ψ_0 (> 0) and $-\psi_0$ (< 0), respectively. Here, ψ_0 is the solution

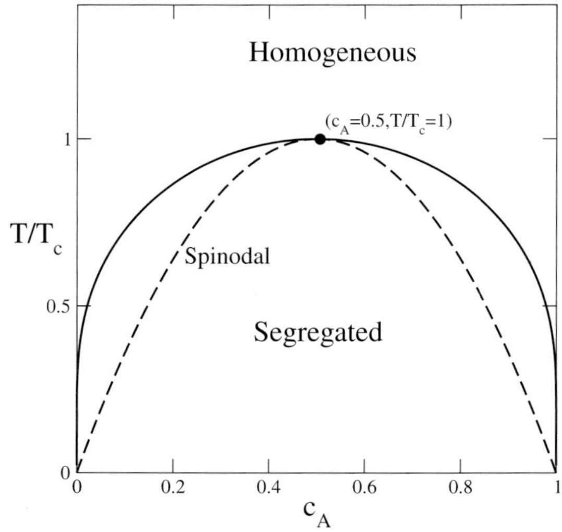

Fig. 2.2. Phase diagram of a binary mixture in the $(c_A, T/T_c)$-plane. The point $(c_A = 0.5, T/T_c = 1)$ is a second-order critical point. Below the coexistence curve (denoted by a solid line), the system segregates into A- and B-rich regions. Their relative proportions are determined by the lever rule. In the region between the spinodal (denoted by a dashed line) and the coexistence curve, an initially homogeneous system is locally stable. This will be discussed further in Chap. 3

of (2.11) with $H = 0$. Of course, the fractions of A-rich (say, x) and B-rich $(1-x)$ regions must satisfy the lever rule, $\psi_h = x\psi_0 + (1-x)(-\psi_0)$. The corresponding phase diagram for a binary mixture in the $(c_A, T/T_c)$-plane is shown in Fig. 2.2.

The above discussion is specific to a binary mixture described by a symmetric free energy $F(T, \psi) = F(T, -\psi)$. For more general functional forms of the free energy, a double-tangent construction [40,41] is employed to determine the equilibrium composition from the following equations:

$$F'(\psi_1) = F'(\psi_2),$$
$$F'(\psi_1) = \frac{F(\psi_1) - F(\psi_2)}{\psi_1 - \psi_2}. \tag{2.13}$$

With this background on the Ising model and its equilibrium properties, the organization of this chapter is as follows. In Sect. 2.3, we introduce the master equation for the Glauber model, and use it to obtain evolution equations for the magnetization and the spin-spin correlation function. In Sect. 2.4, we obtain the corresponding equations for the Kawasaki model. In Sect. 2.5, we formulate a rate theory for atomic ordering, especially in the context of structural phase transitions. In particular, we discuss the role of a *kinetic potential* in driving the ordering process. Finally, Sect. 2.6 discusses coarse-grained models for order-parameter kinetics, and their application to phase ordering dynamics.

2.3 The Spin-Flip Glauber Model

2.3.1 Exact Dynamical Equations

In order to describe dissipative phenomena using the Ising model, the spin system is coupled to an external heat bath. As mentioned earlier, the heat bath induces thermal fluctuations in the system, which approaches an equilibrium state as $t \to \infty$. This state is governed by the Hamiltonian in (2.1) and the temperature T of the heat bath. The analytical framework has already been established in the context of a single spin in Sect. 1.5. The formulation in this section is obtained by direct analogy.

Notice that all the terms involving the spin σ_i in the Hamiltonian in (2.1) can be written as

$$\mathcal{H}_i = -h_i \sigma_i \,, \tag{2.14}$$

where the *local field* h_i at the ith site is defined as

$$h_i = \sum_{j=1}^{N} J_{ij} \sigma_j + H_i \,. \tag{2.15}$$

In the Glauber model, the heat bath induces fluctuations in the system in the form of single-spin-flip processes [42]. We introduce a conditional probability $\mathcal{P}(\{n_i\}, 0|\{m_i\}, t)$, which is the probability that the ith spin is in state m_i ($i = 1 \to N$) at time t, given that it was in state n_i ($i = 1 \to N$) at time $t = 0$. In analogy with (1.56), we now have the master equation:

$$\begin{aligned} \frac{d}{dt}\mathcal{P}(\{m_i\}, t) = &- \sum_{j=1}^{N} W(m_1, ...m_j, ...m_N | m_1, ... - m_j, ...m_N) \mathcal{P}(\{m_i\}, t) \\ &+ \sum_{j=1}^{N} W(m_1, .. - m_j, ...m_N | m_1, ...m_j, ...m_N) \mathcal{P}(\{m_i\}', t) \,, \end{aligned} \tag{2.16}$$

where we suppress the argument $(\{n_i\}, 0|$, for compactness. In the second term on the RHS of (2.16), the configuration $\{m_i\}'$ denotes the state $\{m_i\}$ with $m_j \to -m_j$. Equation (2.16) assumes that the underlying stochastic process is Markovian. The essential physical input is provided by the modeling of the *transition matrix* W.

At a microscopic level, such a system can be described (as in Sect. 1.6) by an interaction Hamiltonian between the Ising system and the heat bath (cf. (1.74)):

$$V = \sum_{j=1}^{N} g_j \hat{b}_j \sigma_{xj} \, . \tag{2.17}$$

As before, the $\hat{b}_j$'s are operators describing heat-bath excitations (e.g., phonons) and the g_j's are coupling constants. We do not address here the issue of how one can *derive* the Glauber master equation by following the approach of Sect. 1.6, and treating V in a suitable perturbation theory for the density matrix [28]. However, we stress that the Glauber model describes *nonconserved* kinetics because the spin-flip processes render the total magnetization $\sum_{i=1}^{N} \langle \sigma_i \rangle$ time-dependent.

To obtain a reasonable model for the transition probability W, we follow the prescription of (1.59)-(1.61). It is easily confirmed that the following choice of transition probability is consistent with the detailed-balance condition [43,44]:

$$\begin{aligned} W(m_1, ... - m_j, ...m_N | m_1, ...m_j, ...m_N) &= \frac{\lambda}{2} \left[1 - \tanh\left(\frac{\beta \Delta \mathcal{H}}{2} \right) \right] \\ &= \frac{\lambda}{2} \left[1 + m_j \tanh\left(\beta h_j \right) \right] , \end{aligned} \tag{2.18}$$

where $j = 1 \to N$, and $\Delta \mathcal{H}$ denotes the change in energy resulting from the spin-flip. In (2.18), h_j is given by (2.15). Equation (2.18) shows that the rate at which the jth spin is flipped depends on the configuration of all other spins through h_j. In the Glauber model, it is assumed that λ^{-1} is a constant which sets the basic time-scale of the nonequilibrium process. This is a simplifying assumption, as is clear from the discussion in Sect. 1.7 (cf. (1.98)), and is reasonable only when the heat bath has no memory.

We do not analyze the complete probability distribution $\mathcal{P}(\{m_i\}, t)$ here, but rather focus upon its first and second moments, which yield the magnetization and the correlation function, respectively. This is achieved by first writing the corresponding results for the single-spin case in a suggestive form. Thus, we have from (1.64) and (1.67),

$$\lambda^{-1} \frac{d}{dt} m_z(t) \equiv \lambda^{-1} \frac{d}{dt} \langle \sigma_z(t) \rangle = -\langle \sigma(t) \rangle + \langle \tanh\left(\beta H \right) \rangle \, , \tag{2.19}$$

$$\begin{aligned} \lambda^{-1} \frac{d}{dt} C(t) &\equiv \lambda^{-1} \frac{d}{dt} \langle \sigma_z(0) \sigma_z(t) \rangle_0 \\ &= -\langle \sigma_z(0) \sigma_z(t) \rangle_0 + \langle \sigma_z(0) \tanh\left(\beta H \right) \rangle_0 \, . \end{aligned} \tag{2.20}$$

Here, the correlation function is calculated in equilibrium, denoted by the subscript "0". Generalizing these results to the N-spin case, we have

$$\lambda^{-1} \frac{d}{dt} \langle \sigma_i(t) \rangle = -\langle \sigma_i(t) \rangle + \left\langle \tanh\left[\beta \left(H_i + \sum_{j=1}^{N} J_{ij} \sigma_j(t) \right) \right] \right\rangle , \tag{2.21}$$

$$\lambda^{-1}\frac{d}{dt}\langle\sigma_i(0)\sigma_j(t)\rangle_0 = -\langle\sigma_i(0)\sigma_j(t)\rangle_0 + \left\langle \sigma_i(0)\tanh\left[\beta\left(H_j + \sum_{k=1}^{N} J_{jk}\sigma_k(t)\right)\right]\right\rangle_0 . \quad (2.22)$$

2.3.2 Mean-Field Approximation

In this section, we restrict our discussion to the ferromagnetic problem, i.e., $J_{ij} > 0$ for all pairs i, j. Then, the Ising model is known to exhibit a phase transition in dimensions $d \geq 2$. At magnetic field $H = 0$, the system becomes spontaneously magnetized below the critical temperature T_c (see Fig. 2.1). As stated earlier, the Ising model is a paradigm for a range of problems in which cooperative interactions lead to phase transitions. Therefore, it is of great interest to understand the kinetics of this model.

Unfortunately, the exact time-dependent equations (2.21) and (2.22) are analytically intractable in $d \geq 2$. (For the $d = 1$ solution, see the work of Glauber [42].) The main obstacle is that the second term on the RHS of (2.21)-(2.22) results in a hierarchical set of correlation functions, as can be seen by expanding the tanh-function. These dynamical equations can be rendered tractable by invoking the MFA, which truncates the hierarchy as described below.

First, notice that the second term on the RHS of (2.22) can be rewritten as

$$\left\langle \sigma_i(0)\tanh\left[\beta\left(H_j + \sum_{k=1}^{N} J_{jk}\sigma_k(t)\right)\right]\right\rangle_0 = \left\langle \tanh\left[\beta\sigma_i(0)\left(H_j + \sum_{k=1}^{N} J_{jk}\sigma_k(t)\right)\right]\right\rangle_0 . \quad (2.23)$$

In (2.21) and (2.22), we now expand the tanh-function, and assume that there is no correlation between different sites, i.e., the average of the product of spin operators can be replaced by the product of their averages. The result of such a random-phase decoupling is that the angular brackets denoting the statistical average can be taken inside the argument of the tanh-function [48,49]. Thus, we obtain

$$\lambda^{-1}\frac{d}{dt}\psi_i(t) = -\psi_i + \tanh\left[\beta\left(H_i + \sum_{j=1}^{N} J_{ij}\psi_j\right)\right] , \quad (2.24)$$

$$\lambda^{-1}\frac{d}{dt}C_{ij}(t) = -C_{ij} + \tanh\left[\beta\left(H_j\psi_i^{\mathrm{eq}} + \sum_{k=1}^{N} J_{jk}C_{ik}\right)\right] , \quad (2.25)$$

where we have introduced the notation

$$\psi_i(t) = \langle \sigma_i(t) \rangle \ , \tag{2.26}$$

$$C_{ij}(t) = \langle \sigma_i(0) \sigma_j(t) \rangle_0 \ . \tag{2.27}$$

Note that, for time-independent effects in equilibrium, the LHS of (2.24)-(2.25) is identically zero. Thus, we have (as $t \to \infty$)

$$\psi_i^{\mathrm{eq}} = \tanh \left[\beta \left(H_i + \sum_{j=1}^{N} J_{ij} \psi_j^{\mathrm{eq}} \right) \right] \ , \tag{2.28}$$

$$C_{ij}(\infty) = \tanh \left[\beta \left(H_j \psi_i^{\mathrm{eq}} + \sum_{k=1}^{N} J_{jk} C_{ik}(\infty) \right) \right] \ . \tag{2.29}$$

We also define $C_{ij}^{\mathrm{eq}} = C_{ij}(0) = \langle \sigma_i(0) \sigma_j(0) \rangle_0$, which can also be similarly obtained as

$$\begin{aligned} C_{ij}^{\mathrm{eq}} = \frac{1}{2} \Bigg\{ & \tanh \left[\beta \left(H_i \psi_j^{\mathrm{eq}} + \sum_{k=1}^{N} J_{ik} C_{jk}^{\mathrm{eq}} \right) \right] \\ & + \tanh \left[\beta \left(H_j \psi_i^{\mathrm{eq}} + \sum_{k=1}^{N} J_{jk} C_{ik}^{\mathrm{eq}} \right) \right] \Bigg\} \ . \end{aligned} \tag{2.30}$$

As (2.28)-(2.30) occur in the *static* version of the MFA, we designate the random-phase approximation which yields the dynamical equations (2.24)-(2.25) as a MFA. However, the resultant equations are still nonlinear because of the presence of the tanh-function, and are only tractable numerically. These equations are often referred to as *mean-field dynamical models* in the literature [50–53]. A further simplification can be effected by expanding the tanh-function and retaining only leading terms as we have done for the Gibbs free energy in (2.9). We will discuss this further in Sect. 2.6 in the context of the dynamical equation for the order parameter. Here, we focus on the equation for the correlation function and demonstrate how one obtains a dynamical version of the Ornstein-Zernike approximation [38].

In (2.25), we set the magnetic field $H_j = 0$, and linearize the tanh-function to obtain

$$\left(1 + \lambda^{-1} \frac{d}{dt} \right) C_{ij}(t) = \beta \sum_{k=1}^{N} J_{jk} C_{ik}(t) \,. \tag{2.31}$$

We assume that J_{ij} only depends on the magnitude of the relative position vector $\boldsymbol{r} = \boldsymbol{r}_i - \boldsymbol{r}_j$ between the lattice points i and j. Then, the correlation function in (2.31) is translationally invariant, and this equation can be solved by introducing the discrete Fourier transform:

$$S(\boldsymbol{k},t) = \sum_{\boldsymbol{r}_j} e^{i\boldsymbol{k}\cdot(\boldsymbol{r}_i-\boldsymbol{r}_j)} C_{ij}(t)\,, \tag{2.32}$$

where $\boldsymbol{k}$ denotes a momentum vector in the reciprocal lattice. The evolution equation for the *structure factor* $S(\boldsymbol{k},t)$ is then obtained as

$$\lambda^{-1}\frac{d}{dt}S(\boldsymbol{k},t) = -\left[1-\beta J(\boldsymbol{k})\right] S(\boldsymbol{k},t)\,, \tag{2.33}$$

where

$$J(\boldsymbol{k}) = \sum_{\boldsymbol{r}_j} e^{i\boldsymbol{k}\cdot(\boldsymbol{r}_i-\boldsymbol{r}_j)} J_{ij}\,. \tag{2.34}$$

The solution of (2.33) is

$$S(\boldsymbol{k},t) = e^{-\lambda t[1-\beta J(\boldsymbol{k})]} S(\boldsymbol{k},0)\,, \tag{2.35}$$

where, by definition,

$$S(\boldsymbol{k},0) = \sum_{\boldsymbol{r}_j} e^{i\boldsymbol{k}\cdot(\boldsymbol{r}_i-\boldsymbol{r}_j)} C_{ij}^{\mathrm{eq}}\,. \tag{2.36}$$

The dynamical Ornstein-Zernike approximation results if we evaluate $S(\boldsymbol{k},0)$ in the same approximation. This yields

$$S(\boldsymbol{k},t) = \frac{e^{-\lambda t[1-\beta J(\boldsymbol{k})]}}{1-\beta J(\boldsymbol{k})}\,. \tag{2.37}$$

Recall that the wave-vector and frequency-dependent susceptibility (per site) is given by [1]

$$\chi(\boldsymbol{k},\omega) = \beta\left[S(\boldsymbol{k},0) - s\tilde{S}(\boldsymbol{k},s)\right]_{s=i\omega}\,, \tag{2.38}$$

where $\tilde{S}(\boldsymbol{k},s)$ is the Laplace transform of $S(\boldsymbol{k},t)$. Therefore, from (2.38), we obtain

$$\chi(\boldsymbol{k},\omega) = \frac{\beta\lambda}{i\omega+\lambda[1-\beta J(\boldsymbol{k})]}\,. \tag{2.39}$$

Finally, we introduce a relaxation time $\tau_{\boldsymbol{k}}$ as

$$\tau_{\boldsymbol{k}}^{-1} = \lambda\left[1-\beta J(\boldsymbol{k})\right]\,. \tag{2.40}$$

Then, (2.39) yields

$$\chi(\boldsymbol{k},\omega) = \frac{\beta}{1-\beta J(\boldsymbol{k})}\,\frac{1}{1+i\omega\tau_{\boldsymbol{k}}}\,. \tag{2.41}$$

This implies that the susceptibility response as a function of ω (for a fixed value of $\boldsymbol{k}$) is a *Debye peak* with width proportional to $\tau_{\boldsymbol{k}}^{-1}$. The width $\tau_{\boldsymbol{k}}^{-1} \to 0$ as $\beta J(\boldsymbol{k}) \to 1$.

2.4 The Spin-Exchange Kawasaki Model

2.4.1 Exact Dynamical Equations

We mentioned earlier that the Glauber model, which assumes single-spin-flip processes, is appropriate for nonconserved kinetics. On the other hand, when the Ising model describes either phase separation ($J_{ij} > 0$) or order-disorder ($J_{ij} < 0$) transitions in a binary (AB) mixture [38–41], the Glauber model is not applicable. For a binary mixture, the Ising spin variable models the presence of an A- or B-atom on a lattice site. Thus, the appropriate microscopic dynamics should involve random exchanges of A- and B-atoms at distinct sites, with their individual numbers being constant. In practice, these jumps are actually mediated by vacancies [54–56] and the system should be described as a ternary (ABV) mixture [52,53,57]. However, when the vacancy concentration is small, it is reasonable to ignore vacancies and assume that the underlying stochastic process is a spin-exchange. This corresponds to the Kawasaki model, which is based on a stationary Markov process involving a spin-exchange mechanism [45,46,48,49]. The resultant master equation is as follows:

$$
\begin{aligned}
&\frac{d}{dt}\mathcal{P}(\{m_i\},t) \\
&= -\sum_{j=1}^{N}\sum_{k\in L_j} W(m_1,...m_j,m_k,...m_N|m_1,...m_k,m_j,...m_N)\mathcal{P}(\{m_i\},t) \\
&\quad +\sum_{j=1}^{N}\sum_{k\in L_j} W(m_1,...m_k,m_j,...m_N|m_1,...m_j,m_k,...m_N)\mathcal{P}(\{m_i\}',t)\,.
\end{aligned}
\tag{2.42}
$$

In this case, the stochastic process involves, e.g., exchange of m_j (or m_k) at site j with m_k (or m_j) at site $k \in L_j$, where L_j denotes a neighboring site of j. In the second term on the RHS of (2.42), $\{m_i\}'$ refers to the configuration obtained from $\{m_i\}$ by interchanging $m_j \leftrightarrow m_k$.

At the microscopic level, the Kawasaki dynamics can be realized by assuming an interaction Hamiltonian between the Ising system and the heat bath as follows (cf. (2.17)):

$$
V = \sum_{j=1}^{N}\sum_{k\in L_j} g_{jk}\left(\hat{b}_j^{+}\hat{b}_k\sigma_j^{-}\sigma_k^{+} + \hat{b}_j\hat{b}_k^{+}\sigma_j^{+}\sigma_k^{-}\right)\,. \tag{2.43}
$$

Here, b_j^{+} (b_j) describe phonon creation (annihilation) operators; $\sigma_j^{+}, \sigma_j^{-}$ are spin-ladder operators [21,22]; and g_{jk}'s are suitable coupling constants. As in the Glauber case, we do not undertake a derivation of the master equation (2.42) using this interaction term, but simply propose a form of the relaxation matrix W, based on the detailed-balance condition. Thus, in analogy with (2.18), we write

$$\begin{aligned} &W(m_1,...m_j,m_k,...m_N|m_1,...m_k,m_j,...m_N) \\ &= \frac{\lambda_K}{2}\left[1-\tanh\left(\frac{\beta\Delta\mathcal{H}}{2}\right)\right] \\ &= \frac{\lambda_K}{2}\left\{1-\tanh\left[\frac{\beta}{2}(m_j-m_k)\left(\sum_{n\in L'_j} J_{jn}m_n - \sum_{n\in L'_k} J_{kn}m_n \right.\right.\right. \\ &\quad \left.\left.\left. + H_j - H_k\right)\right]\right\}, \end{aligned} \tag{2.44}$$

where the subscript K refers to "Kawasaki", and $\Delta\mathcal{H}$ is the energy change due to a Kawasaki spin-exchange. The primes in the summations on the RHS denote constraints as $n \neq k$ (for $n \in L'_j$) and $n \neq j$ (for $n \in L'_k$). Here, we focus on the nearest-neighbor version of the Hamiltonian in (2.1). Notice that $m_j - m_k = 0, \pm 2$, and we can extract this factor from the argument of the tanh-function to obtain

$$\begin{aligned} &W(m_1,...m_j,m_k,...m_N|m_1,...m_k,m_j,...m_N) \\ &= \frac{\lambda_K}{2}\left\{1-\frac{m_j-m_k}{2}\tanh\left[\beta\left(\sum_{n\in L'_j} J_{jn}m_n - \sum_{n\in L'_k} J_{kn}m_n \right.\right.\right. \\ &\quad \left.\left.\left. + H_j - H_k\right)\right]\right\}. \end{aligned} \tag{2.45}$$

Let us focus on the evolution equation for the order parameter $\psi_i(t) = \langle\sigma_i(t)\rangle$. This is obtained by multiplying both sides of (2.42) with m_i and summing over all possible configurations. The evolution equation for $\langle\sigma_i(t)\rangle$ is obtained (after some algebra) as follows [48,49]:

$$\begin{aligned} 2\lambda_K^{-1}\frac{d}{dt}\langle\sigma_i(t)\rangle &= -q\langle\sigma_i\rangle + \sum_{k\in L_i}\langle\sigma_k\rangle \\ &+ \sum_{k\in L_i}\left\langle(1-\sigma_i\sigma_k)\tanh\left[\beta\left(\sum_{n\in L'_i} J_{in}\sigma_n - \sum_{n\in L'_k} J_{kn}\sigma_n + H_i - H_k\right)\right]\right\rangle. \end{aligned} \tag{2.46}$$

In (2.46), q denotes the number of nearest-neighbors (coordination number) of a lattice site. This equation is analogous to (2.21), obtained in the context of Glauber kinetics. Similarly, an evolution equation (analogous to (2.22)) can be obtained for the equilibrium correlation function. We do not discuss this further here, but rather refer the interested reader to [48,49].

Although the Kawasaki model is usually associated with conserved kinetics, we should make a clarifying remark. In the context of binary alloys, a *ferromagnetic* interaction ($J_{ij} > 0$) results in phase separation, i.e., the equilibrium system consists of domains of A- and B-rich phases. The appropriate order parameter is the difference in densities of the A and B species, and is locally conserved by Kawasaki kinetics. The length scale over which the order parameter is conserved increases if we allow *long-ranged* exchanges rather than only *nearest-neighbor* exchanges. In the limit where the spin exchanges are *infinite-ranged*, the Kawasaki model has *global conservation* rather than *local conservation*. In this case, the Kawasaki model is essentially equivalent to the Glauber model [58,59].

It is also of great interest to consider the binary mixture with *antiferromagnetic* interactions, $J_{i_j} < 0$. In this case, there is a phase transition from a high-temperature disordered phase to a low-temperature ordered phase, where the A- and B-atoms order on alternate sub-lattices. The appropriate order parameter is now the *staggered magnetization*, which is the difference between the two sub-lattice *magnetizations*. This quantity is not conserved by Kawasaki kinetics, though the overall concentration is conserved. For the AB alloy with equal fractions of A and B, Kawasaki kinetics in the antiferromagnetic case is equivalent to the ferromagnetic Ising model with Glauber kinetics [60]. For asymmetric compositions, novel features arise due to the conserved concentration variable.

2.4.2 Mean-Field Approximation

In parallel with the discussion in Sect. 2.3.2, we formulate the order-parameter equation in the MFA. The appropriate prescription has already been discussed earlier – we approximate the expectation value of the function on the RHS of (2.46) by the function of the expectation values. This yields

$$2\lambda_K^{-1} \frac{d}{dt}\psi_i(t) = -q\psi_i + \sum_{k \in L_i} \psi_k$$
$$+ \sum_{k \in L_i} (1 - \psi_i\psi_k) \tanh\left[\beta\left(\sum_{n \in L_i} J_{in}\psi_n - \sum_{n \in L_k} J_{kn}\psi_n + H_i - H_k\right)\right] . \tag{2.47}$$

In (2.47), notice that the restrictions on the summations inside the tanh-function have been dropped in the MFA. As stressed by Binder [48,49], this is necessary for (2.47) to contain the correct MF solution in (2.28). The validity of this solution is easily confirmed by direct substitution in the RHS of (2.47), which vanishes identically.

Equation (2.47) is analogous to (2.24), which was obtained in the context of Glauber kinetics. As before, the MFA truncates the infinite hierarchy of equations for spin-spin correlation functions. However, the resultant MF dynamical model is still nonlinear and analytically intractable. In general, (2.24) and (2.47) are usually solved numerically, particularly in the context of phase ordering dynamics. Subsequently, we will also discuss how these MF dynamical models can be used to obtain coarse-grained models for order-parameter kinetics.

2.5 Relaxation Kinetics in Homogeneous Systems

One of the goals of dissipative modeling is to formulate an appropriate rate theory for the approach to equilibrium of a many-body system which is driven away from equilibrium. This is relevant in a variety of physical problems, e.g., chemical reactions, phase ordering dynamics, structural transitions, etc. The kinetic Ising models discussed in Sects. 2.3 and 2.4 are often used to describe nonequilibrium systems. However, even these simple models are analytically intractable, as our earlier discussion has clarified. The MFA yields closed (though nonlinear) dynamical equations, and our subsequent discussion will be in the MF framework.

Equilibrium thermodynamic properties for the Ising model in the MFA are uniquely governed by the Bragg-Williams form of the Gibbs potential in (2.7). It is natural to ask whether there is a corresponding unique *kinetic potential* which describes the evolution of the order parameter. The answer turns out to be in the negative. We will demonstrate that the functional form of the kinetic potential depends on the details of the stochastic processes at the microscopic level. For example, the kinetic potential is different for the Glauber and the Kawasaki models.

Let us focus upon the case where the order parameter ψ is *uniform* and *nonconserved*, so as to clarify the driving force pushing the system towards equilibrium. The rate equation has the following generic form:

$$\frac{d\psi}{dt} = \lambda_0 \beta f(\psi) , \tag{2.48}$$

where λ_0 is a phenomenological rate constant, β is the inverse temperature, and $f(\psi)$ is the driving force. The latter can be assumed to be as follows:

$$f(\psi) = -\frac{\partial G_{\text{kin}}}{\partial \psi} , \tag{2.49}$$

where we refer to G_{kin} as the kinetic Gibbs potential. We expect that $G_{\text{kin}} = G$ in thermodynamic equilibrium because

$$\frac{\partial G}{\partial \psi} = 0 . \tag{2.50}$$

The straightforward generalization is to assume that $G_{\text{kin}} = G(\psi)$ even when the system is out of equilibrium. For simplicity, we consider the Bragg-Williams form of the Gibbs potential in (2.7) with magnetic field $H = 0$. In that case, (2.48) has the following form:

$$\frac{d\psi}{dt} = \lambda_0 \left[\frac{T_c}{T}\psi - \frac{1}{2}\ln\left(\frac{1+\psi}{1-\psi}\right)\right] . \tag{2.51}$$

As expected, (2.51) yields the MF result in equilibrium:

$$\psi_0 = \tanh\left(\frac{T_c}{T}\psi_0\right) . \tag{2.52}$$

Next, let us obtain the kinetic potentials for the Glauber and antiferromagnetic Kawasaki models. Before we proceed, it is relevant to discuss physical situations where these models may be realized. To be specific, we consider examples from materials science. Consider a material which undergoes a structural phase transition with two sites per unit cell on which an A- and B-atom are placed, as shown in Fig. 2.3(a). This is an example of *antiferromagnetic* ordering, but the variable m_i now refers to the ordered configurations in Fig. 2.3(a). Then, the interchange of two atoms in a cell corresponds to a Glauber process. On the other hand, an exchange of configurations in two cells corresponds to a Kawasaki process. Another case is depicted in Fig. 2.3(b) with four atoms per cell. Again, there are intra-cell ordering processes which are Glauber-like and inter-cell exchange processes which are Kawasaki-like [61–64].

The homogeneous version of (2.24) yields the rate equation for the Glauber model as follows:

$$\frac{d\psi}{dt} = -\lambda \left[\psi - \tanh\left(\frac{T_c}{T}\psi\right)\right] , \tag{2.53}$$

which has the same equilibrium solution as (2.51). The corresponding rate equation for the Kawasaki model is obtained from the antiferromagnetic version of (2.47). For uniform antiferromagnetic ordering, $\langle\sigma_i\rangle = \psi$ implies that $\langle\sigma_{L_i}\rangle = -\psi$, because the sites i and L_i belong to two distinct sub-lattices. (We consider the case with equal fractions of A and B in the mixture.) Hence, from (2.47), we obtain

$$\frac{d\psi}{dt} = -\frac{\lambda_K q}{2}\left[2\psi - \left(1+\psi^2\right)\tanh\left(\frac{2T_c}{T}\psi\right)\right] . \tag{2.54}$$

Equation (2.54) also has the correct equilibrium solution. Thus, (2.51), (2.53) and (2.54) constitute three different descriptions of the kinetics of the order parameter, all resulting in the same equilibrium state. To facilitate a comparison of the corresponding kinetic potentials, we rescale the rate constants so that all these models have the same linearized dynamics. This identification leads to

$$\lambda = \lambda_0 ,$$
$$\lambda_K = \frac{\lambda_0}{q} . \tag{2.55}$$

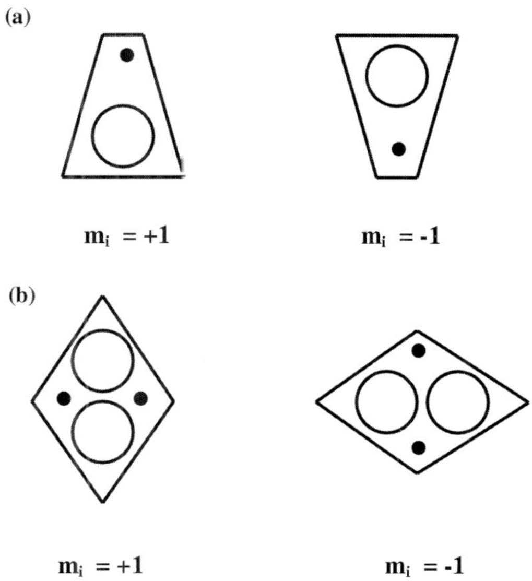

Fig. 2.3. Examples of materials which undergo structural phase transitions. (**a**) Each unit cell consists of an A-atom (denoted by a small filled circle) and a B-atom (denoted by a large open circle), which can be arranged in two different ways. (**b**) Each unit cell consists of four atoms, with a corresponding number of internal degrees of freedom

We then obtain the kinetic potentials as follows:

(i) **Bragg-Williams**

$$G_{\text{kin}}^{BW}(T,\psi) = \frac{k_B T}{2}\left[-\frac{T_c}{T}\psi^2 + (1+\psi)\ln(1+\psi) + (1-\psi)\ln(1-\psi)\right]. \tag{2.56}$$

(ii) **Glauber**

$$G_{\text{kin}}^{G}(T,\psi) = \frac{k_B T}{2}\left\{\psi^2 - \frac{2T}{T_c}\ln\left[\cosh\left(\frac{T_c}{T}\psi\right)\right]\right\}. \tag{2.57}$$

(iii) **Kawasaki**

$$G_{\text{kin}}^{K}(T,\psi) = \frac{k_B T}{2}\left\{\psi^2 - \frac{T}{2T_c}\ln\left[\cosh\left(\frac{2T_c}{T}\psi\right)\right]\right. \\ \left. -\left(\frac{T}{2T_c}\right)^3 \sum_{k=2,4,..}^{\infty} \frac{2^k(2^k-1)B_k}{(k+2)k!}\left(\frac{2T_c}{T}\psi\right)^{k+2}\right\}, \tag{2.58}$$

where B_k are the Bernoulli numbers [65]. In Fig. 2.4, we present plots of $G_{\text{kin}}(T,\psi)/(k_B T_c)$ vs. ψ for the three kinetic potentials. Results are shown

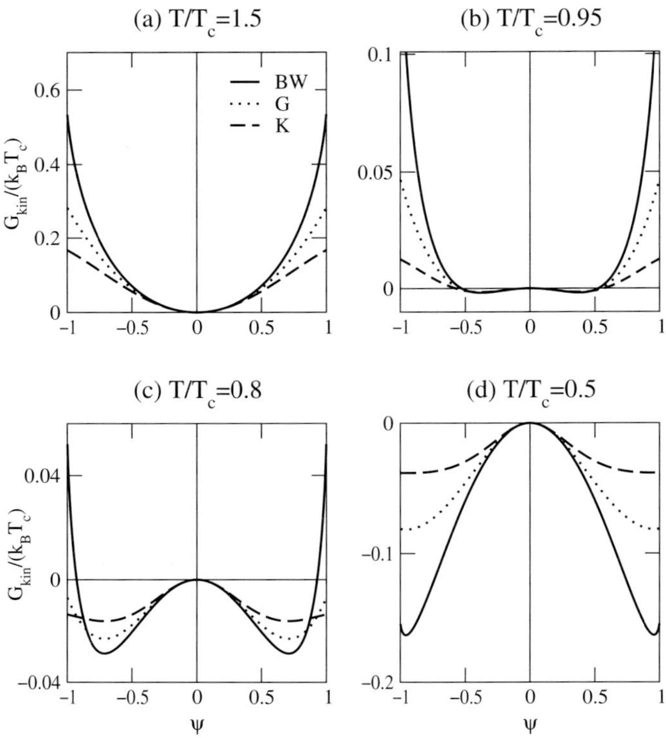

Fig. 2.4a–d. Plots of $G_{\rm kin}(T,\psi)/(k_BT_c)$ vs. ψ for the three kinetic potentials in (2.56)-(2.58). We show results for four different values of T/T_c, as indicated

for four different values of T/T_c. As expected, the distinction between the three kinetic potentials is accentuated in deep-quenching experiments.

In Fig. 2.5, we show the time-dependent evolution of $\psi(t)$ for all three potentials. Again, results are shown for four different values of T/T_c. A comparison of the plots shows that the rate laws are comparable for $T \geq 0.95T_c$. We see significant differences only for $T = 0.5T_c$, with the Kawasaki model giving the slowest dynamics and the Bragg-Williams model the fastest dynamics. This is expected because of the involvement of a pair of spins in each stochastic move for the Kawasaki model.

2.6 Modeling of Order-Parameter Kinetics

2.6.1 Coarse-Grained Models from General Considerations

We have stressed the utility of the order-parameter approach for describing the kinetics of phase transitions. Typically, kinetic processes are comprised of various fluctuating modes, and the order parameter is the slowest one,

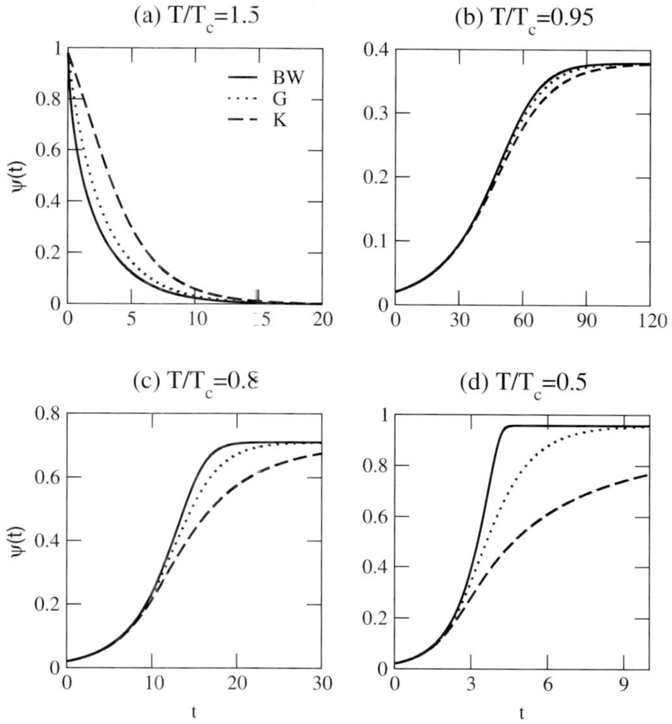

Fig. 2.5. Time-dependent relaxation of the order parameter for the three kinetic potentials in (2.56)-(2.58). The results correspond to four different values of T/T_c, as in Fig. 2.4. The initial values of the order parameter were (**a**) $\psi(0) = 0.98$; (**b–d**) $\psi(0) = 0.02$

both in terms of spatial and temporal modulations. Thus, one needs coarse-grained models for order-parameter kinetics which are valid on length-scales larger than typical lattice spacings and time-scales longer than microscopic relaxation times (such as λ^{-1} in the Glauber model). It is of great relevance to derive such macroscopic models from a microscopic formulation like the kinetic Ising model. Alternatively, we can also use general considerations to obtain such coarse-grained models from the Langevin equation for Brownian motion, discussed in Chap. 1. The discussion in this section focuses on the connection between two different frameworks: (i) Brownian motion, based on *continuous* stochastic processes; and (ii) Kinetic Ising models, based on *discrete* jump processes.

The Langevin equation for phase-space dynamics (cf. (1.17)-(1.18)) reduces to a stochastic differential equation for the particle position in the high-friction limit. When the friction coefficient γ is large, the momentum $P(t)$ rapidly decays to its *stationary* value. We then set $dP/dt \simeq 0$ in (1.17) to obtain

$$P(t) \simeq \frac{1}{\gamma}\left[F(Q)+\theta(t)\right] . \qquad (2.59)$$

Equation (1.18) then yields

$$\frac{dQ}{dt} = \frac{1}{M\gamma}\left[F(Q)+\theta(t)\right] . \qquad (2.60)$$

Further, assuming that the system is conservative (i.e., $F(Q)$ is derivable from the gradient of a potential $U(Q)$), we obtain

$$\frac{dQ}{dt} = -\Gamma\frac{\partial U(Q)}{\partial Q} + \theta'(t) . \qquad (2.61)$$

Here, Γ is the damping coefficient:

$$\Gamma \equiv (M\gamma)^{-1} , \qquad (2.62)$$

and $\theta'(t)$ is the noise term:

$$\theta'(t) = \frac{1}{M\gamma}\theta(t) . \qquad (2.63)$$

In our subsequent discussion, we drop the prime on the stochastic variable $\theta'(t)$, which has the usual white-noise properties (cf. (1.12)):

$$\begin{aligned}\overline{\theta(t)} &= 0 ,\\ \overline{\theta(t')\theta(t'')} &= 2\Gamma k_B T\delta(t'-t'') .\end{aligned} \qquad (2.64)$$

Equation (2.61) is derived in the context of the dynamics of a mechanical particle, but can be generalized to systems which display cooperativity and phase transitions. Thus, interpreting Q as the order parameter ψ measuring the degree of order, and $U(Q)$ as the Gibbs free energy $G(\psi)$ of the system, we can generalize (2.61) as follows:

$$\frac{d\psi}{dt} = -\Gamma\frac{\partial G(\psi)}{\partial \psi} + \theta(t) . \qquad (2.65)$$

The rate equation (2.65) is reasonable because thermodynamic equilibrium is defined by

$$\overline{\frac{\partial G(\psi)}{\partial \psi}} = 0 . \qquad (2.66)$$

Equation (2.65) is usually referred to as the Onsager equation or the Landau-Khalatnikov equation.

Equation (2.65) corresponds to a situation where ordering is uniform, i.e., the order parameter ψ is spatially homogeneous as in Sect. 2.5. However, there are many physical problems where the spatial fluctuations of the order parameter are important, e.g., near the critical point of a phase transition [66] or in phase ordering problems [67]. In such a case, (2.65) is further generalized as follows:

$$\frac{\partial}{\partial t}\psi(\boldsymbol{r},t) = -\Gamma\frac{\delta\mathcal{G}\left[\psi\right]}{\delta\psi} + \theta(\boldsymbol{r},t)\,, \tag{2.67}$$

where $\psi(\boldsymbol{r},t)$ is now a function of space $\boldsymbol{r}$ and time t. In (2.67), $\delta\mathcal{G}/\delta\psi$ denotes the functional derivative of the free-energy functional:

$$\mathcal{G}\left[\psi\right] = \int d\boldsymbol{r}\left[G(\psi) + \frac{1}{2}K(\boldsymbol{\nabla}\psi)^2\right]. \tag{2.68}$$

Typical functional forms of $G(\psi)$ are specified in (2.56)-(2.58). It is usually convenient to consider the Taylor-expanded ψ^4-form of the free energy, as in (2.9). The second term on the RHS of (2.68) accounts for surface tension due to inhomogeneities in the order parameter. The parameter K measures the strength of the surface tension.

The noise term in (2.67) is now space- and time-dependent, and satisfies the fluctuation-dissipation relation:

$$\begin{aligned}\overline{\theta(\boldsymbol{r},t)} &= 0\,,\\ \overline{\theta(\boldsymbol{r}',t')\theta(\boldsymbol{r}'',t'')} &= 2\Gamma k_B T\delta(\boldsymbol{r}'-\boldsymbol{r}'')\delta(t'-t'')\,.\end{aligned} \tag{2.69}$$

Equations (2.67)-(2.69) constitute the so-called Model A of order-parameter kinetics, discussed by Halperin and Hohenberg in the context of dynamical critical phenomena [66]. Alternatively, (2.67) is referred to as the time-dependent Ginzburg-Landau (TDGL) equation [68].

The zero-temperature version of (2.67) is obtained by dropping the noise term as follows:

$$\frac{\partial}{\partial t}\psi(\boldsymbol{r},t) = -\Gamma\frac{\delta\mathcal{G}[\psi]}{\delta\psi}\,. \tag{2.70}$$

Equation (2.70) exhibits purely relaxational dynamics, and is expected to be valid at low temperatures. In the context of phase ordering systems without disorder, the asymptotic behavior of domain-growth kinetics is governed by the zero-temperature limit [69,70].

Thus, we have been able to motivate the TDGL equation from the Langevin equation for Brownian motion. We should stress that the TDGL equation is restricted to *nonconserved* order-parameter kinetics, because

$$\frac{d}{dt}\int d\boldsymbol{r}\psi(\boldsymbol{r},t) \neq 0\,. \tag{2.71}$$

The TDGL equation is appropriate to describe (say) the dynamical evolution of a ferromagnet, e.g., the emergence of spontaneously-magnetized domains when the system is rapidly quenched from $T > T_c$ to $T < T_c$. The resultant far-from-equilibrium evolution of the ferromagnet, and the corresponding pattern formation, is a standard problem in phase ordering dynamics [67].

Next, let us focus upon the case where the order parameter obeys a local conservation law. Consequently, the evolution of the order parameter $\psi(\boldsymbol{r}, t)$ is described by a continuity equation:

$$\frac{\partial}{\partial t}\psi(\boldsymbol{r}, t) = -\boldsymbol{\nabla} \cdot \boldsymbol{J}(\boldsymbol{r}, t) , \tag{2.72}$$

where $\boldsymbol{J}(\boldsymbol{r}, t)$ denotes the current. Further, since the current is driven by concentration fluctuations, we expect

$$\boldsymbol{J}(\boldsymbol{r}, t) = -D\boldsymbol{\nabla}\mu(\boldsymbol{r}, t) , \tag{2.73}$$

where D is the diffusion coefficient and $\mu(\boldsymbol{r}, t)$ is the chemical potential. Finally, the chemical potential is determined as

$$\boldsymbol{\mu}(\boldsymbol{r}, t) = \frac{\delta \mathcal{F}[\psi]}{\delta \psi} , \tag{2.74}$$

where $\mathcal{F}$ refers to the Helmholtz potential, which is the appropriate thermodynamic potential for the binary mixture.

Combining (2.72)-(2.74), we obtain the Cahn–Hilliard (CH) equation [71, 72] for the phase separation of a binary mixture:

$$\frac{\partial}{\partial t}\psi(\boldsymbol{r}, t) = D\nabla^2 \left(\frac{\delta \mathcal{F}[\psi]}{\delta \psi}\right) . \tag{2.75}$$

Equation (2.75) corresponds to the case of a constant diffusion coefficient. There have also been some studies of systems where the diffusion coefficient depends on the local order parameter [73–75].

The effects of thermal fluctuations can be incorporated in the CH equation by including a noise term in the definition of the current in (2.73) [76]. The resultant model is the Cahn–Hilliard–Cook (CHC) equation:

$$\frac{\partial}{\partial t}\psi(\boldsymbol{r}, t) = \boldsymbol{\nabla} \cdot \left\{ D\boldsymbol{\nabla} \left(\frac{\delta \mathcal{F}[\psi]}{\delta \psi}\right) + \boldsymbol{\theta}(\boldsymbol{r}, t) \right\} . \tag{2.76}$$

The noise term is now a d-dimensional vector and satisfies the usual fluctuation-dissipation relation:

$$\begin{aligned} \overline{\boldsymbol{\theta}(\boldsymbol{r}, t)} &= 0 , \\ \overline{\theta_i(\boldsymbol{r}', t')\theta_j(\boldsymbol{r}'', t'')} &= 2Dk_B T\delta_{ij}\delta(\boldsymbol{r}' - \boldsymbol{r}'')\delta(t' - t'') . \end{aligned} \tag{2.77}$$

Equations (2.76)-(2.77) constitute Model B in the classification scheme of Halperin and Hohenberg [66].

The TDGL and CHC equations (or Models A and B) represent the prototypical equations for order-parameter kinetics. In more complex physical situations, there may be multiple order parameters which satisfy different conservation laws. Let us focus on two examples here. First, consider the case of a binary (AB) mixture where one of the components (say, A) has an attached magnetic spin [77]. This system is described by two order parameters – one for the magnetization (which is nonconserved) and the other for the AB density field (which is conserved). The resultant dynamical model is usually referred to as Model C [66] and consists of coupled dynamical equations for the two order parameters. Second, consider the case of a ternary (ABV) mixture [52,53,57]. This system is also described by two order parameters – one for the AV (or BV) density field, and the other for the AB density field. Both these order parameters are conserved, and the corresponding dynamical model is designated as Model D [66,78,79].

2.6.2 Coarse-Grained Models from Kinetic Ising Models

In the previous subsection, we have motivated the TDGL equation for nonconserved kinetics and the CHC equation for conserved kinetics from continuous diffusion processes. We next demonstrate how deterministic versions of these models can be obtained by a suitable coarse-graining of the evolution equations for kinetic Ising models [48,49]. Recall that, in the coarse-grained description, the order parameter is a slowly-varying (continuous) function of space and time. It is convenient to restrict ourselves to the *ferromagnetic* case ($J_{ij} > 0$) of the underlying Ising model. Thus, for nonconserved kinetics, we consider the paramagnetic to ferromagnetic transition in a magnet. For conserved kinetics, the corresponding model describes kinetics of phase separation in a binary mixture.

Let us first consider the MF dynamical model (cf. (2.24)) for the Ising model with Glauber kinetics. For simplicity, we consider the case of zero magnetic field, i.e., $H_i = 0$. In the continuum limit, we define the interaction range R as

$$R^2 = [J(0)]^{-1} \sum_{j=1}^{N} (\boldsymbol{r}_i - \boldsymbol{r}_j)^2 J_{ij} \,, \tag{2.78}$$

where $J(0)$ is defined in (2.8), and $\boldsymbol{r}_i$ is the position vector of site i. We can then expand various terms on the RHS of (2.24) as follows:

$$\sum_{j=1}^{N} J_{ij}\psi_j(t) \simeq J(0) \left[\psi(\boldsymbol{r}_i, t) + \frac{1}{2} R^2 \nabla_i^2 \psi(\boldsymbol{r}_i, t) \right] + \text{other terms} \,. \tag{2.79}$$

Furthermore,

$$\tanh\left[\beta\sum_{j=1}^{N} J_{ij}\psi_j(t)\right]$$
$$\simeq \beta\sum_{j=1}^{N} J_{ij}\psi_j - \frac{1}{3}\left(\beta\sum_{j=1}^{N} J_{ij}\psi_j\right)^3 + \text{ other terms}$$
$$\simeq \frac{T_c}{T}\psi\left(\boldsymbol{r}_i,t\right) - \frac{1}{3}\left(\frac{T_c}{T}\right)^3 \psi\left(\boldsymbol{r}_i,t\right)^3 + \frac{1}{2}\frac{T_c}{T}R^2\nabla_i^2\psi\left(\boldsymbol{r}_i,t\right) + \text{ other terms}\,, \tag{2.80}$$

where we have used (2.79) to obtain the second expression. Therefore, the order-parameter equation for the Glauber-Ising model simplifies as

$$\lambda^{-1}\frac{\partial}{\partial t}\psi(\boldsymbol{r},t) = \left(\frac{T_c}{T}-1\right)\psi(\boldsymbol{r},t) - \frac{1}{3}\left(\frac{T_c}{T}\right)^3\psi(\boldsymbol{r},t)^3 + \frac{T_c}{2T}R^2\nabla^2\psi(\boldsymbol{r},t) + \text{ other terms}\,, \tag{2.81}$$

where we have dropped the subscript i for the position variable.

At this stage a few remarks are in order. Firstly, (2.81) is recognized as the deterministic TDGL equation with the local free energy (cf. (2.68))

$$\beta G(\psi) = -\frac{1}{2}\left(\frac{T_c}{T}-1\right)\psi^2 + \frac{1}{12}\left(\frac{T_c}{T}\right)^3\psi^4\,, \tag{2.82}$$

where we have neglected the higher-order terms in (2.81). The surface-tension parameter K is identified as

$$K = k_B T_c R^2\,. \tag{2.83}$$

Secondly, the approximation of neglecting the higher-order terms in (2.81) is justifiable only for $T \to T_c$, where the order parameter is small. However, it is generally believed that the TDGL equation with the free energy in (2.82) is valid even for deep quenches, at least in terms of containing the correct physics. Thirdly, the noise terms in (2.67) are missing from our present derivation. This is a consequence of neglecting higher spin-spin correlations in the MFA which resulted in (2.24). The noise terms can be reinstated in (2.81) if we model these higher-order correlations as fluctuations.

The above arguments provide a *motivation* for the TDGL equation from microscopic considerations. We deliberately use the term *motivation* rather than *derivation* because of the nature of the approximations involved in obtaining (2.81). Nevertheless, the TDGL equation (2.81) is universally accepted as a reasonable model for a large class of nonconserved ordering problems.

We next consider the kinetic Ising model with Kawasaki spin-exchange (or conserved) kinetics [48,49]. The relevant order-parameter evolution equation in the MFA is (2.47). Again, we focus on the case with $H_i = 0$. We simplify the RHS of (2.47) by using the identity

$$\tanh(X - Y) = \frac{\tanh X - \tanh Y}{1 - \tanh X \tanh Y}, \quad \text{where}$$
$$X = \sum_{n \in L_i} J_{in} \psi_n ,$$
$$Y = \sum_{n \in L_k} J_{kn} \psi_n . \tag{2.84}$$

In equilibrium, we have (cf. (2.28))

$$\psi_i^{\text{eq}} = \tanh \left[\beta \left(\sum_{j=1}^{N} J_{ij} \psi_j^{\text{eq}} \right) \right] . \tag{2.85}$$

We are interested in the late-stage dynamics, where the system has equilibrated locally, so that in (2.47)

$$(1 - \psi_i \psi_k) \left(\frac{\tanh X - \tanh Y}{1 - \tanh X \tanh Y} \right) \simeq \tanh X - \tanh Y . \tag{2.86}$$

Therefore, we can rewrite (2.47) as

$$\begin{aligned} 2\lambda_K^{-1} \frac{d}{dt} \psi_i(t) \simeq & \sum_{k \in L_i} (\psi_k - \psi_i) - \\ & \sum_{k \in L_i} \left\{ \tanh \left[\beta \left(\sum_{n \in L_k} J_{kn} \psi_n \right) \right] - \tanh \left[\beta \left(\sum_{n \in L_i} J_{in} \psi_n \right) \right] \right\} \\ = & \Delta_D \left\{ \psi_i - \tanh \left[\beta \left(\sum_{n \in L_i} J_{in} \psi_n \right) \right] \right\} , \end{aligned} \tag{2.87}$$

where Δ_D denotes the discrete Laplacian operator. We can use the Taylor expansion in (2.80) to obtain the coarse-grained version of (2.87) as

$$\begin{aligned} 2\lambda_K^{-1} \frac{\partial}{\partial t} \psi(\boldsymbol{r}, t) = & -a^2 \nabla^2 \left[\left(\frac{T_c}{T} - 1 \right) \psi - \frac{1}{3} \left(\frac{T_c}{T} \right)^3 \psi^3 + \frac{T_c}{2T} a^2 \nabla^2 \psi \right] \\ & + \text{other terms} , \end{aligned} \tag{2.88}$$

where a is the lattice spacing. Equation (2.88) is identified as the CH equation with the local free energy in (2.82), and diffusion constant $D = \lambda_K \beta a^2 / 2$. The above arguments demonstrate that the CH equation constitutes a reasonable coarse-grained description of the Ising model with Kawasaki kinetics.

3 An Overview of Phase Ordering Dynamics

3.1 Introduction

In Chap. 2, we introduced kinetic Ising models and the corresponding coarse-grained models for order-parameter kinetics. We also discussed the utility of these models in the context of *phase ordering dynamics* or *domain growth kinetics*, viz., the far-from-equilibrium evolution of a homogeneous multi-phase mixture which has been rendered thermodynamically unstable by a rapid change of parameters, e.g., temperature, pressure, etc. These nonequilibrium problems have been the subject of intense experimental, theoretical and numerical investigation in recent years [67,80–82].

This chapter presents an overview of phase ordering dynamics, and has two major objectives. Firstly, we review our theoretical understanding of this subject. This will provide a framework for the material discussed subsequently in Chaps. 4-6. Secondly, we highlight recent developments in this area, with particular attention to the incorporation of experimentally relevant effects in studies of domain growth.

This chapter is organized as follows. In Sect. 3.2, we discuss phase ordering systems with a nonconserved order parameter. The prototypical problem in this class is the ordering of a ferromagnet for $T < T_c$ (see Fig. 2.1). In Sect. 3.3, we discuss phase ordering dynamics with a conserved order parameter. The typical problem in this class is the kinetics of phase separation of a binary (AB) mixture, which has been quenched below the coexistence curve (see Fig. 2.2). Finally, Sect. 3.4 focuses upon the modeling of experimentally relevant effects in phase ordering systems.

3.2 The Case with Nonconserved Order Parameter

3.2.1 Scalar Order Parameter

Consider a ferromagnet, consisting of magnetic spins which energetically prefer to align parallel to each other. The system equilibrium is determined by minimizing the Gibbs potential $G(T, H)$ rather than the internal energy. Therefore, at high temperatures, entropic effects are dominant and the system is disordered. At low temperatures, energetic effects are dominant and

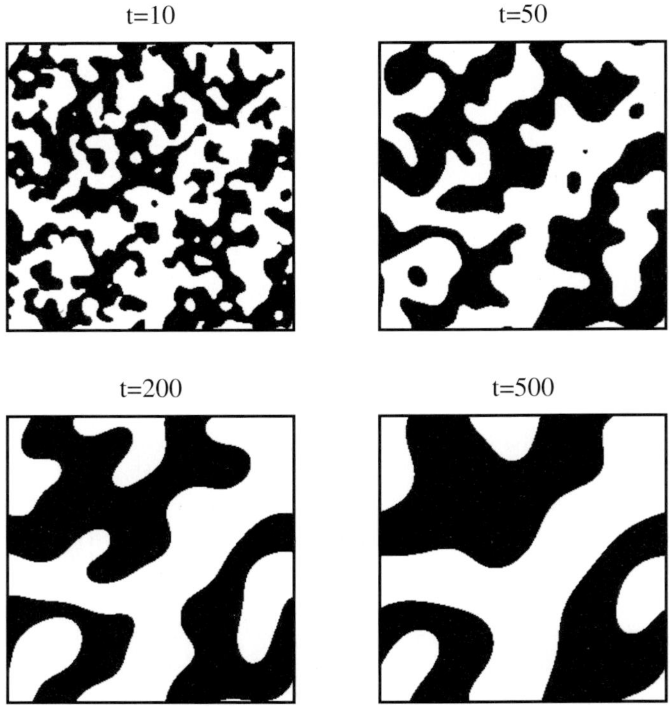

Fig. 3.1. Evolution of a ferromagnet from a disordered initial condition for $T < T_c, H = 0$. Regions with positive magnetization are marked in black, and regions with negative magnetization are unmarked. These pictures were obtained from a simulation of the dimensionless TDGL equation without thermal noise. The Euler-discretization mesh sizes were $\Delta t = 0.1$ and $\Delta x = 1.0$ in time and space, respectively. The lattice size was 256^2. The number labeling each picture corresponds to the dimensionless time after the quench

the system orders with spins parallel to each other. In the absence of an external field ($H = 0$), the spin-up and spin-down states are equivalent. The above considerations are summarized in the phase diagram of a ferromagnet shown in Fig. 2.1.

The corresponding phase ordering problem considers a disordered system at $T > T_c, H = 0$ for time $t < 0$. At $t = 0$, the system is rapidly quenched below the critical temperature. Clearly, the disordered state is no longer the preferred equilibrium state at $T < T_c$. The far-from-equilibrium homogeneous system evolves towards its new equilibrium state by separating into equivalent domains, which are rich in up or down spins, respectively (see Fig. 3.1). These domains coarsen with time and are characterized by a growing length scale $L(t)$. A finite system becomes ordered in either of two equivalent states as $t \to \infty$. At the microscopic level, this evolution can be described by an Ising model with Glauber spin-flip kinetics (see Sect. 2.3). At the coarse-grained level, the appropriate order parameter to describe the dynamics of the system is the local magnetization $\psi(\boldsymbol{r},t)$, introduced in Sect. 2.6.1. The evolution of the magnetization is described by the time-dependent Ginzburg-Landau (TDGL) equation (2.67), which describes nonconserved order-parameter kinetics. For the ψ^4-form of the free energy (cf. (2.9)), the TDGL equation has the form:

$$\frac{\partial}{\partial t}\psi(\boldsymbol{r},t) = \Gamma\left[a(T_c - T)\psi - b\psi^3 + H + K\nabla^2\psi\right] + \theta(\boldsymbol{r},t)\,, \qquad (3.1)$$

where $a, b, K(>0)$ are parameters, and the Gaussian white noise satisfies

$$\begin{aligned}\overline{\theta(\boldsymbol{r},t)} &= 0\,,\\ \overline{\theta(\boldsymbol{r}',t')\theta(\boldsymbol{r}'',t'')} &= 2\Gamma k_B T\delta(\boldsymbol{r}'-\boldsymbol{r}'')\delta(t'-t'')\,.\end{aligned} \qquad (3.2)$$

The parameters in (3.1) can be absorbed into the definitions of space and time by introducing the rescaled variables (for $T < T_c$)

$$\begin{aligned}\psi' &= \frac{\psi}{\psi_0}\,, \qquad \psi_0 = \sqrt{\frac{a(T_c-T)}{b}}\,,\\ t' &= a(T_c-T)\Gamma t\,,\\ \boldsymbol{r}' &= \sqrt{\frac{a(T_c-T)}{K}}\boldsymbol{r}\,,\\ H' &= \frac{H}{a(T_c-T)\psi_0}\,,\\ \theta' &= \frac{\theta}{a(T_c-T)\Gamma\psi_0}\,.\end{aligned} \qquad (3.3)$$

Dropping primes, we obtain the dimensionless TDGL equation:

$$\frac{\partial}{\partial t}\psi(\boldsymbol{r},t) = \psi - \psi^3 + H + \nabla^2\psi + \theta(\boldsymbol{r},t)\,, \qquad (3.4)$$

where

$$\begin{aligned}\overline{\theta(\boldsymbol{r},t)} &= 0\,,\\ \overline{\theta(\boldsymbol{r}',t')\theta(\boldsymbol{r}'',t'')} &= 2\epsilon\delta(\boldsymbol{r}'-\boldsymbol{r}'')\delta(t'-t'')\,,\\ \epsilon &= \frac{k_B T b\left[a(T_c-T)\right]^{(d-4)/2}}{K^{d/2}}\,.\end{aligned} \qquad (3.5)$$

For $H \neq 0$, the system rapidly orders into the preferred state, and the evolution is relatively uninteresting. We will focus on the case with $H = 0$ (shown in Fig. 3.1), where the system evolves into two competing states. The ordering process depicted in Fig. 3.1 is characterized as follows. The domain length scale obeys the Lifshitz-Allen-Cahn (LAC) law, $L(t) \sim t^{1/2}$ for $d \geq 2$ [83]. This growth law can be understood as follows. The bulk domains for (3.4) are defined by order parameter values $\psi = \pm 1$. The interfaces between these domains (i.e., defects) correspond to $\psi = 0$. Allen and Cahn [83] demonstrated that the interfacial velocity in the direction of increasing ψ is determined as $v = -\kappa$, where κ is the local curvature. The LAC growth law is obtained if we identify $v \sim dL/dt$, and $\kappa \sim 1/L$.

The case with $d = 1$ does not obey these general arguments. As a matter of fact, the $d = 1$ nearest-neighbor Ising model is ordered only at $T = 0$. In that case, domain growth is driven by the weak interactions between defect tails, and the corresponding growth law is logarithmic rather than power-law [81].

Furthermore, if the system is characterized by a single length scale, the morphology of the domains does not change with time, apart from the time-dependent scale factor. Therefore, the correlation function for the order parameter exhibits a dynamical-scaling property [84]:

$$\begin{aligned} C(\boldsymbol{r},t) &\equiv \frac{1}{V}\int d\boldsymbol{R}\ \left[\langle\psi(\boldsymbol{R},t)\psi(\boldsymbol{R}+\boldsymbol{r},t)\rangle - \langle\psi(\boldsymbol{R},t)\rangle\langle\psi(\boldsymbol{R}+\boldsymbol{r},t)\rangle\right] \\ &= g\left(\frac{r}{L}\right), \end{aligned} \tag{3.6}$$

where V is the system volume, and the angular brackets denote an averaging over independent initial conditions and thermal fluctuations. In (3.6), $g(x)$ is a time-independent master function. Notice that the equal-time correlation function defined above is a nonequilibrium quantity. It should not be confused with the unequal-time equilibrium correlation function discussed in Chaps. 1 and 2.

Actually, most experiments (e.g., neutron or light scattering [1]) probe the time-dependent structure factor, which is the Fourier transform of the real-space correlation function:

$$S(\boldsymbol{k},t) = \int d\boldsymbol{r} e^{i\boldsymbol{k}\cdot\boldsymbol{r}} C(\boldsymbol{r},t)\,, \tag{3.7}$$

where $\boldsymbol{k}$ is the wave-vector of the scattered beam. Again, we stress that the structure factor defined in (3.7) is a nonequilibrium quantity. The corresponding dynamical-scaling form for $S(\boldsymbol{k},t)$ is

$$S(\boldsymbol{k},t) = L^d f(kL)\,, \tag{3.8}$$

where $f(p)$ is a master function defined as follows:

$$f(p) = \int d\boldsymbol{x} e^{i\boldsymbol{p}\cdot\boldsymbol{x}} g(x)\,. \tag{3.9}$$

The functional forms of the master functions $g(x)$ and $f(p)$ characterize the morphology of the ordering system. Of course, a complete description of the morphology would require knowledge of all higher-order structure factors also, but these have limited experimental relevance.

Let us discuss some general features of the structure factor and correlation function. The presence of sharp interfaces (defects) in the phase ordering system results in a power-law or Porod [85] decay of the structure-factor tail, $f(p) \sim p^{-(d+1)}$ as $p \to \infty$. (This is unusual as one would generally expect a Gaussian decay of the structure factor.) The correlation function exhibits a corresponding non-analytic or singular behavior at short distances. The exact result for the singular behavior due to an interfacial defect is obtained by Bray and Humayun [86] as follows:

$$g(x) = 1 - \frac{2}{\sqrt{\pi}} \frac{\Gamma\left(\frac{d}{2}\right)}{\Gamma\left(\frac{d+1}{2}\right)} x + O(x^3) \,, \tag{3.10}$$

where $\Gamma(y)$ is the gamma function. Notice that the $O(x^2)$-term is missing in the above expansion for $g(x)$. This results in the so-called *Tomita sum rule* [87] for the scaled structure factor:

$$\int_0^\infty dp \left[p^{d+1} f(p) - A\right] = 0 \,, \tag{3.11}$$

where A is the amplitude of the Porod tail, i.e., $f(p) \simeq Ap^{-(d+1)}$ as $p \to \infty$.

Next, let us consider specific results for the nonconserved ordering problem. At the outset, we should stress that thermal noise is asymptotically irrelevant for ordering in systems without quenched disorder. This is because thermal fluctuations only affect the interfacial profile. However, the fixed length scale of the interfacial profile becomes irrelevant in comparison to the diverging domain scale [69]. An equivalent argument is due to Bray [70], who used a renormalization-group (RG) approach to demonstrate that domain growth is driven by a fixed point at $T = 0$.

The classic analytical work in this context is due to Ohta et al. (OJK) [88], who introduced an approximate *nonlinear transformation*, which reduces the ordering problem to the diffusive motion of uncorrelated interfaces. This linearization enabled OJK to analytically solve for the correlation function as follows:

$$C(\boldsymbol{r}, t) = \frac{2}{\pi} \sin^{-1} \gamma \,, \quad \gamma = \exp\left(-\frac{r^2}{8Dt}\right) \,, \tag{3.12}$$

where the diffusion constant is $D = (d-1)/d$. The OJK result has the scaling form in (3.6) with a characteristic length scale $L(t) \sim (8Dt)^{1/2}$. Equation (3.12) is in excellent agreement with results obtained from experiments and numerical simulations. Furthermore, it also explicitly contains general features like the Porod tail and the Tomita sum rule, which were discussed above.

An important extension of the OJK result is due to Oono and Puri (OP) [89], who incorporated the nonuniversal effects of nonzero interfacial thickness into the analytical form for the correlation function. This extension was of considerable experimental and numerical relevance because the nonzero interfacial thickness has a severe impact on the tail of the structure factor. In particular, the power-law decay is replaced by a Gaussian decay for finite times, and the Porod tail is only recovered for $t \to \infty$.

3.2.2 Vector Order Parameter

The vector version of the TDGL equation, with $\psi(\boldsymbol{r},t)$ replaced by an n-component vector $\boldsymbol{\psi}(\boldsymbol{r},t)$, is also of great experimental relevance. This has the following form for $T < T_c$ and $H = 0$ (in dimensionless units):

$$\frac{\partial}{\partial t}\boldsymbol{\psi}(\boldsymbol{r},t) = \boldsymbol{\psi} - |\boldsymbol{\psi}|^2\boldsymbol{\psi} + \nabla^2\boldsymbol{\psi} + \boldsymbol{\theta}(\boldsymbol{r},t)\,, \tag{3.13}$$

where $\boldsymbol{\psi} = (\psi_1, \psi_2, ...\psi_n)$, and

$$\begin{aligned}\overline{\boldsymbol{\theta}(\boldsymbol{r},t)} &= 0\,,\\ \overline{\theta_i(\boldsymbol{r}',t')\theta_j(\boldsymbol{r}'',t'')} &= 2\epsilon\delta_{ij}\delta(\boldsymbol{r}'-\boldsymbol{r}'')\delta(t'-t'')\,.\end{aligned} \tag{3.14}$$

For example, the $n = 2$ case (dynamical XY model) is relevant in the ordering of superconductors, superfluids and liquid crystals. Figure 3.2 shows the evolution of the vector field (ψ_1, ψ_2) from a disordered initial condition for the $d = 2$ XY model. The $n = 3$ case (dynamical Heisenberg model) is also of relevance in the ordering of liquid crystals, and in the evolution dynamics of the early universe.

As in the scalar case, the nature of defects in the ordering system determines some general features of the structure factor and the correlation function. These are defined as the vector generalizations of the corresponding scalar quantities, viz.,

$$\begin{aligned}C(\boldsymbol{r},t) &\equiv \frac{1}{V}\int d\boldsymbol{R}\ \left[\langle\boldsymbol{\psi}(\boldsymbol{R},t)\cdot\boldsymbol{\psi}(\boldsymbol{R}+\boldsymbol{r},t)\rangle - \langle\boldsymbol{\psi}(\boldsymbol{R},t)\rangle\cdot\langle\boldsymbol{\psi}(\boldsymbol{R}+\boldsymbol{r},t)\rangle\right]\,,\\ S(\boldsymbol{k},t) &= \int d\boldsymbol{r}e^{i\boldsymbol{k}\cdot\boldsymbol{r}}C(\boldsymbol{r},t)\,.\end{aligned} \tag{3.15}$$

The presence of n-component defects (e.g., vortices for $n = 2, d = 2$; vortex lines for $n = 2, d = 3$; monopoles for $n = 3, d = 3$, etc.) again yields a power-law or *generalized Porod* tail for the scaled structure factor, i.e., $f(p) \sim p^{-(d+n)}$ as $p \to \infty$ [90,91]. The corresponding exact result for the singular behavior of the scaled correlation function at short distances is [86]

$$g(x) = 1 + \pi^{(n-2)/2}\frac{\Gamma\left(\frac{d}{2}\right)\Gamma\left(\frac{n+1}{2}\right)^2\Gamma\left(-\frac{n}{2}\right)}{\Gamma\left(\frac{d+n}{2}\right)\Gamma\left(\frac{n}{2}\right)}x^n + \text{higher-order terms}\,. \tag{3.16}$$

Notice that the singular behavior for even values of n has to be carefully extracted from the above expression, and involves logarithmic corrections as

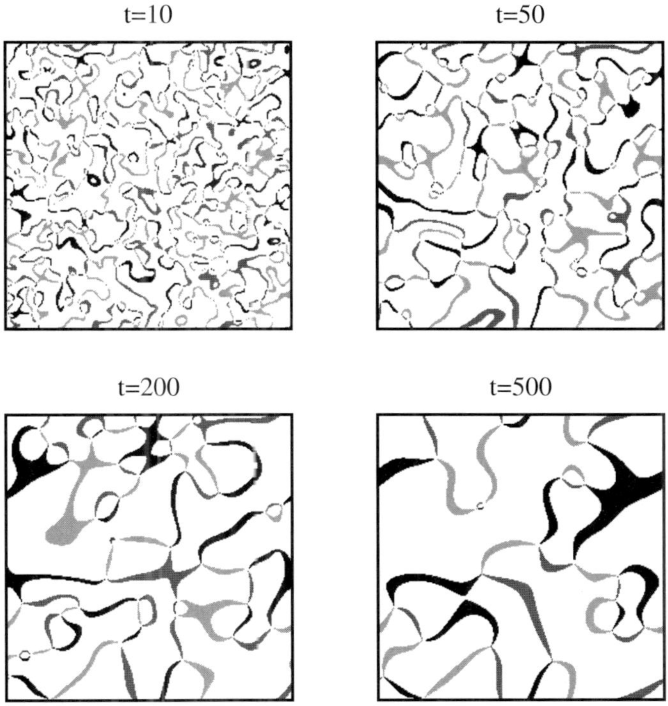

Fig. 3.2. Evolution of the dimensionless XY model from a disordered initial condition for $T < T_c$. The Euler-discretization mesh sizes were $\Delta t = 0.1$ and $\Delta x = 1.0$ in time and space, respectively, and the lattice size was 256^2. The snapshots show regions of constant phase $\theta_\psi = \tan^{-1}(\psi_2/\psi_1)$, measured in radians, with the following color coding: $\theta_\psi \in [1.85, 2.15]$ (black); $\theta_\psi \in [3.85, 4.15]$ (dark grey); $\theta_\psi \in [5.85, 6.15]$ (light grey). Typically, a meeting point of the three colors denotes a vortex defect

$1 - g(x) \sim x^n \ln x$. Furthermore, there is no analog of the Tomita sum rule for the vector ordering problem

Bray and Puri (BP) [90] and (independently) Toyoki (T) [91] used a defect-dynamics approach to obtain an approximate solution for the ordering problem of the n-component TDGL equation in d-dimensional space. The BPT solution is valid for $n \leq d$, corresponding to the case where topological defects are present. These authors demonstrated that the characteristic length scale obeys the LAC growth law, $L(t) \sim t^{1/2}$, in this case also. (A more careful calculation demonstrates that there are logarithmic corrections when $n = d = 2$, and $L(t) \sim (t/\ln t)^{1/2}$ [92].) Furthermore, they also obtained an explicit scaling form for the correlation function:

$$C(\boldsymbol{r}, t) = \frac{n\gamma}{2\pi}\left[B\left(\frac{n+1}{2}, \frac{1}{2}\right)\right]^2 F\left(\frac{1}{2}, \frac{1}{2}; \frac{n+2}{2}; \gamma^2\right), \tag{3.17}$$

where $B(x, y)$ is the beta function; $F(a, b; c; z)$ is the hypergeometric function [65]; and $\gamma = \exp\left(-r^2/L^2\right)$, L being the average defect length-scale. As in the case of the OJK function, (3.17) is valid when the defect core size is identically zero. The nonzero core size in experiments or simulations introduces OP-like corrections in the correlation function and structure factor.

We should stress that the case with $n > d$ is unusual in that there are no topological defects, and it is not possible to describe the evolution of the system in terms of the annealing of defects. As a matter of fact, systems with $n > d$ may not even exhibit dynamical scaling. Some specific results are available for $n = d + 1$, where the system is characterized by the presence of textures [93,94]. In general, systems with textures do not show single-length scaling, as in (3.6). Dynamical scaling is restored in the $n \to \infty$ limit, where the correlation function exhibits a Gaussian decay [81]:

$$C(\boldsymbol{r}, t) = \exp\left(-\frac{r^2}{L^2}\right), \tag{3.18}$$

where $L(t) \sim t^{1/2}$. In general, there are no analytical results available for the correlation function of the n-component TDGL equation with $n > d$.

The results of OJK-OP and BPT have been understood to constitute a complete solution of the dynamics of phase ordering with a nonconserved order parameter. However, the work of Blundell et al. [95] suggests that this may not be the case. Generally, one tests for dynamical scaling by plotting $C(\boldsymbol{r}, t)$ vs. r/L or $S(\boldsymbol{k}, t)L^{-d}$ vs. kL at different times [96]. Equally well, one compares the analytical results with experimental or numerical results using similar plots. However, there is an arbitrariness in the definition of the characteristic length scale which is defined as, e.g., the reciprocal of the first moment of the structure factor, or the distance over which the correlation function falls to half its maximum value. The problem with these definitions, while comparing results from different situations, is that they already build in a high level of agreement on a scaled plot. Blundell et al. [95] have proposed a universal test of dynamical scaling, which does not utilize an internally-defined length scale, but rather uses higher-order structure factors. On such a plot, there is a considerable difference between the OJK/BPT results and the corresponding numerical results. This suggests that our analytical understanding of nonconserved phase ordering may not be so good after all. Subsequently, Mazenko [97] showed that the OJK result is valid for phase ordering systems in the limit $d \to \infty$. There have been attempts to improve on the OJK result for the case of finite d, and these are discussed by Mazenko [98].

3.3 The Case with Conserved Order Parameter

The second class of phase ordering problems we consider involves systems with a conserved order parameter, e.g., phase separation of a binary (AB) mixture. A typical phase diagram for an AB mixture is shown in Fig. 2.2. The appropriate thermodynamic potential to be minimized in equilibrium is now the Helmholtz free energy $F(T, \psi)$, where ψ is the concentration difference of the two species. The energetic considerations are the same for binary mixtures and ferromagnets, Typically, A-A and B-B interactions are attractive ($-E^{AA}, -E^{BB} < 0$ in (2.4)), and A-B interactions are repulsive ($-E^{AB} > 0$ in (2.4)). Thus, it is energetically preferable for the system to separate out into regions rich in A and B, respectively. This is the preferred equilibrium state at low temperatures, i.e., below the coexistence curve in Fig. 2.2. However, at high temperatures, entropic effects cause the system to be disordered (homogeneous). Though the Hamiltonian of the binary mixture is the same as that for the ferromagnet, recall that the phase diagrams in Figs. 2.1 and 2.2 are obtained in different ensembles.

We consider the dynamical evolution resulting due to a quench from above the coexistence curve (homogeneous or disordered phase) to below the coexistence curve (segregated or ordered phase) in Fig. 2.2. The initially homogeneous system separates into regions rich in either A or B. In contrast to the nonconserved case, the evolution in this (conserved) case must satisfy the constraint that numbers of A and B are constant. In the appropriate kinetic Ising model, this was implemented via a spin-exchange Kawasaki mechanism (see Sect. 2.4).

Experimentalists distinguish between shallow quenches (just below the coexistence curve) and deep quenches (far below the coexistence curve). For shallow quenches, in the region between the solid coexistence line and the dashed spinodal lines in Fig. 2.2, the homogeneous system is not spontaneously unstable and decomposes by the *nucleation and growth* of droplets. For deep quenches, into the region below the spinodal lines, the homogeneous system spontaneously decomposes into A-rich and B-rich regions, a process referred to as *spinodal decomposition*. However, there is no sharp physical distinction between the *nucleation and growth* and *spinodal decomposition* regions of the phase diagram [80].

3.3.1 Phase Separation without Hydrodynamic Effects

Let us first focus on the kinetics of phase separation in a binary mixture where hydrodynamic effects are not relevant, e.g., binary alloys. In this case, the primary mechanism for phase separation is diffusion. We have seen in Sect. 2.6.1 that the Cahn–Hilliard–Cook (CHC) equation constitutes a reasonable coarse-grained model for phase separation or conserved kinetics. For the ψ^4-form of the free-energy functional, the CHC equation is as follows:

$$\frac{\partial}{\partial t}\psi(\boldsymbol{r},t) = \boldsymbol{\nabla}\cdot\left\{D\boldsymbol{\nabla}\left[-a(T_c - T)\psi + b\psi^3 - K\nabla^2\psi\right] + \boldsymbol{\theta}(\boldsymbol{r},t)\right\} , \quad (3.19)$$

where $a, b, K (> 0)$ are parameters. The Gaussian white noise satisfies the fluctuation-dissipation relation:

$$\begin{aligned}\overline{\boldsymbol{\theta}(\boldsymbol{r},t)} &= 0\,,\\ \overline{\theta_i(\boldsymbol{r}',t')\theta_j(\boldsymbol{r}'',t'')} &= 2Dk_BT\delta_{ij}\delta(\boldsymbol{r}'-\boldsymbol{r}'')\delta(t'-t'')\,. \end{aligned} \tag{3.20}$$

We rescale variables as follows (for $T < T_c$):

$$\begin{aligned} \psi' &= \frac{\psi}{\psi_0}\,, \quad \psi_0 = \sqrt{\frac{a(T_c-T)}{b}}\,,\\ t' &= \frac{Da^2(T_c-T)^2}{K}t\,,\\ \boldsymbol{r}' &= \sqrt{\frac{a(T_c-T)}{K}}\boldsymbol{r}\,,\\ \boldsymbol{\theta}' &= \frac{\sqrt{bK}}{Da^2(T_c-T)^2}\boldsymbol{\theta}\,. \end{aligned} \tag{3.21}$$

We drop primes to obtain the dimensionless CHC equation

$$\frac{\partial}{\partial t}\psi(\boldsymbol{r},t) = \boldsymbol{\nabla}\cdot\left\{\boldsymbol{\nabla}\left[-\psi+\psi^3-\nabla^2\psi\right]+\boldsymbol{\theta}(\boldsymbol{r},t)\right\}\,, \tag{3.22}$$

where

$$\begin{aligned}\overline{\boldsymbol{\theta}(\boldsymbol{r},t)} &= 0\,,\\ \overline{\theta_i(\boldsymbol{r}',t')\theta_j(\boldsymbol{r}'',t'')} &= 2\epsilon\delta_{ij}\delta(\boldsymbol{r}'-\boldsymbol{r}'')\delta(t'-t'')\,,\\ \epsilon &= \frac{k_BTb\left[a(T_c-T)\right]^{(d-4)/2}}{K^{d/2}}\,. \end{aligned} \tag{3.23}$$

Figure 3.3 shows a typical evolution picture obtained from the dimensionless CHC equation with a random initial condition. The composition of the mixture was 50% A–50% B, i.e., the average value of the order parameter was zero. As in the case of the TDGL equation, thermal noise is asymptotically irrelevant for the CHC equation – the simulation shown in Fig. 3.3 corresponds to the deterministic case with $\epsilon = 0$.

Before we proceed, it is relevant to discuss the applicability of the CHC model to binary alloys. Typically, small lattice parameter mismatches in alloys can set up large strain fields in the intermediate and late stages of phase separation [99–101]. These strain fields drastically modify the somewhat idealized results we quote below, and must be accounted for in any realistic description of phase separation in alloys.

In the absence of such strain effects, the phase-separating system (depicted in Fig. 3.3) is characterized by a unique length scale, $L(t) \sim t^{1/3}$ in $d \geq 2$. This power-law behavior was first derived by Lifshitz and Slyozov (LS) [102] in the context of systems where droplets of one component are growing independently in a homogeneous background, i.e., one of the components is present in a much smaller fraction than the other, corresponding to an *off-critical quench.* Huse [103] demonstrated that the same law is applicable to

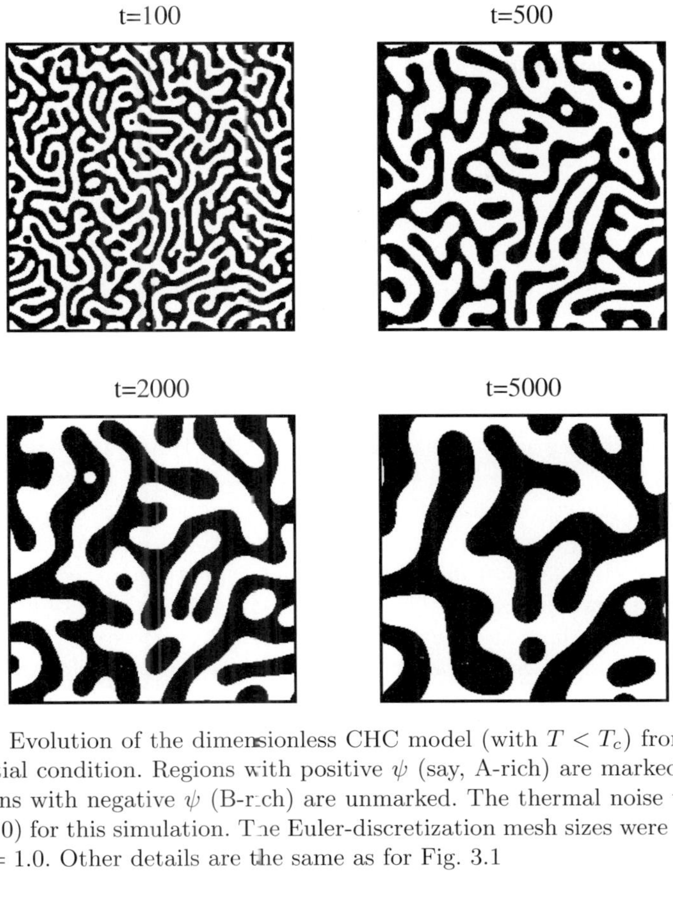

Fig. 3.3. Evolution of the dimensionless CHC model (with $T < T_c$) from a disordered initial condition. Regions with positive ψ (say, A-rich) are marked in black, and regions with negative ψ (B-rich) are unmarked. The thermal noise was set to zero ($\epsilon = 0$) for this simulation. The Euler-discretization mesh sizes were $\Delta t = 0.01$ and $\Delta x = 1.0$. Other details are the same as for Fig. 3.1

spinodal decomposition, where there are approximately equal fractions of the two components and the coarsening structure is bicontinuous. Let us recall the Huse argument here. The chemical potential on the surface of a domain of size L is $\mu \sim \sigma/L$, where σ is the surface tension. The concentration current is obtained as $D|\nabla\mu| \sim D\sigma/L^2$ where D is the diffusion constant. Therefore, the domain size grows as $dL/dt \sim D\sigma/L^2$, or $L(t) \sim (D\sigma t)^{1/3}$.

As in the nonconserved case, the relevant experimental quantities which characterize the evolution morphology in Fig. 3.3 are the correlation function and the structure factor. As before, the existence of a characteristic length scale ensures the dynamical scaling of these quantities. Furthermore, the presence of interfacial defects results in a Porod tail for the scaled structure factor, $f(p) \sim p^{-(d+1)}$ as $p \to \infty$, and a singular short-distance behavior of the correlation function as in (3.10). Additionally, the structure factor obeys the Tomita sum rule in (3.11).

The above properties are common to the nonconserved and conserved cases. There are some additional features due to the conservation law, which we now discuss. For example, the conservation constraint dictates the sum rule:

$$\int d\boldsymbol{r} C(\boldsymbol{r}, t) = 0 \,, \quad \text{or}$$
$$S(0, t) = 0 \,. \tag{3.24}$$

Furthermore, the conservation law also determines the $p \to 0$ behavior of the scaled structure factor as $f(p) \sim p^4$ [104,105].

We have a good understanding of various general features of the morphology. There have also been extensive numerical simulations of phase separation, which have obtained detailed results for late-time behavior. These studies are based on *cell dynamical system* (CDS) models [96,106]; discrete simulations of the CHC equation [107]; and Monte Carlo (MC) simulations of the spin-exchange kinetic Ising model [108,109]. However, inspite of many attempts [110–112], there is still no comprehensive theory for the entire functional form of the correlation function or structure factor. A major analytical obstacle in this regard is the strongly correlated motion of interfaces, resulting from the conservation law. In particular, *Gaussian closure* techniques, which involve linearization of the dynamical equations via a nonlinear transformation [88,90], have not worked well in the conserved case [113].

There have also been some studies of the case with vector conserved order parameter. For brevity, we only mention some representative results here. In the case where topological defects are present ($n \leq d$), Puri et al. [114] demonstrated that Gaussian closure techniques work better than in the case with scalar order parameter. They used these techniques to obtain approximate analytical results for the correlation functions of the conserved XY and Heisenberg models. In the limit $n \to \infty$, an important result is due to Coniglio and Zannetti [115], who showed that the structure factor exhibits *multiple-length* scaling rather than single-length scaling. However, Bray and Humayun [116] have demonstrated that multi-scaling is a singular property of the $n = \infty$ case.

3.3.2 Phase Separation with Hydrodynamic Effects

Next, let us consider the phase separation of immiscible binary fluids. In this case, the hydrodynamic flow field provides an additional mechanism for transport of the segregating components. The appropriate equation for the order-parameter evolution in this case is [66]

$$\frac{\partial}{\partial t}\psi(\boldsymbol{r}, t) = \boldsymbol{\nabla} \cdot \left\{ D\boldsymbol{\nabla} \left(\frac{\delta \mathcal{F}[\psi]}{\delta \psi} \right) - \psi \boldsymbol{v} + \boldsymbol{\theta}(\boldsymbol{r}, t) \right\} , \tag{3.25}$$

where $\boldsymbol{v}(\boldsymbol{r}, t)$ denotes the fluid velocity field. We assume that the fluid is incompressible (with constant density ρ). This imposes a constraint on the velocity field as $\boldsymbol{\nabla} \cdot \boldsymbol{v} = 0$.

The corresponding equation for the velocity field is the usual Navier-Stokes equation with an "external" force:

$$\frac{\partial}{\partial t}\boldsymbol{v}(\boldsymbol{r},t) = \eta\nabla^2\boldsymbol{v} - \left[\psi\boldsymbol{\nabla}\left(\frac{\delta\mathcal{F}[\psi]}{\delta\psi}\right)\right]_\perp + [\boldsymbol{\zeta}(\boldsymbol{r},t)]_\perp\,, \tag{3.26}$$

where the pressure has been eliminated using the incompressibility condition, and η is the viscosity. Furthermore, $[\boldsymbol{X}(\boldsymbol{r})]_\perp$ denotes the transverse part of the vector $\boldsymbol{X}(\boldsymbol{r})$. In momentum space, this is computed as

$$[\boldsymbol{X}(\boldsymbol{k})]_\perp = \boldsymbol{X}(\boldsymbol{k}) - \frac{\boldsymbol{k}\cdot\boldsymbol{X}(\boldsymbol{k})}{k^2}\boldsymbol{k}\,. \tag{3.27}$$

The Gaussian white noises in (3.25)-(3.26) satisfy the fluctuation-dissipation relations:

$$\begin{aligned} \overline{\boldsymbol{\theta}(\boldsymbol{r},t)} &= 0\,, \\ \overline{\theta_i(\boldsymbol{r}',t')\theta_j(\boldsymbol{r}'',t'')} &= 2Dk_BT\delta_{ij}\delta(\boldsymbol{r}'-\boldsymbol{r}'')\delta(t'-t'')\,, \end{aligned} \tag{3.28}$$

and

$$\begin{aligned} \overline{\boldsymbol{\zeta}(\boldsymbol{r},t)} &= 0\,, \\ \overline{\zeta_i(\boldsymbol{r}',t')\zeta_j(\boldsymbol{r}'',t'')} &= -2\eta k_BT\delta_{ij}\nabla^2\delta(\boldsymbol{r}'-\boldsymbol{r}'')\delta(t'-t'')\,. \end{aligned} \tag{3.29}$$

Equations (3.25)-(3.29) are referred to as Model H in the classification scheme of Hohenberg and Halperin [66]. Let us focus on the deterministic ($T=0$) case. The relaxation of the velocity field is much faster than that of the order-parameter field. Thus, we set $\partial\boldsymbol{v}/\partial t = 0$ on the LHS of (3.26). The resultant equation is easily solved in Fourier space for $\boldsymbol{v}$ as

$$\begin{aligned} v_i(\boldsymbol{k},t) &= -T_{ij}(\boldsymbol{k})X_j(\boldsymbol{k},t) \\ &= -\frac{1}{\eta k^2}\left(\delta_{ij} - \frac{k_ik_j}{k^2}\right)X_j(\boldsymbol{k},t)\,, \\ \boldsymbol{X}(\boldsymbol{k},t) &= \int d\boldsymbol{r}e^{i\boldsymbol{k}\cdot\boldsymbol{r}}\psi\boldsymbol{\nabla}\left(\frac{\delta\mathcal{F}[\psi]}{\delta\psi}\right)\,. \end{aligned} \tag{3.30}$$

In (3.30), repeated indices are summed over and we have introduced the Oseen tensor $T_{ij}(\boldsymbol{k})$. In $d=3$, the real-space Oseen tensor is

$$T_{ij}(\boldsymbol{r}) = \frac{1}{8\pi\eta r}\left(\delta_{ij} + \frac{r_ir_j}{r^2}\right)\,. \tag{3.31}$$

Replacing the expression for the velocity in the deterministic version of the order-parameter equation, we obtain the closed evolution equation

$$\frac{\partial}{\partial t}\psi(\boldsymbol{r},t) = \boldsymbol{\nabla}\cdot\left\{D\boldsymbol{\nabla}\left(\frac{\delta\mathcal{F}[\psi]}{\delta\psi}\right) + \psi\int d\boldsymbol{r}'T_{ij}(\boldsymbol{r}-\boldsymbol{r}')X_j(\boldsymbol{r}',t)\right\}\,. \tag{3.32}$$

Equation (3.32) is an appropriate starting point for understanding domain growth laws for coarsening in binary fluids. At early times, growth is diffusion-limited, as in the case of binary alloys. However, there is a crossover to a hydrodynamic growth regime, where material is rapidly transported along domain boundaries by advection [117,118]. The growth laws for different regimes are summarized by Bray [81], and we quote the relevant results here:

$$\begin{aligned} L(t) &\sim (D\sigma t)^{1/3}, \quad L \ll (D\eta)^{1/2}, \\ &\sim \frac{\sigma t}{\eta}, \quad (D\eta)^{1/2} \ll L \ll \frac{\eta^2}{\rho\sigma}, \\ &\sim \left(\frac{\sigma}{\rho}\right)^{1/3} t^{2/3}, \quad \frac{\eta^2}{\rho\sigma} \ll L. \end{aligned} \tag{3.33}$$

The above regimes correspond to *diffusive growth*, *viscous hydrodynamic growth*, and *inertial hydrodynamic growth*, respectively.

Recently, Grant and Elder [119] have argued that scaling requires that the asymptotic growth exponent $\phi \leq 1/2$. Otherwise, the system enters a turbulent regime and would undergo remixing. Alternatively, they suggest that a faster growth law than $L(t) \sim t^{1/2}$ can be sustained if the system does not exhibit dynamical scaling. However, the Grant-Elder arguments have been criticized by Kendon et al. [120]. These authors present results from extensive lattice-Boltzmann simulations [121,122] of phase separation in binary fluids. Their results support the inertial hydrodynamic growth law, $L(t) \sim t^{2/3}$ [118].

We should stress that domain connectivity plays a crucial role in enabling hydrodynamic transport. In highly off-critical quenches, the morphology consists of droplets of a minority phase in a matrix of the majority phase. Then, the hydrodynamic mechanism is disabled and domain growth is analogous to that for binary alloys. Furthermore, thermal fluctuations can drive the Brownian motion and coalescence of droplets. This also gives rise to a LS-like growth law, $L(t) \sim (k_B T t/\eta)^{1/3}$ [117], though the physical mechanism is quite different.

Finally, we should remark on the morphology, as characterized by the correlation function or the structure factor. These statistical properties and the domain growth laws have been investigated in many numerical studies. These include CDS simulations [106,123,124]; molecular dynamics (MD) studies [125,126]; lattice-Boltzmann simulations [120–122], etc. As regards analytical results, the general morphological features discussed in Sect. 3.3.1 arise in the present context also. However, as in the previous case, there is no satisfactory theory for the entire functional form of the structure factor. This remains one of the major unsolved problems in the area of phase ordering dynamics.

3.4 Incorporation of Experimentally Relevant Effects

We have discussed in some detail the two prototypical problems of phase ordering dynamics, viz., the cases with nonconserved and conserved order parameter. These two simple problems provide the basis for a range of further studies, as we briefly elucidate here. Recall that the discussions in Sects. 3.2 and 3.3 were in the context of pure and isotropic systems. Of course, real experimental systems are neither pure nor isotropic. Recent research in this area has attempted to incorporate and study various experimentally relevant effects in phase ordering systems. Let us discuss some of these recent research directions here.

Phase ordering systems typically contain disorder, which can be either quenched or annealed. Quenched (or immobile) disorder is in the form of large impurities, which act as pinning centers for domain interfaces. At early times, domains are small and the effects of disorder are negligible. However, at later times, the interfaces are trapped by disorder sites, and domains can grow only by thermally-activated hopping over disorder traps. The presence of quenched disorder drastically changes the nature of the asymptotic domain growth law, but does not appreciably alter the domain morphology of the evolving system [127–131]. Domain growth with quenched disorder has received considerable attention in the literature, but there are still many issues to be clarified in this context. Chapter 4 will examine some of these issues.

Another class of important problems concerns the role of annealed (or mobile) disorder. Let us consider two particularly important classes of annealed disorder, viz., surfactants and vacancies. Surfactants are amphiphilic molecules which reduce the surface tension between two immiscible fluids (e.g., oil and water) and promote mixing. Consider a phase-separating binary fluid with a small concentration of surfactants. (In general, this would constitute a ternary or three-component mixture.) Typically, the surfactants migrate rapidly to interfacial regions – diminishing the surface tension and, thereby, the drive to segregate. Thus, a binary fluid with surfactants can exhibit a range of fascinating meso-scale structures [132,133]. Vacancies in binary alloys also play a similar role as surfactants [52,56,57,134]. However, we should stress that phase separation in binary alloys is mediated by vacancies rather than direct A-B interchanges, which are energetically expensive. Therefore, vacancies are necessary for the phase separation of binary alloys, and not just a complicating feature in the phase diagram.

Next, we consider the role of anisotropies in phase ordering systems. Many physical situations give rise to anisotropic phase ordering dynamics. This anisotropy can result from external fields. For example, a gravitational field distinguishes between the lighter and heavier components of a mixture, resulting in the growth of elongated domains in the direction of gravity [135–139]. Alternatively, a surface may preferentially attract one of the components of a binary mixture, resulting in an interplay between the dynamics of wetting at the surface and kinetics of phase separation [140,141]. Chapter 5 is devoted to a discussion of this problem of *surface-directed spinodal decomposition*.

Anisotropies in phase ordering systems can result from internal fields also, as mentioned earlier. In real binary alloys, strain fields are invariably set up at interfaces between A-rich and B-rich domains due to lattice parameter mismatches. These long-ranged strain fields induce directional anisotropies, which strongly affect the intermediate and late stages of phase separation in binary alloys [99–101].

The examples quoted above merely constitute the tip of the iceberg. There are a large number of experimentally relevant effects in phase ordering systems, e.g., viscoelasticity and gelation [142–146], shear flows [147–150], multiple components [151–153], etc. These effects are being actively investigated by various groups. As yet, we do not even have a comprehensive experimental or numerical understanding of the asymptotic behavior of phase ordering dynamics in most of the above situations – leave alone an analytical understanding. Therefore, this promises to be a fertile research direction.

We also stress that the general principles of phase ordering kinetics and defect dynamics have rather broad applicability. In Chap. 6, we will discuss how pattern formation in the complex Ginzburg-Landau (CGL) equation can be characterized by using some of the ideas presented here.

4 Domain Growth in Systems with Quenched Disorder

4.1 Introduction

The phase ordering systems described in Chap. 3 were typically pure and isotropic. However, real experimental systems contain disorder in the form of impurities, which may be either mobile or immobile. This chapter focuses on domain growth in systems with immobile or quenched disorder. In Chap. 3, we have already summarized the growth laws and scaling functions which characterize domain growth in pure systems. In general, sites of quenched disorder act as traps for coarsening domains. Once a domain wall is trapped, it can only move by thermal activation over the corresponding energy barrier. There are various classes of disorder studied in the context of experimental systems, e.g., random exchange interactions, random fields, etc.

To be more specific, consider a nearest-neighbor spin-1/2 Ising model with spin variables σ_i on a set of sites labeled by i. In the pure case, recall that the Hamiltonian of this system is (cf. (2.6))

$$\mathcal{H} = -J \sum_{\langle ij \rangle} \sigma_i \sigma_j - H \sum_{i=1}^{N} \sigma_i , \quad \sigma_i = \pm 1 . \tag{4.1}$$

The phase diagram of the ferromagnet is well-understood and is shown in Fig. 2.1. The effect of disorder can be modeled by introducing randomness in either the exchange interaction or the local magnetic field, depending on the physical context. The nearest-neighbor Ising Hamiltonian with disorder is as follows (cf. (2.1)):

$$\mathcal{H} = -\sum_{\langle ij \rangle} J_{ij} \sigma_i \sigma_j - \sum_{i=1}^{N} H_i \sigma_i . \tag{4.2}$$

Here, J_{ij} is a random-bond disorder depending on the pair $\langle ij \rangle$, and H_i is a random-field disorder depending on the site i. The phase diagram of the disordered Ising model has been extensively discussed in the literature [154–160].

One can consider cases where the disorder amplitude has various distributions, e.g., Gaussian, uniform, bimodal, etc. with either zero or nonzero mean. A particularly interesting class of bond disorder in ferromagnets arises from dilution, i.e., replacement of a magnetic atom by a non-magnetic impurity. Then, $J_{ij} = Jp_ip_j$, where p_i takes values 1 (with probability p) or 0 (with probability $1-p$), depending on whether or not the site i is occupied by a magnetic atom. Here, we are primarily interested in the nonequilibrium dynamics of disordered systems and will refer to relevant equilibrium results only when necessary. Chapter 7 contains a more detailed discussion of the equilibrium properties of disordered systems in the context of *quantum glasses*.

The coarse-grained free-energy functional which is analogous to the microscopic disordered Hamiltonian in (4.2) is obtained as a generalization of the pure ψ^4-functional in (2.68) [161]:

$$\mathcal{G}[\psi] = \int d\boldsymbol{r} \left[-\frac{\tau(\boldsymbol{r})}{2}\psi^2 + \frac{b}{4}\psi^4 - H(\boldsymbol{r})\psi + \frac{1}{2}K(\boldsymbol{r})(\boldsymbol{\nabla}\psi)^2 \right] . \tag{4.3}$$

In (4.3), the presence of disorder is modeled via a randomization of the phenomenological coefficient $\tau(\boldsymbol{r})$ (corresponding to the local critical temperature), the magnetic field $H(\boldsymbol{r})$, and the surface tension $K(\boldsymbol{r})$. We are interested in the physical problem of ordering in a ferromagnet or binary alloy with quenched disorder, and will consider the case where $\tau(\boldsymbol{r}), K(\boldsymbol{r}) > 0$ everywhere. Thus, the system is ferromagnetic but there is a random fluctuation in the critical temperature.

This chapter is organized as follows. In Sect. 4.2, we summarize analytical arguments for phase ordering in systems with quenched disorder. In Sect. 4.3, we describe experimental studies of this problem. Finally, Sect. 4.4 describes various models used for numerical studies of this problem, and discusses representative results from these studies.

4.2 Phase Ordering Systems with Quenched Disorder

Let us briefly summarize theoretical arguments for domain growth in systems with quenched disorder. In this context, a useful classification scheme for domain growth laws is due to Lai et al. (LMV) [162]. The LMV scheme is summarized in Sect. 4.2.1. Subsequently, we focus on two separate limits of the Hamiltonian in (4.2). In Sect. 4.2.2, we discuss the case with $H_i = 0$ and random J_{ij}, corresponding to the random-exchange Ising model (REIM). In Sect. 4.2.3, we discuss the case with $J_{ij} = J$ and random fields H_i, corresponding to the random-field Ising model (RFIM).

4.2.1 Classification Scheme for Domain Growth Laws

Typically, a phase ordering system evolves from an initial nonequilibrium configuration towards a final state by going through a number of intermediate configurations. The evolution of the system can be understood in terms of paths in state space. The probability associated with a particular path is determined by the statistical weight of intervening configurations. At low temperatures, paths which encounter the lowest energy barriers will clearly dominate. At higher temperatures, the entropic factor arising from multiplicity of paths should also be accounted for. Therefore, the evolution problem is equivalent to one of diffusion in a high-dimensional configuration space with a complex energy landscape.

Let us elucidate this point with a familiar example, i.e., the pure $d = 2$ Ising model with zero magnetic field. After a quench, the system starts ordering locally and soon develops a number of competing up and down domains (see Fig. 3.1). As time goes on, the number of domains decreases due to growth in the domain size. To understand this coarsening, it is instructive to consider shrinking of a single square domain of side L. The energy cost to flip a spin inside a domain is $8J$. On the boundary, this cost is $4J$, whereas it is 0 at the corners of the domain. Since a corner spin can be flipped without energy cost, the most probable path for domain shrinking begins with the flipping of one of the corner spins, which creates two corners on the edges. This flip is followed by a random walk of the corner along an edge, each step of which has no energy cost. Thus, an edge of size L is eliminated in a time proportional to L^2. This time can be computed accurately if one considers a random walk with a probability p to jump right or left, and a probability q of not jumping. The resultant expression for the first-passage probability of covering a distance L in t steps is as follows [163]:

$$P(L,t) = 1 - \frac{2}{2L-1} \sum_{k=0}^{L-1} \frac{\cos^2 \phi_k}{\sin \phi_k} \left\{1 - [q + 2p\cos(2\phi_k)]^t\right\} , \tag{4.4}$$

where $\phi_k = \pi(2k+1)/(4L-2)$. The first-passage time (upto a prefactor) is obtained by setting $P(L,t)$ to a fixed value, yielding a $t \sim L^2$ behavior even at rather small values of L [57].

The above considerations again lead us to the LCA growth law $L(t) \sim t^{1/2}$, providing an equivalent way of interpreting curvature-driven growth in the context of discrete lattice models. Furthermore, since no energy barrier occurs on this path (i.e., corner performing random walk on an edge), such coarsening occurs even at $T = 0$

An important step towards a general understanding of domain growth laws is due to Lai et al. [162], who provided a classification of growth laws for nonconserved systems – both with and without disorder. LMV proposed four different classes of systems, determined by the dependence of the energy barrier to coarsening on the characteristic scale. Recall the equation for curvature-driven growth in Sect. 3.2.1:

$$\frac{dL}{dt} = \frac{a(L,T)}{L} , \tag{4.5}$$

where we have explicitly included the diffusion coefficient $a(L,T)$. In general, this quantity depends on the length scale and temperature. The LMV classification can be stated in terms of the behavior of $a(L,T)$:

1. *Class 1* systems are those for which $a(L,T)$ is independent of L and remains non-zero as $T \to 0$, i.e., $a(L,t) = a_0$. In these systems, there exist relaxation paths which have no energy barrier, and the nature of relaxation does not depend on the coarsening length scale. The pure Ising model discussed above belongs to this class. In general, $L(t) \sim t^{1/2}$ for this class of systems.
2. *Class 2* systems are those where the minimum energy barriers are again independent of L, but the elementary diffusion process (e.g., a corner moving along an edge) has an energy barrier E_B, so that $a(L,T) = a_0 \exp(-\beta E_B)$. In this case, we again obtain a $t^{1/2}$-law, but with a time-scale $t_0 \sim \exp(\beta E_B)/a_0$. In this case, domain growth will be slow with L constant for times $t \ll t_0$. For large times the growth law is $L(t) \sim (t/t_0)^{1/2}$.
3. For *Class 3* systems, the activation energy grows linearly with L. An example of this is a system with quenched impurities which trap the domain walls with a fixed energy (say ϵ) per unit length. In that case, we have a short-time growth law $L(t) \sim t^{1/2}$, which crosses over to a logarithmic growth law at late times, viz., $L(t) \sim T\epsilon^{-1}\ln(t/t_0)$ with $t_0 = k_B T/(a_0 \epsilon)$.
4. Finally, *Class 4* systems are those for which the energy barriers grow as $E_B \sim \epsilon L^{\psi}$ where $\psi \neq 1$. This situation also arises in the context of systems with quenched disorder [127], and will be discussed subsequently. Again, the short-time growth law is $L(t) \sim t^{1/2}$, which crosses over to a "logarithmic" growth law, viz., $L(t) \sim (T\epsilon^{-1})^{1/\psi}\left[\ln(t/t_0)\right]^{1/\psi}$ with $t_0 = k_B T/(\psi a_0 \epsilon)$. Clearly, systems in Class 3 are just a special case of Class 4 systems.

4.2.2 Random-Exchange Ising Model (REIM)

The appropriate Hamiltonian for the REIM is

$$\mathcal{H} = -\sum_{\langle ij \rangle} J_{ij} \sigma_i \sigma_i \, . \tag{4.6}$$

The corresponding coarse-grained Hamiltonian is obtained by setting $H(\boldsymbol{r}) = 0$ in (4.3). One can consider various possible distributions for the exchange interaction in (4.6):

1. *Gaussian*

$$P(J_{ij}) = \frac{1}{\sqrt{2\pi \overline{J}^2}} \exp\left[-\frac{(J_{ij} - J)^2}{2\overline{J}^2}\right] , \tag{4.7}$$

where J is the mean and $\overline{J}^2$ is the variance.

2. *Uniform*

$$P(J_{ij}) = \frac{1}{J_2 - J_1}, \quad J_1 < J_{ij} < J_2 ,$$
$$= 0 , \quad \text{otherwise} , \tag{4.8}$$

where J_1 and J_2 are the minimum and maximum values of J_{ij}.

3. *Bimodal*

$$P(J_{ij}) = \frac{1}{2}\left[\delta\left(J_{ij} - J_1\right) + \delta\left(J_{ij} - J_2\right)\right] , \tag{4.9}$$

where the bonds have strengths of either J_1 or J_2 with equal probability.

As stated earlier, we will consider the case where $J_{ij} \geq 0$, so that the system is locally ferromagnetic everywhere. For example, one could consider the case where J_{ij} varies uniformly between $(1-\epsilon)J$ and J. Here, ϵ quantifies the degree of disorder. The limit $\epsilon = 0$ corresponds to the pure case with a critical temperature $T_c(\epsilon = 0) \simeq 2.269J/k_B$ for a $d = 2$ square lattice, etc. [38,39]. Further, $\epsilon = 1$ corresponds to the maximally disordered case with $T_c(1) \simeq T_c(0)/2$. Alternatively, one could consider the dilute Ising model (DIM), where magnetic spins are present with probability p. This corresponds to the bimodal distribution with $J_1 = 0, J_2 = J$, though with different weights for the two values. In the DIM, an important characteristic is the *percolation probability* p_c, which depends on both dimensionality and the lattice structure [164,165]. In this case, sufficient dilution ($p < p_c$) results in the breakdown of long-range ferromagnetic order as domains cannot form percolating clusters. In the vicinity of the critical point $p = p_c, T = 0$, the critical line is determined as $k_B T_c(p) \sim -J/\ln(p - p_c)$ in all dimensions [157,166]. We will consider the physical situation where $p > p_c$ and there is a non-zero critical temperature $T_c(p)$. Thus, when the system is quenched below its critical temperature, it is driven towards an asymptotically ordered state.

The physical situation where J_{ij} can be both ferromagnetic and antiferromagnetic is also of great interest in the context of *spin glasses*. The equilibrium and nonequilibrium properties of spin glasses will be discussed in Chap. 7. In this chapter, we will briefly discuss domain growth in spin glasses at an appropriate point.

An important theoretical investigation of nonconserved ordering in the REIM is due to Huse and Henley (HH) [127], whose arguments we briefly present here. In general, quenched disorder tends to trap moving interfaces at locally favorable sites. At temperature $T = 0$, a trapped interface cannot move further and domain growth freezes. However, at non-zero T, the interfaces are thermally activated over the free-energy barriers and continue to move, albeit rather slowly. As discussed earlier, barrier hopping occurs on an Arrhenius time-scale

$$t \sim a_0^{-1} e^{\beta E_B} , \tag{4.10}$$

where a_0^{-1} is a characteristic time-scale, and E_B is the barrier height. HH argue that the energy barriers scale as a power of the characteristic domain size, $E_B(L) \sim \epsilon L^{\psi}$, where ϵ sets the disorder scale. The exponent ψ is related to the roughening exponent ζ and the pinning exponent χ as $\psi = \chi/(2-\zeta)$. Furthermore, the roughening and pinning exponents are related as $\chi = 2\zeta + d - 3$. In the LMV classification of Sect. 4.2.1, the REIM corresponds to a Class 4 system, and the domain growth law is

$$L(t) \sim \left[\frac{T}{\epsilon} \ln\left(\frac{t}{t_0}\right)\right]^{1/\psi} . \tag{4.11}$$

For $d = 2$, the exponents are known to be $\zeta = 2/3$ and $\chi = 1/3$ [167,168], yielding $\psi = 1/4$. For $d = 3$, a perturbative calculation gives $\psi \simeq 0.55$ [127]. Subsequently, we will discuss experiments and simulations conducted to examine the veracity of the HH growth law. However, it should be stressed that (4.11) only represents the asymptotic domain growth law. During the early stages of ordering, the domains are not hindered by the energy barriers and exhibit domain growth similar to that for the pure case. As the domain scale increases, the energy barriers become appreciable and there is a crossover from power-law growth to the HH regime, which is earlier for higher amplitudes of disorder. The crossover time (and length-scale) can be obtained by comparing the length scales for early-time and asymptotic growth in Class 4 systems (see Sect. 4.2.1).

Recall that one of the standard means of randomizing exchanges is by dilution with probability $1-p$. For $p > p_c$, the HH arguments are still applicable. However, unusual physical features are expected at $p = p_c$. Henley [169] and Rammal and Benoit [170] have studied the critical dynamics of the Ising model with Glauber kinetics at $p = p_c, T = 0$. They find that the fractal nature of percolation clusters results in a barrier-scaling form $E_B(L) \sim \ln L$ rather than $E_B(L) \sim L^{\psi}$. The corresponding scaling form for the relaxation time-scale τ_r in the vicinity of the critical point is

$$\tau_r \sim \xi_b^z \equiv \xi_b^{\beta J} , \tag{4.12}$$

where ξ_b is the correlation length, and the dynamical exponent $z = \beta J$ is now temperature-dependent. This should be contrasted with the usual dynamical-scaling scenario for pure systems, where z is a constant [66].

Next, consider the case with $p \to p_c^+$ and $T_c(p) > 0$. In this case, the disordered system undergoes phase ordering if it is quenched below $T_c(p)$. Then, the form of the barrier energy results in the novel growth law

$$L(t) \sim t^{cT/J} , \tag{4.13}$$

where c is a constant. Thus, one expects to recover power-law growth just above the percolation threshold but with a temperature-dependent exponent [171]. This novel growth law would apply for extended periods of time ($\to \infty$ as $p \to p_c^+$) before a crossover to the asymptotic HH regime.

It is also relevant to study the effect of disorder on the scaling forms of the correlation function or structure factor, which characterize the morphology of the evolving system. The typical transverse displacement of interfaces due to disorder roughening is $L^{\zeta/(2-\zeta)}$ [127]. At late times, one has $L \gg L^{\zeta/(2-\zeta)}$, because $\zeta < 1$ above the lower critical dimension. (If $\zeta > 1$, disorder-induced roughening would destroy long-range order or LRO in the system.) Thus, in the asymptotic regime, the roughness is irrelevant compared to the typical domain size, and domain growth is driven by the curvature-reduction mechanism but with a renormalized diffusion constant. Therefore, the evolution morphologies and their statistical properties should be independent of disorder in the asymptotic regime. This has been confirmed in various numerical simulations [129,130], as we will discuss in Sect. 4.4.3.

The above discussion has focused on ordering in ferromagnets, where the system evolves towards an asymptotic state with LRO. Let us briefly discuss the corresponding problem for nearest-neighbor spin glasses, where the exchange interactions can take both positive and negative values, e.g., $J = 0$ in the Gaussian distribution of (4.7). This results in local frustration and a multiplicity of metastable states, with a corresponding slowing-down of dynamical processes in the spin-glass state [172–174]. An important study of nonequilibrium dynamics in spin glasses is due to Fisher and Huse (FH) [128]. The FH arguments are based on the so-called *droplet model*, where the spin glass is presumed to approach either of two equivalent ground states. These states are disordered and are related by a simple symmetry transformation $\sigma_i \to -\sigma_i$. The FH droplet model should be contrasted with the *replica-symmetry breaking (RSB) model* of Parisi [175]. The RSB scenario argues that there are infinitely many ground states, and is verified only for the infinite-ranged Sherrington-Kirkpatrick (SK) model [176].

In the context of the droplet model, one can again investigate domain growth subsequent to a quench, except that the ordering occurs in either of two disordered-looking ground states. In this case, FH use barrier-hopping arguments analogous to those for the REIM, and find that the domain scale grows as (cf. (4.11))

$$L(t) \sim \left[\frac{T}{\Delta(T)} \ln\left(\frac{t}{t_0}\right)\right]^{1/\overline{\psi}}, \tag{4.14}$$

where $\Delta(T)$ sets the disorder scale which is temperature-dependent. FH did not determine the value of the barrier-scaling exponent $\overline{\psi}$, and only suggested the weak bounds, $0 \le \overline{\psi} \le d - 1$.

The spatial structure of coarsening spin glasses has been tested in Monte Carlo (MC) simulations in $d = 2, 3$ by Rieger and others [177,178]. These authors examined the length scales associated with the appropriate correlation function. They found that their length-scale data is equally well-described by the logarithmic law in (4.14) with $\overline{\psi} \simeq 0.71$ (in $d = 3$) and $\overline{\psi} \simeq 0.61$ (in $d = 2$), or power-law growth with a temperature-dependent exponent, $L(t) \sim t^{\phi(T)}$. As yet, there is no conclusive data which confirms the logarithmic growth law in (4.14). As a matter of fact, there continues to be vigorous debate over the applicability of the droplet model or the RSB model to short-ranged spin glasses. Many relevant features of nonequilibrium dynamics in spin glasses have been reviewed by Bouchaud et al. [179] and Crisanti and Ritort [180].

So far, we have been discussing nonconserved domain growth in ferromagnets (or spin glasses) with quenched disorder. It is also relevant to study the phase separation of binary mixtures with quenched disorder. The appropriate microscopic model for this problem is the Hamiltonian in (4.6) in conjunction with Kawasaki spin-exchange kinetics (see Sect. 2.4). Notice that the arguments leading to the domain growth law in (4.11) do not refer to the nature of the microscopic dynamics. The relevant time-scale is the barrier hopping time, which is unaffected (except for the coefficient) by whether or not the order parameter is conserved. Thus, we expect the same asymptotic growth law to apply for the nonconserved and conserved cases [130]. As far as the structure-factor scaling function is concerned, the previous argument suggests that the functional form should not be affected by the presence of disorder. Bray [81] makes the even stronger claim, based on RG arguments, that the asymptotic scaling function for the conserved case is the same as that for the nonconserved case. He argues that the conservation law is irrelevant for disordered systems and, therefore, the conserved and nonconserved systems belong to the same universality class. As we will see later, available results do not substantiate this claim.

4.2.3 Random-Field Ising Model (RFIM)

Another important class of problems with quenched disorder involves the RFIM, which is described by the microscopic Hamiltonian

$$\mathcal{H} = -J \sum_{\langle ij \rangle} \sigma_i \sigma_j - \sum_{i=1}^{N} H_i \sigma_i \,, \tag{4.15}$$

where the field H_i is now a site-dependent random variable. The typical distributions for the random field are analogous to those described in (4.7)-(4.9). The corresponding coarse-grained Hamiltonian is obtained from (4.3) by setting $\tau(\boldsymbol{r}) = \tau, K(\boldsymbol{r}) = K$. As in the case of the REIM, there have been many studies of the equilibrium phase diagram of the RFIM [158–160,181]. For $d \geq 3$, the RFIM is ordered at sufficiently low temperatures for a range of field strengths. However, there is no LRO for $d = 2$, even for infinitesimally small random fields. Thus, the lower critical dimension for the RFIM is $d_c = 2$ [182,183].

Let us next consider the nonconserved ordering dynamics of the RFIM. The broad scenario is similar to that for the REIM or Class 4 systems in Sect. 4.2.1. For early times, the ordering domains are unaffected by the random field and coarsen according to the LAC law. However, once the domains become larger than a critical size, the random-field pinning becomes the growth-determining factor. Subsequent growth proceeds by thermally-activated hopping over free-energy barriers.

The dynamics of hopping was first investigated by Villain [184], Grinstein and Fernandez (GF) [185], and others [186,187]. Villain [184] considered a coarse-grained version of the RFIM and estimated the energy-barrier scaling as $E_B(L) \sim \overline{H}^2 L/J$, where $\overline{H}^2$ is the variance of the random field which has zero average. The corresponding domain growth law in the asymptotic regime is (see Sect. 4.2.1)

$$L(t) \sim \frac{JT}{\overline{H}^2} \ln\left(\frac{t}{t_0}\right) , \qquad (4.16)$$

where $t_0 = k_B TJ/(a_0 \overline{H}^2)$. Equation (4.16) is valid at all times subsequent to the crossover for $d \geq 3$, and at pre-asymptotic times for $d = 2$, before there is a saturation of domain growth due to the absence of LRO. GF [185] arrived at the same asymptotic result in the context of the lattice RFIM. However, for intermediate times, GF argued that domain growth proceeds by the diffusion of corner sites on domain boundaries. The corresponding barriers scale as $E_B(L) \sim \overline{H} L^{1/2}$, and the growth law is

$$L(t) \sim \left[\frac{T}{\overline{H}} \ln\left(\frac{t}{t_0}\right)\right]^2 . \qquad (4.17)$$

As in the case of the REIM, the crossover time to the disorder-affected regime depends inversely on the strength of the random field. An expression for the crossover time can be obtained by comparing the early-time and disorder-affected growth laws for Class 4 systems.

Apart from the domain growth law, the other experimentally relevant quantity is the scaling form of the structure factor. The arguments for the REIM in this regard are applicable in the present context also, with appropriate modifications. Thus, disorder-induced interfacial roughness is asymptotically irrelevant, and we expect dynamical scaling to hold for domain growth in the RFIM with $d \geq 3$. Furthermore, the scaling function should be independent of the random-field amplitude. As we will see later, numerical simulations for the $d = 3$ case are consistent with this prediction.

However, in the $d = 2$ case, we expect a breakdown of dynamical scaling due to the absence of LRO [188]. In physical terms, the asymptotic scaling regime is accessed when $L(t) \gg \xi_D$, the disorder-induced interfacial roughness. The presence of strong disorder delays the onset of asymptotic behavior, because of the correspondingly higher value of ξ_D. Equally well, in the absence of LRO as in $d = 2$, the domain scale saturates at an equilibrium value $L_s(\to \infty$ as $\overline{H} \to 0)$, making it difficult for the above condition to be satisfied. Thus, for *weak* fields in the $d = 2$ case, L_s is large and the RFIM appears to exhibit scaling over simulation time-scales, though it does not access the true scaling regime where $\xi_D/L_s \to 0$. Furthermore, the scaling function is approximately independent of the random-field amplitude. On the other hand, for *strong* fields in the $d = 2$ case, there is a clear breakdown of dynamical scaling and this has been observed in various numerical studies. We will summarize some of these results in Sect. 4.4.4.

Finally, we consider the case of the RFIM with conserved dynamics, which is an appropriate description of a binary mixture with impurities. As far as the asymptotic growth law is concerned, one does not expect any difference between the conserved and nonconserved cases for reasons we have elaborated in Sect. 4.2.2. With regard to structure-factor scaling, we again expect the functional form of the scaling function to be independent of the random field for $d \geq 3$. Furthermore, the $d = 2$ case is expected to exhibit a breakdown of dynamical scaling due to the absence of LRO.

Before concluding this section, we should stress that logarithmic growth laws are not uniquely associated with disordered systems. An important study in this context is due to Shore and others [189,190], who formulated arguments for logarithmic domain growth in Ising models with competing nearest-neighbor and next-nearest-neighbor interactions. Furthermore, as discussed in Sect. 3.2.1, domain growth in $d = 1$ systems is often characterized by logarithmic growth laws. For example, Nagai and Kawasaki [191] have studied the $d = 1$ TDGL equation, where coarsening is driven by an exponentially-decaying interaction between defect tails. This weak interaction again gives rise to logarithmic domain growth.

4.3 Experimental Studies of Domain Growth with Disorder

In principle, all experimental systems contain disorder and should therefore manifest the effects discussed in Sect. 4.2. Clearly, a generic experimental system will contain both quenched and annealed disorders. However, the effects of disorder may not always be seen due to extremely low concentrations. In this section, we discuss representative experiments which are designed to mimic ordering dynamics in the REIM and RFIM. The discussion here is not an exhaustive survey of domain growth in systems with quenched disorder. Rather, we discuss specific experimental studies in the context of the analytical arguments presented in Sect. 4.2.

4.3.1 Domain Growth in the REIM

An important experimental realization of the REIM is the dilute Ising antiferromagnet (AF) $Rb_2Co_pMg_{1-p}F_4$, where the magnetic ion Co is diluted by the introduction of Mg impurities. The dynamics of ordering in this compound has been studied by Ikeda et al. [192]. In the absence of an external magnetic field, the nonconserved ordering of an AF is equivalent to that of a ferromagnet. Of course, the appropriate order parameter for the AF is the *staggered magnetization* (see Sect. 2.5). When the pure AF is placed in a uniform magnetic field, the system is analogous to a ferromagnet in an alternating magnetic field. Therefore, the dilute AF in a uniform magnetic field is analogous to the REIM with a random magnetic field. As stated earlier, the RFIM does not exhibit LRO in the $d = 2$ case. This has been used by Ikeda et al. to investigate the kinetics of disordering for the REIM in a random magnetic field [193,194].

Let us focus on results from the ordering experiments of Ikeda et al. [192], which were done on a thin film of $Rb_2Co_pMg_{1-p}F_4$ with $p = 0.6$. The LRO in this system at low temperatures was destroyed by application of a strong uniform magnetic field, giving rise to a disordered system. The magnetic field was then turned off, and the ordering dynamics of the system was studied by neutron scattering and direct magnetization measurements. At the outset, we should stress that $p = 0.6$ is not a convenient dilution value to investigate many of the theoretical predictions of Sect. 4.2.2. This value is dangerously close to the percolation threshold ($p_c = 0.593$) for the square lattice [164,165], where one expects power-law domain growth for extended periods of time (see Sect. 4.2.2). Regardless of this, Ikeda et al. attempted to fit their experimental data to the logarithmic law in (4.11). They found that a reasonable fit could only be obtained if the exponent $1/\psi$ was taken to be temperature-dependent, which is at variance with the HH arguments. As a matter of fact, even with a temperature-dependent growth exponent, Ikeda et al. do not obtain a particularly good fit to their experimental data. This is not intended to detract from the important work of these authors, but rather to highlight the difficulty of performing clean experiments (or simulations) on disordered systems. Ikeda et al. did not investigate the disorder-independence of the scaling function which characterizes the morphology of domain growth.

Another experimental study on the compound $Rb_2Cu_pCo_{1-p}F_4$ is due to Schins et al. [195]. This compound has a random distribution of ferromagnetic and antiferromagnetic bonds, and is effectively 2-dimensional because of its layered structure. Depending on the value of p, the system can be in either of three states at low temperatures, i.e., antiferromagnetic, spin glass or ferromagnetic. Schins et al. studied ordering dynamics in both the spin glass ($p = 0.67$) and ferromagnetic ($p = 0.89$) phases. They investigated only the asymptotic domain growth law, and did not examine the behavior of the scaling function. Schins et al. found that their data for domain growth in the ferromagnetic phase can be fitted by a logarithmic growth law as in (4.11) with $\psi \simeq 0.20$. This result is consistent with the HH prediction but we should remark that it was obtained only for a limited time-range.

In more recent work, Shenoy et al. [196] have studied the effects of quenched disorder on the coarsening of *reverse tilt domains* in nematic liquid crystals confined to a thin film ($d = 2$). The system of tilt domains can be mapped onto an Ising model, and heterogeneity in the film plays the role of disorder. Shenoy et al. examined the time-dependence of the domain perimeter, and fitted their data to the theoretical growth laws for the pure case ($L(t) \sim t^{1/2}$), the REIM ($L(t) \sim [\ln(t/t_0)]^4$ in $d = 2$), and the RFIM ($L(t) \sim \ln(t/t_0)$). On the basis of the fits obtained, Shenoy et al. concluded that their coarsening system was consistent with the HH scenario for the REIM.

Finally, we discuss the experiments of Likodimos et al. [197,198], who studied domain growth in $d = 2$ *ferroelectrics* (e.g., triglycine sulfate). In this system, the growth law for the pure system is approximately $L(t) \sim t^{1/3}$, which is associated with the presence of long-ranged Coulombic interactions rather than a local conservation law [199,200]. For the disordered system, Likodimos et al. found that the length-scale data is consistent with both the HH law and power-law growth with a temperature-dependent exponent. They also confirmed the dynamical scaling of the correlation function, but did not study the disorder-independence of the scaling function.

There appear to be no experimental studies which specifically investigate the effects of random-exchange quenched disorder in phase-separating binary mixtures. It is easy to introduce heavy (immobile) impurities into binary alloys and one might imagine that these would constitute convenient experimental systems. However, phase separation in binary alloys is often drastically affected by long-ranged strain fields due to lattice parameter mismatches [99–101]. These strain-field effects are dominant in the intermediate to late stages of phase separation in binary alloys. In the context of binary fluids, it is more relevant to discuss the role of mobile disorder. Of course, it is possible to introduce sites of quenched disorder into a binary fluid but it is not apparent how effectively these sites would trap coarsening domains. The presence of hydrodynamic flows would facilitate the evolution of a domain around an obstacle, obviating the need for thermal activation over a free-energy barrier.

4.3.2 Domain Growth in the RFIM

Next, we consider experimental studies of ordering dynamics in the RFIM. We first consider the case with nonconserved dynamics. As discussed in Sect. 4.3.1, a convenient realization of the RFIM is a dilute AF in a uniform magnetic field. The Hamiltonian for this system has been mapped onto that of the RFIM by Fishman and Aharony [201] and Cardy [202]. Other realizations of the RFIM include adsorbed monolayers on a heterogeneous substrate [203].

There are two classes of experiments on ordering dynamics in dilute antiferromagnets in a uniform magnetic field. The first class involves zero-field cooling (ZFC), i.e., the system is prepared in a low-temperature ordered state at zero field, and then the field is turned on. The subsequent evolution and asymptotic state of the system are determined by the amplitude of the random field and the system dimensionality. The second class of experiments involves field cooling (FC), i.e., the system is prepared in a disordered state at high temperatures in a magnetic field, and is then quenched below the ordering temperature.

Wong and Cable [204] conducted early experiments on time-dependent phenomena in the dilute AF $Fe_pMg_{1-p}Cl_2$ (with $p = 0.7$) in a uniform magnetic field. This system constitutes a $d = 3$ realization of the RFIM. In particular, Wong and Cable were interested in the hysteretic behavior arising from different experimental protocols, i.e., the path taken in the $(T, \overline{H})$-plane. On the time-scales of their experiments, they found that ZFC paths resulted in states with LRO, whereas FC paths did not. Similar results were obtained by Cowley and others [205,206] on $Mn_pZn_{1-p}F_2$ and $Fe_pZn_{1-p}F_2$, and Shapira et al. [207] on $Mn_pZn_{1-p}F_2$ (with $p = 0.75$). These hysteretic effects result from the presence of many metastable states in the corresponding free-energy landscape. As we have discussed in Sect. 4.2, trapping in these metastable states gives rise to slow logarithmic growth as in (4.16).

More quantitative studies of the dynamical behavior of the RFIM are due to Lederman et al. [208], who studied $Fe_pZn_{1-p}F_2$ (with $p = 0.46$) over a wide range of temperatures and magnetic fields in the ordered state. These authors studied domain growth subsequent to a quench, and summarized their results in a *dynamical phase diagram* in $(T, \overline{H})$-space. Essentially, they identified three regions of parameter space:

1. *Random-field dynamics* occurs for values of $\overline{H}$ and T in a region just below the paramagnetic-ferromagnetic transition line. Here, domain growth is consistent with the logarithmic law in (4.16).
2. *Random-exchange dynamics* occurs for values of $\overline{H} \to 0$ and moderate values of T, where the effects of random-field disorder are not seen on experimental time-scales.
3. *Frozen states* occur for low values of $\overline{H}$ and T.

Finally, we mention the study of Feng et al. [209] on domain growth in $Fe_pZn_{1-p}F_2$ (with $p = 0.5$). They obtained results consistent with (4.16) for moderate field values. Feng et al. also observed an extended regime of square-logarithmic growth as in (4.17) for small field values, where the onset of the asymptotic regime of (4.16) is delayed considerably.

We have stated earlier that adsorbed monolayers on impure substrates also constitute an experimental realization of the RFIM, though with dimensionality $d = 2$. Zuo et al. [210] have studied the ordering of oxygen adsorbed on a W(112) surface at half-monolayer coverage. They found that the initial growth regime was consistent with the LAC law. Subsequently, there was a crossover to logarithmic growth as in (4.16), with the crossover being earlier for higher levels of randomness. Similar results were obtained from quenching experiments in a rather different system, i.e., the kinetics of structural transitions in the high-temperature superconductor $YBa_2Cu_3O_{7-x}$ (with $x = 0.5$). This was studied by Schleger et al. [211], who demonstrated that the ordering to the Ortho-II state of YBCO was characterized by logarithmic dynamics.

As in the case of the REIM, experiments on RFIM systems have primarily focused on domain growth laws. There have been few results concerning the nature of the structure-factor scaling function, and its dependence on disorder. Recall that we expect the evolution morphology to be independent of the disorder amplitude for $d \geq 3$, where the RFIM exhibits LRO.

Next, let us consider the problem of phase separation in the RFIM. It was first suggested by Brochard and de Gennes [212,213] that phase-separating binary fluids in porous media constitute a realization of the RFIM with conserved kinetics. They argued that the porous medium gives rise to a local field because of its preference for one of the components of the mixture. The density of the porous medium has a random spatial variation, so the local field has both systematic and random parts. However, this argument has been criticized in the context of low-porosity media like Vycor glass [214–216]. In this case, it is believed that a single-pore model without any randomness can be used to understand domain growth, i.e., effects of geometrical confinement determine the nature of ordering in low-porosity media. We do not wish to enter into details of this controversy here – a good review is given by Liu and others [214,215]. Rather, we confine ourselves to the statement that binary fluids in high-porosity media (e.g., silica gels) constitute a reasonable realization of phase separation in the RFIM. Of course, the phase separation of binary fluids is more complicated than that of binary alloys as one has to account for hydrodynamic flows in the former case.

As important set of experiments on critical phenomena in binary fluids (lutidine-water [217] and isobutyric acid-water [218,219]) in silica gels were conducted by Frisken, Cannell and co-workers. These early works primarily focused on the static aspects of phase separation, and the critical exponents associated with the phase transition between the homogeneous and phase-separated states.

Subsequently, Frisken et al. [220] and Bailey et al. [221] have also studied dynamical aspects of phase separation of lutidine-D_2O and isobutyric acid-water mixtures in silica gels. Their results demonstrate that domain growth occurs slower than in pure mixtures, but the domain size becomes considerably larger than the distance over which fluctuations in the silica concentration field are correlated. These authors did not quantify the asymptotic domain growth law or the functional form of the time-dependent structure factor.

This concludes our review of representative experimental studies of domain growth in the REIM and RFIM. Most of these studies have focused on identifying the domain growth law in disordered systems. In general, the evidence seems to support the HH growth law (4.11) for the REIM, and the logarithmic growth law (4.16) for the RFIM. However, most of the fits to experimental data have been guided by the analytical predictions, and it is not clear as to what other functional forms might be reasonable. Furthermore, the experimental data is usually available over restricted time-scales, whereas the unambiguous confirmation or contradiction of the logarithmic laws discussed in Sect. 4.2 would require quite a few decades in time.

4.4 Numerical Studies of Domain Growth with Disorder

Clearly, numerical simulations of appropriate models are relevant in the present context. In this section, we will discuss numerical results for ordering dynamics in systems with quenched disorder. We will also use this opportunity to provide details of simulational approaches to domain growth problems.

We can study domain growth in disordered systems through MC simulations of the Ising model with stochastic dynamics, e.g., Glauber spin-flip or Kawasaki spin-exchange kinetics. The MC approach will be discussed in Sect. 4.4.1 below. In general, MC simulations of kinetic Ising models in the pure case have not proved convenient for accessing the late stages of domain growth, especially in the context of conserved kinetics. The quench temperature cannot be too low, as this results in freezing in metastable states. However, for moderately high temperatures, there are large microscopic fluctuations which give rise to ragged interfaces. Recall that the asymptotic regime arises when the interface thickness is irrelevant compared to the characteristic domain size. Thus, the rough interfaces in MC simulations of ordering dynamics delay the onset of the asymptotic regime.

An alternative approach for studies of domain growth problems is based on simulations of the coarse-grained models discussed in Sects. 3.2 and 3.3, i.e., the time-dependent Ginzburg-Landau (TDGL) equation for the nonconserved case, and the Cahn-Hilliard-Cook (CHC) equation for the conserved case. In Sect. 4.4.2, we will discuss how these equations can be solved numerically. Apart from their numerical utility, these coarse-grained models are also convenient starting points for analytical approaches to phase ordering dynamics.

In the specific context of disordered systems, one can use either of the above approaches. It is clear from the discussion in Sects. 4.2 and 4.3 that the broad issues to be clarified are as follows:

1. Is asymptotic domain growth characterized by a unique length scale? If yes, what is the time-dependence of this length-scale?
2. Is the scaling form of the correlation function or the structure factor independent of the disorder amplitude?

3. In the case of random exchange arising from dilution of magnetic spins, are there any unusual effects associated with the percolation threshold for the dilute magnet?

4.4.1 Monte Carlo Simulations

Let us start by briefly discussing MC simulation techniques. The MC approach to statistical physics has been extensively discussed by Binder and Heermann [44] and Binder [222]. Here, we will only focus upon MC techniques for domain growth problems. The MC approach has the advantage that the microscopic model is almost obvious, but suffers from the disadvantage that inherent fluctuations hinder the extraction of asymptotic results. The appropriate microscopic Hamiltonian for disordered systems is given by (4.2). In Chap. 2, we had stressed that the Ising model has no intrinsic dynamics. Therefore, we associate stochastic dynamics with the Ising model, induced through contact with an external heat bath. The spin configuration $\{\sigma_i\}$ or $\{m_i\}$ is evolved according to a transition probability $W(\{m_i\}|\{m_i\}')$, which depends on the energy change and the heat-bath temperature. This transition probability must satisfy the detailed-balance condition so that the evolution drives the system towards equilibrium.

Typical forms of the transition probability used in the MC literature [44] are as follows (cf. (2.18) with $\lambda = 1$):

$$W(\{m_i\}|\{m_i\}') = \frac{1}{2}\left[1 - \tanh\left(\frac{\beta \Delta\mathcal{H}}{2}\right)\right], \tag{4.18}$$

or

$$\begin{aligned} W(\{m_i\}|\{m_i\}') &= e^{-\beta\Delta\mathcal{H}} \quad \text{if} \quad \Delta\mathcal{H} > 0\,, \\ &= 1 \quad \text{if} \quad \Delta\mathcal{H} \leq 0\,. \end{aligned} \tag{4.19}$$

In (4.18)-(4.19), $\Delta\mathcal{H}$ is the change in energy associated with the change in configuration:

$$\Delta\mathcal{H} = \mathcal{H}(\{m_i\}') - \mathcal{H}(\{m_i\})\,, \tag{4.20}$$

where $\mathcal{H}(\{m_i\})$ is obtained from (4.2) by replacing σ_i with m_i.

A typical MC simulation of domain growth starts from a random configuration of spins, mimicking the high-temperature disordered state before the quench. Depending on the physical context, this spin configuration is then evolved via an appropriate stochastic kinetics at a temperature below the ordering temperature. The subsequent evolution mimics the far-from-equilibrium evolution of a phase ordering system. Finally, the system relaxes to equilibrium (for a finite system) and then wanders around in an equilibrium ensemble. We should stress that the MC dynamical procedure differs from the equilibrium prescription in three important respects:

1. The initial ensemble $\mathcal{P}(\{m_i\}, 0)$ is carefully chosen to reflect the far-from-equilibrium state of interest. For studies of phase ordering dynamics, one considers the initially disordered state as corresponding to a configuration in which spins are randomly oriented with no spatial correlation between different spins.
2. We must be careful about how we make random changes to the configuration $\{m_i\}$ as the stochastic dynamics must reflect the underlying physics of the problem. Of course, in equilibrium MC, we can use a wide range of stochastic moves to change the configuration because we are interested only in the equilibrium ensemble, which is independent of the dynamics of configurational change.
3. In the nonequilibrium case, it is appropriate to work with an ensemble of systems and evolve them separately according to the MC prescription. The probability distribution of systems in the ensemble is time-dependent and the expectation value (also time-dependent) of an operator $A(\{m_i\})$ can be obtained by averaging over this distribution as

$$\langle A(t) \rangle = \sum_{\{m_i\}} A(\{m_i\}) \mathcal{P}(\{m_i\}, t) . \tag{4.21}$$

This should be contrasted with the equilibrium case, where one can invoke ergodicity of the system (which holds in most cases of interest) to compute the expectation value of an operator as the time-average of a single system.

4.4.2 Simulations of Coarse-Grained Models

Next, let us consider simulations of coarse-grained models for domain growth with quenched disorder. In this context, one possibility is to study mean-field (MF) dynamical models for kinetic versions of the Ising model in (4.2). For the spin-flip case, the relevant MF dynamical model is (2.24), with appropriate distributions for J_{ij} and H_i. The corresponding model for the spin-exchange case is (2.47). Equations (2.24) and (2.47) are obtained by averaging over configurations and noise, in conjunction with the MF approximation. This eliminates the fluctuations present at the microscopic level, and it is appropriate to think of (2.24) and (2.47) as macroscopic equations, even though they are defined on a discrete lattice. These MF dynamical models constitute a convenient method of accessing the late stages of phase ordering dynamics [52,53]. Furthermore, they are directly defined in terms of a continuum spin variable $\psi_i = \langle \sigma_i \rangle$, eliminating the need for identifying an appropriate order parameter, which may not always be an easy task.

Before proceeding, it is relevant to discuss the disadvantages of this approach. As a result of the MF approximation, the disorder is smeared out and geometrical properties (e.g., percolation) are not present. Moreover, the macroscopic models in (2.24) and (2.47) have to be supplemented with thermal fluctuations for them to be useful in the context of ordering in systems with quenched disorder. Recall that domain growth in these systems proceeds via thermally-activated hopping of domains over disorder barriers.

Alternatively, we can investigate phenomenological models of domain growth like the TDGL equation in Sect. 3.2, and the CHC equation in Sect. 3.3. Recall the TDGL equation from (2.67):

$$\frac{\partial}{\partial t}\psi(\boldsymbol{r},t) = -\Gamma\frac{\delta\mathcal{G}[\psi]}{\delta\psi} + \theta(\boldsymbol{r},t)\,. \tag{4.22}$$

The free-energy functional in (4.3) yields the evolution equation:

$$\frac{\partial}{\partial t}\psi(\boldsymbol{r},t) = \Gamma\left[\tau(\boldsymbol{r})\psi - b\psi^3 + H(\boldsymbol{r}) + K\nabla^2\psi\right] + \theta(\boldsymbol{r},t)\,, \tag{4.23}$$

where we have neglected the (irrelevant) spatial variation of the surface tension, i.e., $K(\boldsymbol{r}) = K$. Typically, the parameters $\tau(\boldsymbol{r})$ and $H(\boldsymbol{r})$ exhibit fluctuations about an average value, i.e., $\tau(\boldsymbol{r}) = \tau_0 + \delta\tau(\boldsymbol{r})$, $H(\boldsymbol{r}) = H_0 + \delta H(\boldsymbol{r})$. In (4.22)-(4.23), the Gaussian white noise $\theta(\boldsymbol{r},t)$ has zero mean and obeys the fluctuation-dissipation relation:

$$\overline{\theta(\boldsymbol{r}',t')\theta(\boldsymbol{r}'',t'')} = 2\Gamma k_B T\delta(\boldsymbol{r}'-\boldsymbol{r}'')\delta(t'-t'')\,. \tag{4.24}$$

As stated earlier, we focus on the case where $\tau(\boldsymbol{r}) > 0$ for all $\boldsymbol{r}$ so that the homogeneous system is thermodynamically unstable everywhere. Of course, the local minima of the ψ^4-functional will now have a spatial variation whose amplitude depends upon the amount of disorder.

As in Sect. 3.2, it is convenient to introduce dimensionless variables so as to focus only on relevant parameters. The appropriate rescaling is as follows (cf. (3.3)):

$$\begin{aligned}
\psi' &= \frac{\psi}{\psi_0}\,, \qquad \psi_0 = \sqrt{\frac{\tau_0}{b}}\,,\\
t' &= \tau_0\Gamma t\,,\\
\boldsymbol{r}' &= \sqrt{\frac{\tau_0}{K}}\boldsymbol{r}\,,\\
H_0' &= \frac{1}{\tau_0\psi_0}H_0\,,\\
\theta' &= \frac{1}{\tau_0\Gamma\psi_0}\theta\,.
\end{aligned} \tag{4.25}$$

The corresponding dimensionless forms of (4.23)-(4.24) are

$$\frac{\partial}{\partial t}\psi(\boldsymbol{r},t) = [1+\delta a(\boldsymbol{r})]\psi - \psi^3 + H_0 + \delta b(\boldsymbol{r}) + \nabla^2\psi + \theta(\boldsymbol{r},t)\,, \tag{4.26}$$

$$\overline{\theta(\boldsymbol{r}',t')\theta(\boldsymbol{r}'',t'')} = 2\epsilon\delta(\boldsymbol{r}'-\boldsymbol{r}'')\delta(t'-t'')\,, \tag{4.27}$$

where the primes have been dropped. In (4.26), $\delta a(\boldsymbol{r}) = \delta\tau(\boldsymbol{r})/\tau_0$ and $\delta b(\boldsymbol{r}) = \delta H(\boldsymbol{r})/(\tau_0\psi_0)$ are random fluctuations with zero mean. The amplitude of these fluctuations characterizes the amount of disorder. Furthermore, the noise amplitude is

$$\epsilon = \frac{k_B T b \tau_0^{(d-4)/2}}{K^{d/2}} . \tag{4.28}$$

A simple method of solving (4.26) numerically is by implementing an Euler-discretized version on a lattice. There are two important features which must be respected by such a discretization. First, the spatial mesh size Δx must be sufficiently small to resolve the interface region. The equilibrium kink solution for the dimensionless TDGL equation at $T = 0$ and without disorder is

$$\psi_s(z) = \tanh\left(\frac{z - z_0}{\sqrt{2}}\right) , \tag{4.29}$$

where z_0 denotes the position of the interface. Therefore, the typical interface width $\simeq 2\sqrt{2}$. Second, the time mesh size Δt must be small enough to avoid a subharmonic bifurcation in the bulk domains where $\psi = \pm 1$ [107,223]. We know that bulk domains are stable for the TDGL equation, and this property must be respected by the Euler discretization. The corresponding constraint on Δt for the disorder-free TDGL equation is

$$\Delta t < \frac{\Delta x^2}{\Delta x^2 + 2d} . \tag{4.30}$$

An Euler scheme which respects these conditions will give the correct physics of domain growth. Of course, one can always improve the simulation by using more sophisticated schemes to reduce the numerical error and weaken the constraint on Δt. It is not our intention to discuss these schemes in detail here.

An alternative route is due to Oono and Puri [96], who formulated cell dynamical system (CDS) models of phase ordering dynamics. The principle underlying CDS modeling is the direct implementation of the relaxational continuous-time dynamics of the order parameter at the discrete level by introduction of a suitable local mapping. Again, the guiding principle is the replication of correct physics, rather than a numerically accurate (and possibly useless) simulation of the TDGL or CHC equations.

Next, focus on the case with conserved order parameter. The appropriate evolution equation for the case without disorder is the CHC equation (3.19). Following the rescaling in Sect. 3.3.1, we obtain the dimensionless CHC equation for the case with disorder as

$$\frac{\partial}{\partial t}\psi(\boldsymbol{r},t) = \boldsymbol{\nabla}\cdot\left\{\boldsymbol{\nabla}\left[-(1+\delta a(\boldsymbol{r}))\psi + \psi^3 - \delta b(\boldsymbol{r}) - \nabla^2\psi\right] + \boldsymbol{\theta}(\boldsymbol{r},t)\right\} , \tag{4.31}$$

where

$$\overline{\theta_i(\boldsymbol{r}',t')\theta_j(\boldsymbol{r}'',t'')} = 2\epsilon\delta_{ij}\delta(\boldsymbol{r}'-\boldsymbol{r}'')\delta(t'-t'') . \tag{4.32}$$

The corresponding Euler-discretized version must satisfy similar conditions as for the TDGL equation. In the disorder-free case, the constraint on the time mesh size Δt is

$$\Delta t < \frac{\Delta x^4}{4d(\Delta x^2 + 2d)} . \tag{4.33}$$

This condition is considerably stronger than (4.30) because of the additional Laplacian operator in the CHC equation.

Let us next discuss specific numerical results obtained for domain growth in disordered systems using the above techniques.

4.4.3 Domain Growth in the REIM

Case with Nonconserved Kinetics

The relevant experimental system in this case is the random-exchange ferromagnet (or antiferromagnet), where quenched disorder is introduced in the form of immobile non-magnetic impurities. The appropriate microscopic Hamiltonian is (4.6), and the corresponding coarse-grained Hamiltonian is (4.3) with $H(\boldsymbol{r}) = 0$.

First, we focus on MC simulations of the REIM with Glauber kinetics. Early MC simulations of ordering in random magnets are due to Grest and Srolovitz, who studied domain growth in $d = 2$ for the DIM [224] and the dilute Potts model [225]. These authors considered quenches to both $T = 0$ and $T > 0$. They found that domain growth in the $T = 0$ case was pinned by impurities with a saturation length $L_s \sim (1-p)^{-1/2}$, where p is the concentration of magnetic spins. For $T > 0$, Grest and Srolovitz observed that domain growth was nonalgebraic in time but were unable to quantify the asymptotic growth law. Subsequent MC studies of the DIM were done by Chowdhury and others [226,227]. These studies found the same broad features, viz., a crossover from an early-time LAC growth regime to a regime in which domain growth was anomalously slow and did not exhibit power-law behavior. The crossover to the nonalgebraic regime was faster for higher disorder amplitudes and lower temperatures. These early studies were of a somewhat qualitative nature and did not clarify the nature of the growth law in the asymptotic regime. Furthermore, these studies did not investigate the scaling form of the time-dependent structure factor.

In related work, Oh and Choi [228] undertook MC studies of the nonconserved REIM in $d = 2$ with a Gaussian distribution of the disorder. They focused on the time for shrinkage of initially circular domains, which is directly related to the corresponding domain growth law, as discussed in Sect. 4.2.1.

Oh and Choi argued that the shrinkage time t_s for a domain of initial size L obeyed the power law $L \sim t_s^{\phi(T)}$, with the exponent being smaller for larger disorder amplitudes and lower temperatures.

Later MC studies have attempted to quantify the nature of domain growth in the REIM. Bray and Humayun [229] conducted a comprehensive $d = 2$ MC study of domain growth in the REIM with a uniform distribution of J_{ij}. Of course, one does not expect the precise distribution of disorder to materially affect asymptotic results for this problem. The results of Bray and Humayun were in conformity with earlier MC results regarding the nonalgebraic nature of the domain growth law. An important feature which was confirmed by them is the disorder-independence of the scaled correlation function. (The same result was simultaneously obtained in CDS simulations of domain growth in random magnets [129], which we will discuss shortly.) The Bray-Humayun study is more ambiguous about the form of the asymptotic domain growth law and neither confirms nor contradicts the HH law.

Before proceeding, it is relevant to discuss the difficulties associated with unambiguous numerical observation of a logarithmic growth law of the HH form in (4.11). Power-law growth in pure systems is relatively easy to confirm – data over 2-3 decades in time usually suffices to fix the exponent ϕ in a growth law of the form $L(t) \sim t^{\phi}$. However, a logarithmic growth law of the HH form would require at least 9-10 decades in time to get a clear indication of the growth exponent $1/\psi$. Unfortunately, the available data ranges over only a few decades in time. Let us elaborate this point further. For a weak disorder amplitude, the crossover to disorder-affected growth is delayed and there is a possibility of late-stage data being unreliable due to finite-size effects. Therefore, one needs to focus on the strong disorder regime. However, in the case of dilution disorder, the amplitude of disorder must not be so strong that the system is near the percolation threshold. As discussed in Sect. 4.2.2, unusual effects are expected in the vicinity of percolation. Now, in the strong disorder regime, there is a rapid crossover from power-law growth to slow growth. However, the system exhibits freezing in the intermediate to late stages, limiting the time-regime over which there is any appreciable domain growth. In principle, it may be argued that this is perfectly reasonable and this *frozen* regime actually corresponds to the slow logarithmic growth. However, in practice, the frozen regime shows no domain growth at all, complicating the extraction of an unambiguous growth law. These and similar problems have plagued the conclusive confirmation of logarithmic domain growth in disordered systems.

Next, consider the $d = 2$ MC studies of Chowdhury and Biswal [171], who investigated the effects of percolation on domain growth in the DIM. Recall that one expects a power-law growth with a temperature-dependent exponent over extended time-scales when $p \simeq p_c$. The MC results of Chowdhury and Biswal were obtained for $p = 0.75$ and 0.65, which is considerably above the $d = 2$ percolation threshold, i.e., $p_c = 0.593$. In this case, one expects to see the effects of percolation over a time-scale which diverges as $p \to p_c^+$. Chowdhury and Biswal found results consistent with the Henley-Rammal-Benoit arguments in Sect. 4.2.2. However, these results are not conclusive because of considerable scatter in the numerical data. Chowdhury and Biswal did not examine the scaling function for domain growth. The morphology of coarsening domains is substantially different on a fractal percolation cluster [230] and signals of this difference should be seen even for $p > p_c$. Clearly, it is desirable that further simulations be conducted to elucidate the nature of domain growth in dilute magnets for $p \simeq p_c$.

In recent work, Jain [231] has focused upon *persistence exponents* [232, 233] in the $d = 2$ DIM with both weak and strong dilution. In pure systems, the probability that a spin has not flipped (or persists) upto time t decays as $P(t) \sim t^{-\theta_p}$, where θ_p is the persistence exponent. Jain found that the persistence probability for disordered systems showed a crossover from this algebraic form to a logarithmic form, but did not quantify the exponents associated with the logarithmic decay.

Next, consider results obtained from simulations of coarse-grained models. Puri et al. (PCP) [129] undertook detailed studies of $d = 2$ CDS models for ordering in random magnets for a wide range of disorder amplitudes. Of course, it should be kept in mind that percolation effects are not present in these coarse-grained models. PCP demonstrated that the scaled structure factors were independent of the disorder amplitude. However, their results for the domain growth laws were inconclusive, as with previous studies. For example, PCP obtained domain growth consistent with the HH prediction for a range of disorder amplitudes and time-regimes. However, they were unable to describe all their numerical data by a single universal law.

Let us present some typical numerical results for domain growth in disordered ferromagnets. Figure 4.1 shows evolution snapshots and order-parameter profiles obtained from the TDGL equation in (4.26) with zero magnetic field. The snapshots on the LHS correspond to the same time for different disorder amplitudes, and should be compared with the snapshots for the pure ferromagnet in Fig. 3.1. Clearly, the morphology for the disordered case is similar to that for the pure case. The frames on the RHS of Fig. 4.1 show the variation of the order parameter along a typical cross-section. The bulk domain structure is considerably ragged due to the random variation of the local critical temperature.

Figure 4.2 shows the scaled structure factors for the evolution depicted in Fig. 4.1. In Fig. 4.2(a), we plot $S(k,t)L^{-2}$ vs. kL, which is the scaling suggested by (3.8). Notice that $S(k,t)$ is obtained by spherically averaging the vector function $S(\boldsymbol{k},t)$. Data is shown for different disorder amplitudes, as well as for the pure case (denoted by a solid line). Here, the characteristic length scale is defined from the structure factor, $L(t) = \langle k \rangle^{-1}$, where $\langle k \rangle$ is the first moment of the spherically-averaged structure factor:

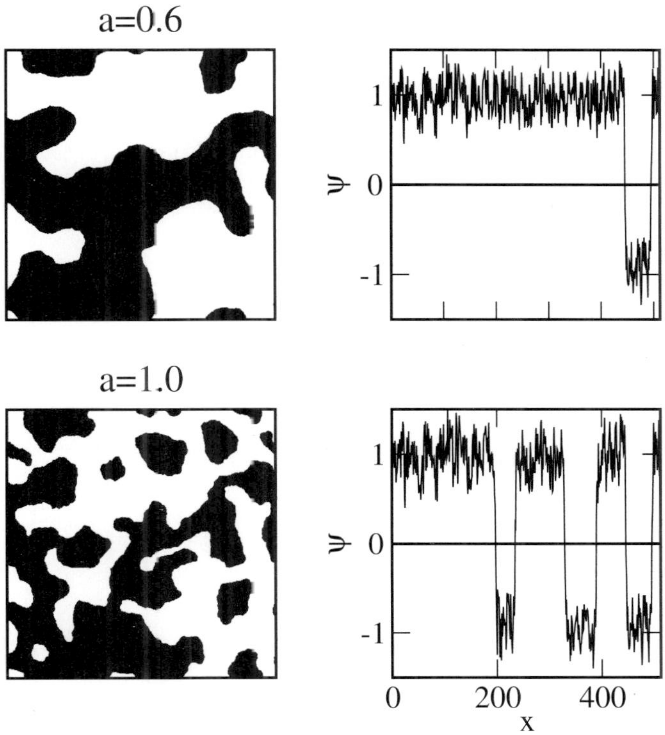

Fig. 4.1. Evolution pictures and order-parameter profiles for the REIM with nonconserved kinetics. These pictures were obtained from an Euler-discretized version of the TDGL equation (4.26) with zero field. The discretization mesh sizes were $\Delta t = 0.1$ and $\Delta x = 1.0$, and the lattice size was 512^2. The noise amplitude was $\epsilon = 0.2$. The disorder field $\delta a(\boldsymbol{r})$ is uniformly distributed in the interval $[-a, a]$. The snapshots on the LHS correspond to $t = 10000$ for different disorder amplitudes, $a = 0.6$ and $a = 1.0$. The black regions have order parameter $\psi > 0$. The frames on the RHS show the variation of the order parameter along a horizontal cross-section in the middle of the snapshots

$$\langle k \rangle = \frac{\int_0^{k_m} dk k S(k,t)}{\int_0^{k_m} dk S(k,t)} . \tag{4.34}$$

In (4.34), k_m is an upper cut-off on the momentum integral. Of course, the characteristic length scale can be defined in a variety of ways, e.g., from higher moments of $S(k,t)$, or zero-crossings of the correlation function $C(r,t)$. All these definitions are equivalent in the dynamical-scaling regime.

The scaled structure factors are seen to be disorder-independent in Fig. 4.2(a). This universality extends into the tail region, as is seen in Fig. 4.2(b), which is a log-log plot of Fig. 4.2(a). Notice that the structure factor data exhibits the Porod tail $S(k,t) \sim k^{-3}$, which characterizes scat-

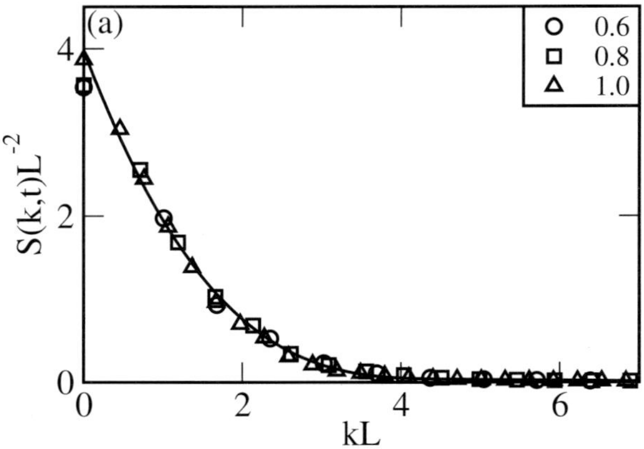

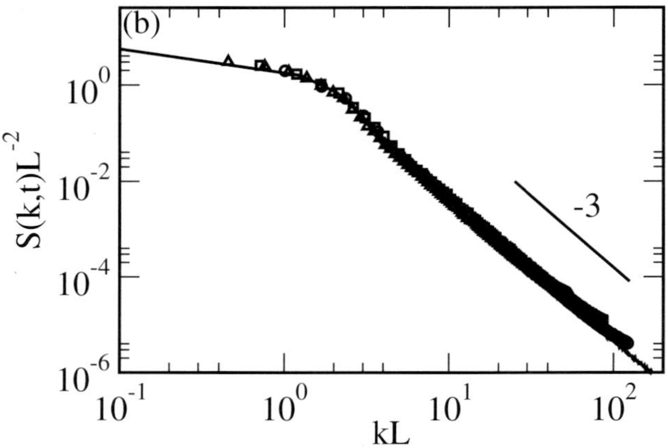

Fig. 4.2. (**a**) Plot of scaled structure factors, $S(k,t)L^{-2}$ vs. kL, for three disorder amplitudes, $a = 0.6, 0.8, 1.0$. The simulation details are the same as for Fig. 4.1. The data corresponds to $t = 10000$, and is obtained as an average over 10 independent runs on systems of size 1024^2. The solid line is the scaled structure factor for the pure TDGL equation ($a = 0.0$), also at $t = 10000$. (**b**) Log-log plot of the data in (a). The line with slope -3 corresponds to Porod's law, $S(k,t) \sim k^{-(d+1)}$ with $d = 2$

tering off sharp interfaces (see Sect. 3.2.1). This universality of the scaling function has been clarified further by Biswal et al. [234]. These authors used MF dynamical models to demonstrate that universality holds in an even more stringent test of dynamical scaling [95], which does not involve the somewhat arbitrary definition of a length scale from the structure factor.

Next, we focus on the time-dependence of the length scale. Figure 4.3 plots $\ln[L(t)]$ vs. $\ln t$ for the pure case and for various disorder amplitudes. The pure case obeys the LAC growth law, as expected. However, the disordered case does not exhibit power-law growth over extended time-regimes. One could

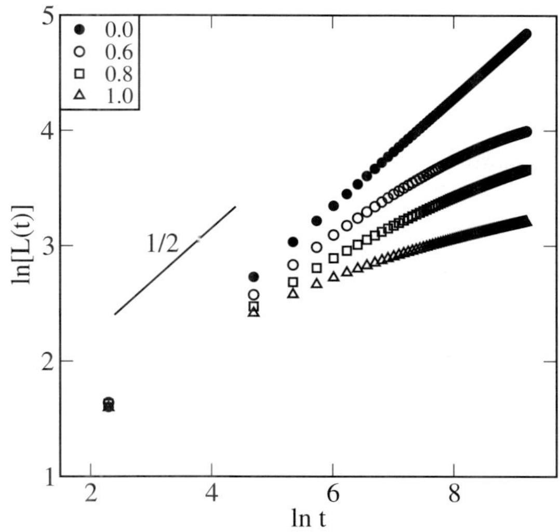

Fig. 4.3. Time-dependence of length scales, $\ln[L(t)]$ vs. $\ln t$, for the pure case ($a = 0.0$) and $a = 0.6, 0.8, 1.0$. The line of slope 1/2 corresponds to the LCA growth law

test for the HH growth law by attempting the power-law fit $\ln t = c_1 L^{\psi} + c_2$, where c_1, ψ and c_2 are fit parameters. The results of this can be summarized as follows:

1. The HH growth law does not give a good fit to the data.
2. For the available fits, the exponent ψ exhibits a strong dependence on the disorder amplitude.

In Fig. 4.4, we attempt a rather biased test of the HH law by plotting $L^{1/4}$ vs. $\ln t$. Recall that $\psi = 1/4$ in $d = 2$, according to the HH prediction. However, the data does not exhibit linear behavior over any extended regime, which would correspond to HH-type growth.

Finally, we mention a set of CDS simulations due to Hayakawa and Iwai [235,236], who found results at variance with those described above. The CDS model used by them introduces disorder via *missing bonds* at a coarse-grained level. This randomizes the local critical temperature, and the discrete Laplacian operator, where some of the bonds are eliminated. Hayakawa and Iwai find that the scaled structure factor has a strong dependence on the disorder amplitude, in contrast to analytical expectations and the results discussed above. It is appropriate to make two remarks in this context. Firstly, the relevance of introducing a *bond-dilution* disorder in the coarse-grained model is unclear. Generally, microscopic dilution disorder results in a randomization of the coefficients of free-energy terms (cf. (4.3)), and does not affect the form of the Laplacian operator. Secondly, macroscopic bond-dilution introduces effects of connectivity (or lack thereof) at the coarse-grained level.

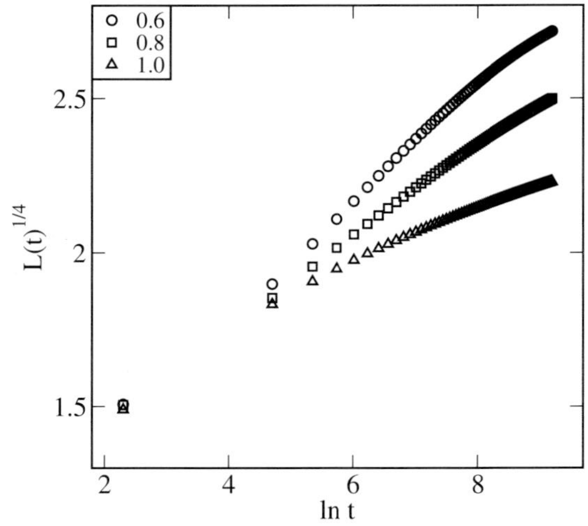

Fig. 4.4. Plot of $L(t)^{1/4}$ vs. $\ln t$ for the data from Fig. 4.3, excluding the pure case. The HH growth law would be linear on this plot

It is not clear how these connectivity effects alter the late stages of domain growth. Thus, a further clarification of the results of Hayakawa and Iwai seems necessary.

Case with Conserved Kinetics

Next, consider the REIM with conserved kinetics. The appropriate coarse-grained model is the CHC equation (4.31) with zero field. Puri and Parekh [130] have studied a $d = 2$ CDS model which is equivalent to the CHC equation. Again, their results are similar to those obtained for random magnets. For example, the scaled structure factor exhibits dynamical scaling, and the scaling function is independent of disorder. Furthermore, the domain growth law exhibits a crossover from LS behavior ($L(t) \sim t^{1/3}$) to nonalgebraic behavior in the late stages of phase separation. As in the nonconserved case, it was not possible to conclusively ascertain the asymptotic domain growth law, or even whether there is a universal law valid for all disorder amplitudes and temperatures.

A more recent study of the CHC equation is due to Gyure et al. [131], who studied an Euler-discretized version of (4.31) in $d = 2$. Gyure et al. considered quenches to both $T = 0$ and $T > 0$. Their study confirmed the disorder-independence of the structure-factor scaling function. For quenches to $T = 0$, the evolving system freezes into a metastable state with mesoscale segregation. Gyure et al. found that the saturation length scale $L_s \sim a^{-1}$, where a is the disorder amplitude. Furthermore, the crossover time from LS growth to disorder-affected growth scales as $t_c \sim a^{-3}$. However, the situation is not so clear regarding the domain growth law for the case with $T > 0$. For

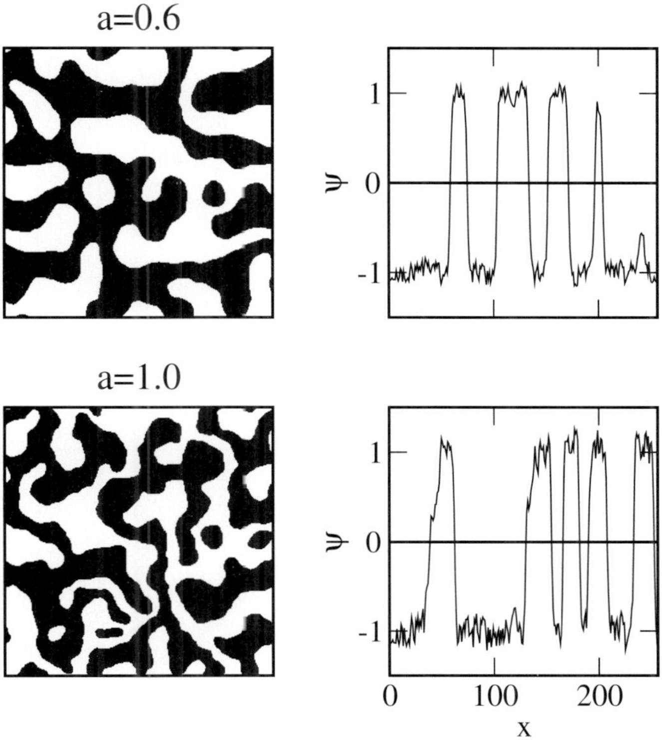

Fig. 4.5. Evolution pictures and order-parameter profiles for phase separation with quenched disorder. These pictures were obtained from a numerical solution of the CHC equation (4.31) with zero field. The discretization mesh sizes were $\Delta t = 0.01$ and $\Delta x = 1.0$, and the lattice size was 256^2. Other details are the same as for Fig. 4.1. The snapshots and profiles correspond to $t = 10000$

example, Gyure et al. found that fits to the HH form yielded disorder- and temperature-dependent exponents.

At this stage, it is relevant to present typical results for phase separation in disordered systems. Figure 4.5 shows evolution pictures and order-parameter profiles obtained from a simulation of the CHC equation with disorder. The snaphots on the LHS should be compared with the snapshots in Fig. 3.3 for the disorder-free CHC equation. As in the nonconserved case, there appears to be little effect of the disorder on the evolution morphology. This is confirmed in Fig. 4.6, where the scaled structure factors for different disorder amplitudes are superposed. Again, the scaling function is independent of the disorder amplitude, and comparable to that for the pure case. Notice that there is no numerical evidence for a shift of the scaling function towards that for the nonconserved case, as suggested by Bray [81].

Finally, focus on the time-dependence of the domain length scale. Figure 4.7 plots $\ln[L(t)]$ vs. $\ln t$ for various disorder amplitudes, including the

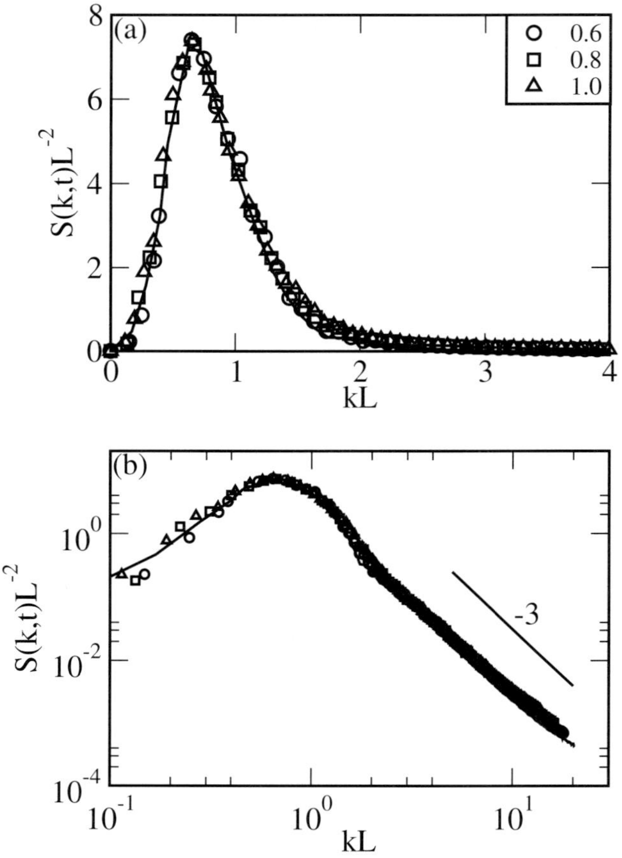

Fig. 4.6a,b. Analogous to Fig. 4.2, but for the case with conserved order parameter

pure case with $a = 0.0$. The domain growth law exhibits a crossover from LS growth to nonalgebraic growth. However, asymptotic growth is not consistent with the HH growth law (see Fig. 4.8). To date, there is no clear confirmation of the logarithmic growth law in this context.

We had earlier mentioned the work of Hayakawa and Iwai [235,236] in the context of nonconserved domain growth. These authors also used their CDS models to study the problem of phase separation. As in the nonconserved case, Hayakawa and Iwai found results at variance with the above results, i.e., the scaling function depends on the disorder amplitude. As stated earlier, these studies need further clarification, particularly with regard to the physical interpretation of missing "bonds" in their coarse-grained model. Their model is reminiscent of phase separation in a porous medium, where diffusive transport is not permitted in locally random directions.

Before proceeding, it is appropriate to summarize numerical results for domain growth in the REIM. On the basis of analytical arguments and nu-

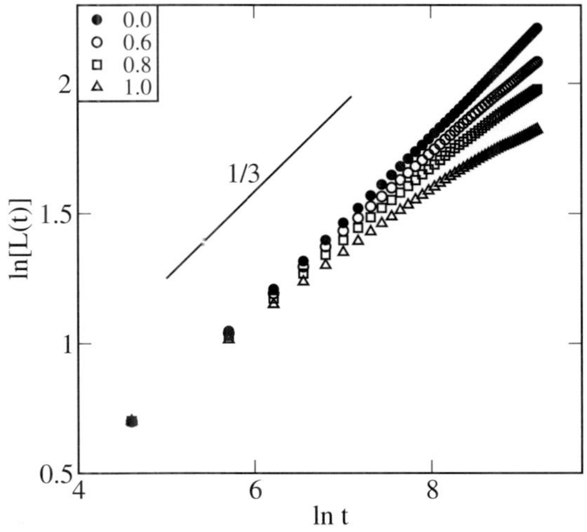

Fig. 4.7. Analogous to Fig. 4.3, but for the case with conserved order parameter

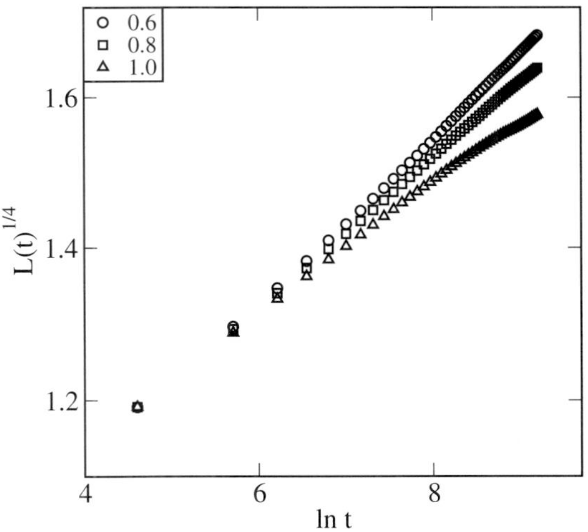

Fig. 4.8. Analogous to Fig. 4.4, but for the case with conserved order parameter

merical studies, it is clear that the scaling functions which characterize the evolution morphology are universal. However, the nature of the asymptotic domain growth law is ambiguous. Thus, it is clear that there is a crossover from the LCA or LS law to nonalgebraic growth, and the crossover time is inversely proportional to the disorder amplitude. However, the nature (or even existence) of the asymptotic growth law remains unclear.

4.4.4 Domain Growth in the RFIM

Case with Nonconserved Kinetics

There have also been many numerical studies of domain growth in the RFIM, and we discuss some representative results here. Early MC studies of the RFIM with Glauber kinetics are due to Gawlinski et al. [237,238] and Pytte and Fernandez [239]. These studies observed slowing-down of domain growth, but were unable to quantify the asymptotic domain growth law. As regards structure-factor scaling, recall that we expect important differences between $d = 2$ and $d = 3$ (see Sect. 4.2.3). In their MC simulations of the $d = 2$ RFIM, Gawlinski et al. [237,238] found that the scaled structure factor did not exhibit dynamical scaling. As discussed earlier, this is due to the absence of LRO in $d = 2$. Clearly, for sufficiently large systems and long simulation times, the $d = 2$ RFIM will show nonuniversal effects, regardless of the disorder amplitude.

A more detailed MC study of the $d = 2$ RFIM with Glauber kinetics is due to Anderson [240], who presented results for a range of quench temperatures and random-field amplitudes. Anderson focused on the domain growth law and undertook a careful test of various theoretical predictions (see Sect. 4.2.3). Essentially, his results are consistent with the logarithmic growth law in (4.16), before a crossover to an equilibrium length scale. For temperatures $T > 0$, Binder [241] predicted the following scaling behavior for the equilibrium length:

$$\ln L_s \sim \left(\frac{k_B T}{J}\right)^{1/3} e^{2\beta J/3} \left(\frac{\overline{H}}{J}\right)^{-4/3} . \tag{4.35}$$

In the absence of fluctuations ($T = 0$), the corresponding equilibrium length is predicted to behave as [203,242]

$$\ln L_s \sim \left(\frac{\overline{H}}{J}\right)^{-2} . \tag{4.36}$$

Anderson's numerical results for $T > 0$ are consistent with Binder's prediction. He also performed simulations for $T = 0$, where the system invariably freezes into a metastable state and does not evolve to equilibrium.

Oguz et al. [243] studied the $d = 2$ TDGL equation with random fields, viz., (4.26) with $\delta a(\boldsymbol{r}) = 0$ and $H_0 = 0$. They observed a logarithmic growth law over limited time-windows. Oguz et al. also observed a breakdown of dynamical scaling for the structure factor. Similar results were reported by Puri and Parekh [244] from their simulations of $d = 2$ CDS models.

Next, let us discuss some studies of the $d = 3$ case, where the RFIM exhibits LRO, and domain growth does not saturate. Rao and Chakrabarti [245] have numerically solved the corresponding TDGL equation, and found a clear confirmation of the growth law (4.16). They also demonstrated that the scaled correlation function is independent of the disorder amplitude. Their results have been confirmed by Oguz [246] in $d = 3$ MC simulations of the RFIM with Glauber kinetics.

Case with Conserved Kinetics

Finally, we consider domain growth in the RFIM with conserved kinetics. Recall that we expect the same growth law as in the nonconserved case. The first study of this problem is due to Puri and Parekh [244], who undertook CDS simulations in $d = 2$. Their results are similar to those for the nonconserved RFIM. Thus, domain growth is consistent with logarithmic growth but only over limited time-windows. Furthermore, the structure factor does not exhibit dynamical scaling. A similar study has been done by Rao and Chakrabarti [247], who numerically solved the CHC equation with random fields. Their results are analogous to those of Puri and Parekh.

Before concluding, it is appropriate to summarize numerical results for domain growth in the RFIM. The situation for the RFIM appears to be considerably clearer than that for the REIM. Thus, there are many studies which confirm logarithmic growth as in (4.16). Of course, this growth law is pre-asymptotic for $d = 2$, where there is an equilibration of the length scale due to the absence of LRO. In the $d = 2$ case, there is a breakdown of dynamical scaling because there are two relevant length scales, viz., the disorder-enhanced interfacial width ξ_D, and the characteristic domain size. In contrast with usual domain growth problems, ξ_D does not become asymptotically irrelevant in this case. The disorder-independence of the morphology is recovered for $d > 2$, where domain growth proceeds for all times.

5 Surface-Directed Spinodal Decomposition and Surface Enrichment

5.1 Introduction

In Chaps. 2 and 3, we described models for the kinetics of phase separation of a binary mixture. We also summarized available results for growth kinetics in the case of pure and isotropic systems. In Chap. 4, we examined the effects of *quenched disorder* (or *impurities*) on phase ordering systems, and found that disorder can drastically alter the nature of domain-growth kinetics. In this chapter, we study another experimentally important effect, viz., the presence of surfaces with a preferential attraction for one of the components of the phase-separating mixture.

There has been much interest in the properties of immiscible binary mixtures AB in contact with a surface S. Let γ_A, γ_B and σ denote the surface tension between AS, BS and AB, respectively. Assume that $\gamma_A < \gamma_B$, so that A is the preferred component. Then, the contact angle between the AB interface and the surface S is determined by Young's condition $\sigma \cos\theta = \gamma_B - \gamma_A$ [248]. This equation has no solution if $\gamma_B - \gamma_A > \sigma$, which corresponds to a situation where the component A completely wets the surface and B has no contact with the surface. This is referred to as the *completely wet* (CW) morphology. For $\gamma_B - \gamma_A < \sigma$, both A and B are in contact with the surface in a *partially wet* (PW) morphology. The equilibrium transitions between the PW and CW morphologies have been extensively studied [249–254].

In the vicinity of the critical point of the binary mixture, we have $\sigma \simeq \sigma_0 \tau^{\mu}$ (with $\tau = (T_c - T)/T_c$ and $\mu \simeq 1.26$) and $\Delta\gamma = \gamma_B - \gamma_A \simeq \Delta\gamma_0 \tau^{\delta}$ (with $\delta \simeq 0.34$). The τ-dependence of both σ and $\Delta\gamma$ are related to the change in equilibrium composition of the mixture as $\tau \to 0^+$ or $T \to T_c^-$. Then, Young's condition becomes $\cos\theta \simeq \tau^{-0.92} \Delta\gamma_0/\sigma_0$, i.e., sufficiently close to the critical temperature, one always has complete wetting in equilibrium [249].

The effects of geometry and composition can be incorporated as follows [214]. Consider a binary mixture with A and B present in fractions ϕ_A and ϕ_B $(= 1-\phi_A)$, respectively. Suppose that the mixture is confined in a *thin-film* geometry or a $d = 2$ *capillary* with identical surfaces. The crossover condition between CW and PW morphologies can be estimated by comparing the free energies per unit area for the 2 cases, $F_{CW} = 2(\gamma_A+\sigma)$ and $F_{PW} = 2(\phi_A\gamma_A+\phi_B\gamma_B)$. Note that the interface energy between A and B has been neglected in the PW case. This yields the crossover composition $\phi_{B,c} = \sigma/(\gamma_B - \gamma_A)$ such that $\phi_B > \phi_{B,c}$ or $\phi_B < \phi_{B,c}$ yields CW or PW morphologies, respectively. For a *cylindrical* geometry or a $d = 1$ *capillary*, the corresponding crossover composition is $\phi_{B,c} = [\sigma/(\gamma_B - \gamma_A)]^2$.

With this background, we consider the following generalization of the bulk phase separation problem. A homogeneous mixture AB is in contact with a surface which prefers A. At time $t = 0$, the system is quenched deep below the coexistence curve (see Fig. 2.2). As discussed in Chap. 3, the system undergoes phase separation. In addition, there is an accretion of the preferred component (or wetting) at the surface. This chapter will focus on the dynamical interplay of phase separation and wetting kinetics. There are a number of variants on the above theme, many of which are experimentally and technologically important. For example, the system may be confined between two parallel surfaces, both of which have a preferential attraction for A. Alternatively, one of the surfaces may attract A while the other attracts B, etc.

The first experimental study of this problem is due to Jones et al. [255], in the context of phase-separating polymer mixtures in a *semi-infinite* geometry. We will provide a brief overview of this and other experiments subsequently. A detailed review of experimental techniques and some of the earlier experiments is due to Krausch [140]. Reviews of analytical and numerical approaches to this problem have been undertaken by Puri and Frisch [141] and Binder [256].

A large part of this chapter will focus on the above problem of *surface-directed phase separation* or *surface-directed spinodal decomposition (SDSD)*. In Sect. 5.2, we provide an overview of experimental results to motivate the subsequent discussion. In Sect. 5.3, we discuss the theoretical modeling of SDSD. Section 5.4 contains an overview of analytical and numerical results for this problem. Finally, in Sects. 5.5 and 5.6, we will turn our attention to the *kinetics of surface enrichment*, viz., the accretion of the preferred component from a stable homogeneous mixture (at $T > T_c$).

5.2 Overview of Experimental Results

As stated earlier, the first relevant experimental study of SDSD is due to Jones et al. [255], who studied the segregation of critical mixtures of poly(ethylenepropylene) (PEP) and perdeuterated poly(ethylenepropylene) (d-PEP) in contact with a surface having a preferential attraction for d-PEP. Jones et al. observed that the surface became the origin of composition waves with wave-vectors normal to the surface and propagating into the bulk, as shown in Fig. 5.1. They used the nomenclature *SDSD waves* to describe this phenomenon. Apart from these qualitative observations, the initial experiments of Jones et al. [255] did not provide quantitative results for the temporal evolution of these waves and associated domain-growth kinetics.

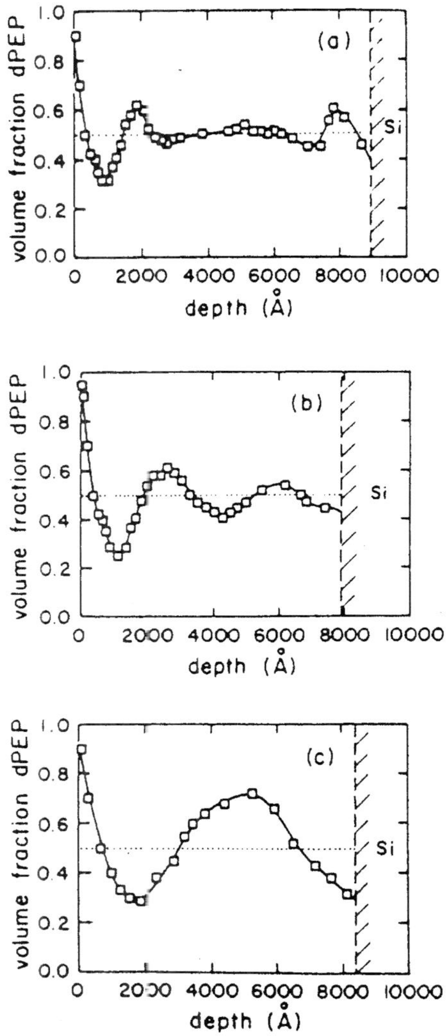

Fig. 5.1. Evolution of SDSD profiles for an unstable polymer mixture (PEP and d-PEP) in contact with a wetting surface [255]. The frames show the depth-dependence of the laterally averaged density of d-PEP at (**a**) 19200 s, (**b**) 64440 s, and (**c**) 172800 s, after the quench

Krausch et al. [257] improved on these experiments, studying a similar experimental system but with better techniques. Their results indicated that the first zero crossing of the SDSD waves ($R_1(t)$) obeyed the LS growth law $R_1(t) \sim t^{1/3}$ (see Sect. 3.3.1). The experimental results of Krausch et al. are shown in Fig. 5.2. Their parameters correspond to a situation in which the

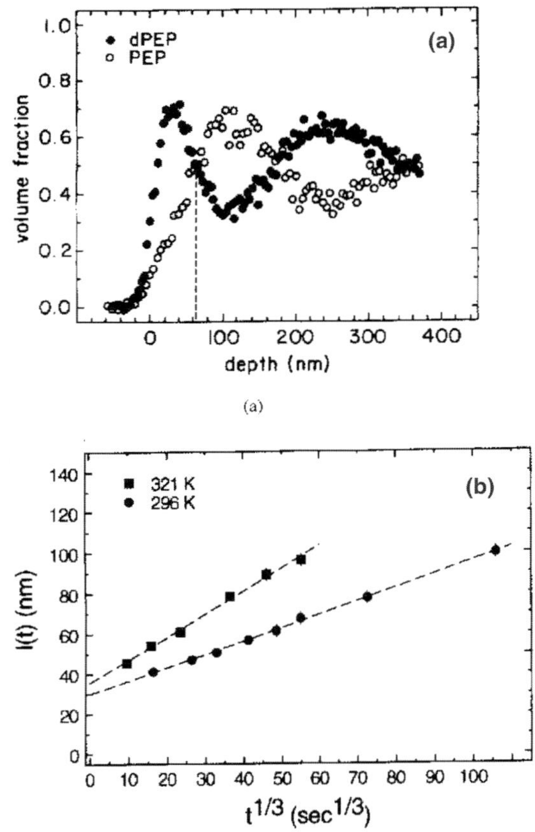

Fig. 5.2. (**a**) Analogous to Fig. 5.1, but from the experiments of Krausch et al. [257]. Density vs. depth profiles are shown for both polymers in the mixture (PEP and d-PEP) at 14400 s after the quench. To compare these profiles with those in Fig. 5.1, the pre-surface region of $\sim$ 20 nm should be discarded. (**b**) Plot of wetting-layer thickness vs. $t^{1/3}$ for two different quench depths [257]. The dashed lines denote the best linear fits to the experimental data sets

surface field is weak, and the surface is only partially wetted by the droplets of the preferred phase.

A similar physical situation was also studied by Bruder, Brenn and others [258,259] using critical polymer mixtures of deuterated polystyrene (dPS) and partially-brominated polystyrene (PBr_xS). These authors found that the preferred phase initially formed a plated layer at the surface. This grew out for a while before decomposing into a PW morphology. The layer thickness (defined from the first zero of the laterally averaged profiles) again obeyed the LS growth law, $R_1(t) \sim t^{1/3}$.

In a related experiment, Krausch et al. [260] studied the evolution of an initial state consisting of a multilayer of the coexisting phases of a polymer mixture (viz., dPS and PBr_xS) in contact with an unstable bulk. For early and intermediate times, they found that coarsening near the surface occurred by the dissolution of alternating layers of the initial structure. At late times, they again found results similar to those in [257].

One can also consider the opposite limit of a strong surface field, where the surface is completely wet. In this situation, a multilayered structure rapidly forms at the surface, and consists of alternating layers of the preferred and non-preferred phases. For extended times, the growth of the wetting layer is determined by the surface potential. The experiments by Geoghegan et al. [261,262] are relevant to this limit. These authors studied SDSD in blends of dPS and poly(α-methylstyrene) (PαMS). They considered a variety of quench depths, and observed an LS growth of the wetting layer for deep quenches, and much slower growth for shallow quenches. They associated this difference with the crossover from a PW to a CW surface morphology.

Recent interest has also focused on SDSD in other contexts. For example, various groups have studied SDSD on patterned substrates [263–265]. In these experiments, the surfaces are specially prepared (e.g., in a *checkerboard* pattern) so that different regions are preferentially wetted by different components. This provides interesting possibilities for technological applications involving special preparation of materials. Furthermore, Aichmayer et al. [266] have reported the first observation of SDSD in solid mixtures. They studied cylindrical samples of austenitic stainless steel, which exhibits macroscale segregation into concentric regions enriched in ferrite and austenite, respectively. Figure 5.3 shows an evolution picture from their experiments.

As far as domain growth in the direction parallel to the surface is concerned, there have been a number of apparently conflicting observations. Again, it is appropriate to distinguish between two different physical situations. In the PW case, where domains of both phases are in contact with the surface, it is reasonable to examine the lateral domain growth at the surface itself. On the other hand, in the CW case, where the surface is completely covered by a layer of the preferred phase, it is reasonable to consider lateral domain growth in the region outside the multilayered structure. Straub et al. [259] found that the lateral size of surface domains (say, $L_{\parallel}(t)$) formed in the late stages of SDSD in the PW case obeys the LS growth law.

These results should be contrasted with the work of Wiltzius, Cumming and others [267–270], who reported observations of a *fast mode* of surface domain growth in their experiments on SDSD in both polymer and fluid mixtures. They observed that the structure factor of the coarsening system exhibited two peaks, rather than the single peak usually seen in bulk spinodal decomposition. The inverse position of the second peak (located at larger wave-vectors or smaller length-scales) obeyed the appropriate bulk growth law, i.e., $L(t) \sim t$ or $L(t) \sim t^{1/3}$, depending on whether or not hydrodynamics is relevant. However, the position of the first peak (located at smaller wave-vectors or larger length-scales) exhibited an anomalously fast growth law

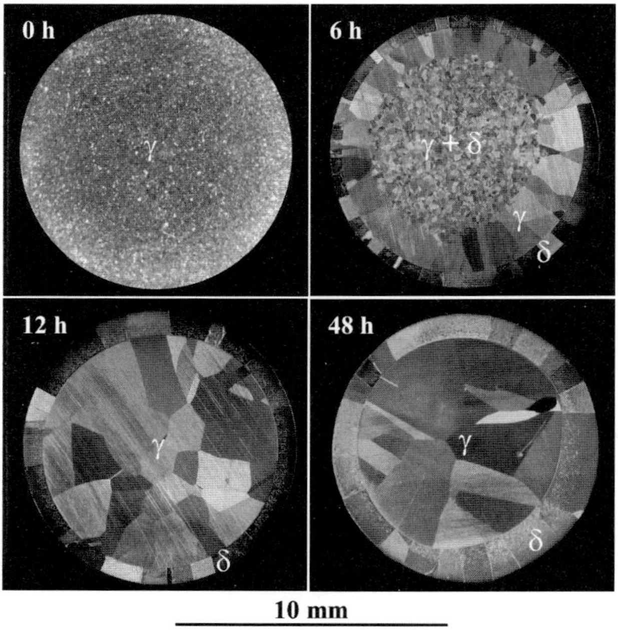

Fig. 5.3. Evolution of a cylindrical sample of austenitic stainless steel, quenched below its coexistence temperature. The snapshots show radial cross-sections for different annealing times: 0 h (only solution treated), 6 h, 12 h and 48 h. The annealing temperature was 1325° C. The ferrite and austenite phases are denoted as δ and γ, respectively

$L(t) \sim t^{\phi}$, with $\phi \sim 1.1 - 1.5$. These authors interpreted the fast growth as an early-time regime in which a droplet morphology evolves to one where the surface is completely covered by a continuous layer, which grows into the bulk at later times. This interpretation has been supported by the molecular dynamics (MD) simulations of Ma et al. [271,272]. However, Troian [273] has argued that this fast mode is the consequence of coalescence of domains of the non-wetting phase in contact with a continuous layer of the wetting phase at the surface.

The results of [267–270] are not at variance with the results of Straub et al. [259] because they correspond to different time-regimes and physical situations. The asymptotic state in [259] corresponds to a partly wet surface with domains of both phases in contact with the wall. This should be contrasted with the experiments of [267–270], where the fast mode precedes the formation of a continuous layer at the wall.

Guenoun et al. [274] also investigated the interplay of wetting and phase separation in binary fluid mixtures (cyclohexane and methanol) at critical concentrations. The time-regime they study corresponds to a situation in which the surface is already coated with a layer of the preferred phase. They found that the wetting layer grew as $R_1(t) \sim t^{\phi}$, with $\phi \simeq 0.56$. Guenoun et al. reported that domain growth perpendicular to the surface was consistent with the bulk growth law for fluids, viz., $L(t) \sim t$. However, in contrast to the experiments of [267–270], they found that domain growth parallel to the surface was actually suppressed and was characterized by a growth exponent $\phi \sim 0.5 - 0.7$.

Finally, Tanaka [275] has also conducted experiments on polymer mixtures phase-separating in $d = 1, 2$ capillaries. Most of Tanaka's experiments were conducted on mixtures of polyvinylmethylether (PVME) with water, or oligomer mixtures. His primary focus was the interaction between phase separation and the wetting layer growing from the surface. In particular, he focused on the late stages of phase separation, where hydrodynamic effects dominate the segregation dynamics. Tanaka found a strong dependence of the evolution morphology on the composition of the mixture. A recent review by Tanaka [276] summarizes results from his experiments, and provides useful scaling arguments for domain growth laws in fluid mixtures near a surface. We will reproduce some of these arguments in Sect. 5.4.6.

Of course, we have only surveyed a small fraction of the experimental results available for SDSD. Nevertheless, these representative experiments provide guidance to the relevant theoretical questions. Before proceeding, it is appropriate to summarize here the experimentally relevant features which must be manifested by any reasonable model for SDSD:

1. The attracting surface is rapidly enriched in the preferred component and becomes the source of anisotropic spinodal decomposition waves which propagate into the bulk perpendicular to the surface.
2. The morphology of these waves and their evolution depends critically on the relative strengths of the surface field, surface tension and thermal noise, i.e., on whether the surface potential results in a partial or complete wetting of the surface.
3. Domain growth parallel to the surface (but outside the wetting layer, if any) obeys the same growth law as that in the bulk. However, there may be an enhancement of the prefactor near the surface due to the orientational effect of wetting layers formed at the surface.

The subsequent discussion primarily focuses on the case where phase separation occurs via diffusion. At appropriate points, we will highlight the modifications which arise when hydrodynamic effects are incorporated.

5.3 Theoretical Modeling

5.3.1 Early Studies of Surface-Directed Spinodal Decomposition

One of the earliest studies of the effect of surfaces on spinodal decomposition is due to Xiong and Gong [277], who considered a semi-infinite system with a surface at $z = 0$. In their model, the order parameter obeys the usual CHC equation (cf. (3.22)):

$$\frac{\partial}{\partial t}\psi(\boldsymbol{\rho}, z, t) = -\boldsymbol{\nabla} \cdot \left\{\boldsymbol{\nabla}\left[\psi - \psi^3 + \nabla^2\psi\right] + \boldsymbol{\theta}\right\}, \quad z > 0. \tag{5.1}$$

Here, $\psi(\boldsymbol{\rho}, z, t)$ is the order parameter at space-point $\boldsymbol{r} \equiv (\boldsymbol{\rho}, z)$ and time t, with $\boldsymbol{\rho}$ and z denoting coordinates parallel and perpendicular to the surface, respectively. In (5.1), $\boldsymbol{\theta}$ is a Gaussian white noise which obeys the appropriate fluctuation-dissipation relation in (3.23). Furthermore, it is assumed that the surface gives rise to a delta-function (or short-ranged) potential, which does not penetrate into the bulk of the system and only affects the boundary conditions. Xiong and Gong supplemented the CHC equation with a boundary condition at the surface position $z = 0$:

$$\left.\frac{\partial}{\partial z}\psi(\boldsymbol{\rho}, z, t)\right|_{z=0} = C\psi(\boldsymbol{\rho}, 0, t), \tag{5.2}$$

where C measures the strength of the surface potential.

Following Langer et al. [73], Xiong and Gong used this model as the basis of an approximate theory for the time-dependent structure factor in the surface layer. Their results suggested that phase separation in the boundary layer is faster than that in the bulk, but they did not quantify the nature of domain growth near the surface. Unfortunately, the modeling of Xiong and Gong is open to criticism for two reasons. Firstly, their model is incomplete without a second boundary condition at the surface, which should be of the *no-flux* or *zero-current* type:

$$\left.\left\{\frac{\partial}{\partial z}\left[\psi - \psi^3 + \nabla^2\psi\right] + \theta_z\right\}\right|_{z=0} = 0. \tag{5.3}$$

Furthermore, the form of the surface free energy used by Xiong and Gong does not result in a preferential attraction of either of the components to the surface, and is inappropriate in the context of SDSD.

Jiang and Ebner [278] studied the effect of surfaces on phase separation through Monte Carlo (MC) simulations of a semi-infinite Ising model (discussed in Sect. 5.3.2) with Kawasaki spin-exchange kinetics (see Sect. 2.4). In their modeling, the surface preferentially attracted one of the components of the mixture with either a short-ranged or long-ranged (power-law) surface potential. They found that the thickness of the surface layer in their simulations always obeyed the LS growth law, regardless of the nature of the potential. However, their numerical results show a large degree of scatter and it is difficult to conclusively ascertain an asymptotic growth exponent from their data.

The next relevant study is due to Ball and Essery [279], who modeled surface-affected spinodal decomposition using (5.1)-(5.3). As we have remarked earlier, the boundary condition in (5.2) does not lead to SDSD for critical quenches because the corresponding free energy does not discriminate between the two phases. For off-critical quenches, this model does exhibit symmetry breaking but in a somewhat artificial fashion. Nevertheless, the pioneering work of Ball and Essery provided guidance to the initial experiments of Jones et al. [255] on SDSD. Furthermore, the initial study of Ball and Essery already recognized the effect of surface morphology on the evolution dynamics of the "wetting" layer.

In passing, we should mention that Ball and Essery also attempted to model effects due to temperature gradients, which have also been experimentally important. This was done by directly coupling the temperature-dependent coefficient in the linear term of the CHC equation (3.19) with a diffusion equation for the temperature field. Their description of this problem is also incomplete as there are terms coupling temperature and order-parameter gradients, which are missing in their model.

5.3.2 Model Hamiltonian and Static Formalism

In Sect. 2.4, we saw that the phase separation of an AB mixture in the bulk is described by an Ising model in which the constituents A and B interchange directly via Kawasaki spin-exchange kinetics. This is not a microscopically realistic description of phase separation in solid alloys which involves a vacancy-mediated mechanism [54,55]. It is also not accurate for the intermediate and late stages of phase separation in binary fluids, where hydrodynamic effects are crucial (see Sect. 3.3.2). However, the kinetic Ising model is a convenient model for MC simulations. Furthermore, we have seen in Sect. 2.6.2 that it can also be used to motivate the coarse-grained CHC model. Therefore, it is useful to study the relevant spin Hamiltonian in the present case, and understand the role played by various parameters.

Consider an AB mixture in contact with a surface S. There are pair-wise interactions $-E_{ij}^{\alpha\beta}$ between atoms α and β at sites i and j (see Sect. 2.1). In terms of local concentration variables, $n_i^A = 1 - n_i^B = 1$ or 0, if site i is occupied by an A or B atom. The corresponding Hamiltonian is (cf. (2.4))

$$\begin{aligned}\mathcal{H} = &- \sum_{i>j=1}^{N} \left[E_{ij}^{AA} n_i^A n_j^A + E_{ij}^{BB} n_i^B n_j^B + E_{ij}^{AB} \left(n_i^A n_j^B + n_i^B n_j^A\right)\right] \\ &+ \sum_{i=1}^{N} \left[V_A(\boldsymbol{r}_i) n_i^A + V_B(\boldsymbol{r}_i) n_i^B\right] ,\end{aligned} \tag{5.4}$$

where lattice sites now exist in the positive half-space ($z \geq 0$) only. In (5.4), $V_A(\boldsymbol{r}_i)$ and $V_B(\boldsymbol{r}_i)$ are potentials on A or B atoms at site $\boldsymbol{r}_i$ due to the hard wall at $z = 0$. In the case of a free surface (in contact with vacuum or air) at $z = 0$, it would seem reasonable to set $V_A(\boldsymbol{r}_i) = V_B(\boldsymbol{r}_i) = 0$. However, this ignores any intrinsic roughness of the interface between the mixture and the *vapor phase* at $z < 0$. For solids above their roughening transition temperature T_R [280,281] (and for fluid mixtures, where the fluid-gas interface is always rough), this assumption is only reasonable on length scales much larger than the scale of atomic roughness. Furthermore, in general, we expect that the pair-wise interactions $-E_{ij}^{\alpha\beta}$ depend not only on the relative distance $|\boldsymbol{r}_i - \boldsymbol{r}_j|$, but also on $\boldsymbol{r}_i, \boldsymbol{r}_j$ separately. For example, different interactions occur if both sites i, j are in the surface layer (labeled as $i_z = 0$) [251,282].

The correct description of atomistic detail is not relevant here. Rather, one wishes to use the Hamiltonian in (5.4) to motivate reasonable coarse-grained models. The validity of these models is then expected to be substantially greater than the "derivation" suggests. Therefore, we simplify the Hamiltonian by assuming that there are only nearest-neighbor interactions. Furthermore, the interactions are taken to be independent of the sites i and j, except when they both lie in the surface layer. As regards the potential, this usually depends only on the distance from the surface. One is interested in both short-ranged potentials ($V(z) \sim \delta(z)$ or $V(z) \sim \exp(-z/z_0)$) and long-ranged power-law potentials ($V(z) \sim z^{-n}$). Notice that there are significant physical differences between wetting by short-ranged and long-ranged potentials [254].

We introduce Ising variables through the usual transformation $\sigma_i = 2n_i^A - 1 = 1 - 2n_i^B$. The corresponding semi-infinite Ising Hamiltonian is (cf. (2.4)-(2.5))

$$\mathcal{H} = -\sum_{\langle ij \rangle} J_{ij}\sigma_i\sigma_j - H\sum_{i=1}^{N}\sigma_i - H_1\sum_{i_z=0}\sigma_i + \sum_{i_z \neq 0} V(z_i)\sigma_i + \mathcal{H}_0\,, \quad (5.5)$$

where the subscript $\langle ij \rangle$ denotes a sum over nearest-neighbor pairs. Furthermore, the subscript $i_z = 0$ denotes sites i lying in the surface layer, and $\mathcal{H}_0$ is a constant which only affects the energy scale. The pair-wise exchange interaction J_{ij} is identified as

$$J_{ij} = J = \frac{E^{AA} + E^{BB} - 2E^{AB}}{4}\,, \quad (5.6)$$

when at least one of the sites i, j is not in the surface layer. If both sites are in the surface layer,

$$J_{ij} = J_s = \frac{E_s^{AA} + E_s^{BB} - 2E_s^{AB}}{4}\,, \quad (5.7)$$

where the subscript s denotes the pair-wise interaction potential in the surface layer. The bulk "magnetic field" H is

$$H = \frac{q}{4}\left(E^{AA} - E^{BB}\right)\,, \quad (5.8)$$

where q is the coordination number of a site. For sites in the surface layer, the magnetic field (for a cubic lattice) is

$$H + H_1 = \frac{q-2}{4}\left(E_s^{AA} - E_s^{BB}\right) + \frac{1}{4}\left(E^{AA} - E^{BB}\right) - \frac{1}{2}\left[V_A(0) - V_B(0)\right] . \tag{5.9}$$

Finally, we identify the potential as $V(z_i) = [V_A(z_i) - V_B(z_i)]/2$. Notice that we have absorbed the potential contribution at the surface into the definition of H_1. For the specific case of a delta-function surface potential which only acts in the surface layer, the potential term $V(z_i)$ in (5.5) is absent.

A non-zero surface field H_1 arises even for the case where $V_A = V_B = 0$, and interactions are unchanged at the surface, i.e., $E_s^{AA} = E^{AA}$ and $E_s^{BB} = E^{BB}$, provided $E^{AA} - E^{BB} \neq 0$. This is a consequence of the missing neighbors of sites in the first layer. Of course, this field vanishes if we consider the symmetric case with $E^{AA} = E^{BB}$. However, this case has little physical relevance for real binary mixtures. Therefore, in general, a surface field manifests itself whenever a surface is introduced.

The bulk field in (5.5) is irrelevant because we work in a fixed-concentration ensemble. However, the surface field H_1 and the potential term remain as parameters in the problem. These terms are responsible for both surface enrichment and wetting phenomena in mixtures [251–254]. The generalization of the Hamiltonian in (5.5) to any other geometry is obvious. For example, in a $d = 2$ capillary of thickness D, the RHS of (5.5) would contain additional terms arising from the introduction of a surface at $z = D$.

Let us next discuss the framework for characterization of equilibrium critical phenomena for confined binary mixtures. We focus on the case of a short-ranged surface field, where the potential only appears in the surface field H_1. Then, surface critical phenomena [251,253,254] and wetting [254,283] can be discussed in terms of the singular behavior of the surface excess free-energy density $G_s(T, H, H_1)$. This is defined from the free-energy density $G(T, H, H_1)$ of a film of thickness D with two equivalent surfaces as follows:

$$G(T, H, H_1) = G_b(T, H) + \frac{2}{D} G_s(T, H, H_1), \quad D \longrightarrow \infty . \tag{5.10}$$

Here, $G_b(T, H)$ is the bulk free-energy density of the system at temperature T and bulk field H. (In this discussion, we explicitly use the subscript b to denote the bulk.) The physical quantities of interest are the response functions, e.g., the surface susceptibility χ_s defined as

$$\begin{aligned} \chi_s &= -\frac{\partial^2}{\partial H^2} G_s(T, H, H_1)\bigg|_{T,H_1} \\ &= \sum_n (\chi_n - \chi_b) \\ &= \int_0^\infty dz [\chi(z) - \chi_b] , \end{aligned} \tag{5.11}$$

where the index n labels the layers of the lattice parallel to the surface. Here, the bulk susceptibility is $\chi_b = -\partial^2 G_b(T,H)/\partial H^2|_T$.

The surface layer susceptibility is defined as

$$\begin{aligned}\chi_0 &= -\frac{\partial^2}{\partial H \partial H_1} G_s(T,H,H_1)\bigg|_T \\ &= \frac{\partial \psi_0}{\partial H}\bigg|_{T,H_1} \\ &= \beta \sum_j \left(\langle \sigma_i \sigma_j \rangle_0 - \langle \sigma_i \rangle_0 \langle \sigma_j \rangle_0\right) ,\end{aligned} \tag{5.12}$$

where $\psi_0 = \langle \sigma_i \rangle_0$, with i lying in the surface layer ($i_z = 0$). We can also define a susceptibility with respect to the surface field H_1 as

$$\begin{aligned}\chi_{00} &= -\frac{\partial^2}{\partial H_1^2} G_s(T,H,H_1)\bigg|_{T,H} \\ &= \beta \sum_{j_z=0} \left(\langle \sigma_i \sigma_j \rangle_0 - \langle \sigma_i \rangle_0 \langle \sigma_j \rangle_0\right) .\end{aligned} \tag{5.13}$$

The layer magnetization ψ_n and the layer susceptibilities χ_n, χ_{nn} are obvious generalizations to the case where a field H_n acts on spins in layer n:

$$\begin{aligned}\chi_n &= \frac{\partial \psi_n}{\partial H}\bigg|_{T,H_n} \\ &= \beta \sum_j \left(\langle \sigma_i \sigma_j \rangle_0 - \langle \sigma_i \rangle_0 \langle \sigma_j \rangle_0\right), \quad i_z = n ,\end{aligned} \tag{5.14}$$

and

$$\begin{aligned}\chi_{nn} &= \frac{\partial \psi_n}{\partial H_n}\bigg|_{T,H} \\ &= \beta \sum_{j_z=n} \left(\langle \sigma_i \sigma_j \rangle_0 - \langle \sigma_i \rangle_0 \langle \sigma_j \rangle_0\right), \quad i_z = n .\end{aligned} \tag{5.15}$$

The correlation functions on the RHS of (5.14)-(5.15) are related to the long-wavelength limits of the corresponding structure factors:

$$S_n(\boldsymbol{k}) = \sum_j e^{i\boldsymbol{k}\cdot(\boldsymbol{r}_i - \boldsymbol{r}_j)} \left(\langle \sigma_i \sigma_j \rangle_0 - \langle \sigma_i \rangle_0 \langle \sigma_j \rangle_0\right), \quad i_z = n , \tag{5.16}$$

and

$$S_{nn}(\boldsymbol{k}_\parallel) = \sum_{j_z=n} e^{i\boldsymbol{k}_\parallel\cdot(\boldsymbol{r}_i - \boldsymbol{r}_j)} \left(\langle \sigma_i \sigma_j \rangle_0 - \langle \sigma_i \rangle_0 \langle \sigma_j \rangle_0\right), \quad i_z = n , \tag{5.17}$$

where $\boldsymbol{k}_\parallel$ is a wave-vector lying in a plane parallel to the surface.

In analogy with (5.10), it is reasonable to consider the surface excess $S_s(\boldsymbol{k})$ of the total scattering intensity $S(\boldsymbol{k})$ of a film:

$$S(\boldsymbol{k}) = S_b(\boldsymbol{k}) + \frac{2}{D} S_s(\boldsymbol{k}), \quad D \to \infty\,, \tag{5.18}$$

where

$$\begin{aligned} S_s(\boldsymbol{k}) &= \sum_n [S_n(\boldsymbol{k}) - S_b(\boldsymbol{k})] \\ &= \int_0^\infty dz [S(\boldsymbol{k}, z) - S_b(\boldsymbol{k})]\,. \end{aligned} \tag{5.19}$$

In equilibrium, the small-k behavior of the structure factors defined via (5.16)-(5.19) characterizes the typical length scales of the problem. For surface critical phenomena, the bulk correlation length ξ_b determines these length scales for both the ordinary and special surface transitions [251,253]. However, for critical wetting, one has to consider separate correlation lengths $\xi_{\|}, \xi_{\perp}$ in directions parallel and perpendicular to the surface [254].

We conclude this discussion by formulating the coarse-grained free-energy functional corresponding to the Hamiltonian in (5.5). We introduce the space-dependent order parameter $\psi(\boldsymbol{r}_i) = \langle \sigma_i \rangle$. In terms of this, the free-energy functional can be obtained by identifying $\psi(\boldsymbol{r}_j) = \psi(\boldsymbol{r}_i + \boldsymbol{r}_j - \boldsymbol{r}_i)$, and Taylor-expanding around the point $\boldsymbol{r}_i$. The resultant expression is as follows:

$$\begin{aligned} \mathcal{F}[\psi] &= \mathcal{H} - TS \\ &\simeq \int d\boldsymbol{r} \left[-\frac{1}{2} k_B (T_c - T) \psi^2 + \frac{k_B T}{12} \psi^4 + \frac{J}{2} (\boldsymbol{\nabla}\psi)^2 + V(z)\psi \right] \\ &\quad + \int d\boldsymbol{\rho} \left\{ -\frac{1}{2} [(q-2)J_s + J - k_B T] \psi(\boldsymbol{\rho}, 0)^2 - H_1 \psi(\boldsymbol{\rho}, 0) \right. \\ &\quad \left. + \frac{J_s}{2} \left[\boldsymbol{\nabla}_{\|} \psi(\boldsymbol{\rho}, 0) \right]^2 - \frac{J}{2} \psi(\boldsymbol{\rho}, 0) \left. \frac{\partial \psi}{\partial z} \right|_{z=0} \right\}\,. \end{aligned} \tag{5.20}$$

The first term on the RHS of (5.20) corresponds to the usual bulk energy, supplemented by a surface potential term. The bulk critical temperature is identified as $T_c = qJ$. The second term on the RHS of (5.20) is the surface contribution. Here, the different value of the exchange interaction in the surface layer has been explicitly accounted for. Furthermore, the term $\partial\psi/\partial z|_{z=0}$ appears because of the missing neighbors for $z < 0$. The expansion which results in (5.20) is only justifiable near criticality, where the order-parameter amplitude is small. However, we will also use this coarse-grained free energy for parameter values far from criticality. As usual, the justification for this lies in the ability of the resultant model to describe experimental phenomenology. The appropriate minimal model for description of surface critical phenomena has been discussed extensively in the literature [254].

5.3.3 Coarse-Grained Dynamical Model

As stated earlier, one can associate Kawasaki spin-exchange kinetics with the Ising Hamiltonian in (5.5). This constitutes a good microscopic model for SDSD, and has been used by some authors, as we will discuss shortly. However, the coarse-grained counterpart of the kinetic Ising model is expected to be more amenable to theoretical analysis, analogous to Models A and B in Sect. 2.6. In early work, Binder and Frisch [284] used the master-equation approach described in Chap. 2 to obtain a coarse-grained equivalent of the kinetic Ising model for the case of a delta-function surface potential. The Binder-Frisch model consisted of the CH equation in the bulk supplemented by two boundary conditions representing the surface. One of these boundary conditions was later modified by Puri and Binder [285], who explicitly incorporated a no-flux boundary condition into the model. The Puri-Binder model was the first successful coarse-grained model for SDSD. It is convenient to motivate it directly from the free-energy functional in (5.20), as was done for the CHC equation in Sect. 2.6.1.

The bulk order-parameter equation is obtained from (2.76) with $\mathcal{F}[\psi]$ from (5.20). The resultant CHC equation is then

$$\begin{aligned}\frac{\partial}{\partial t}\psi(\boldsymbol{r},t) &= -\boldsymbol{\nabla}\cdot\boldsymbol{J}(\boldsymbol{r},t)\\ &= \boldsymbol{\nabla}\cdot\{D\boldsymbol{\nabla}\mu(\boldsymbol{r},t)+\boldsymbol{\theta}(\boldsymbol{r},t)\}\\ &= \boldsymbol{\nabla}\cdot\left\{D\boldsymbol{\nabla}\left[-k_B(T_c-T)\psi+\frac{k_BT}{3}\psi^3-J\nabla^2\psi+V(z)\right]+\boldsymbol{\theta}\right\},\end{aligned} \tag{5.21}$$

where D is the diffusion coefficient. As usual, the quantities $\boldsymbol{J}$ and μ in (5.21) denote the current and chemical-potential difference between A and B, respectively. For convenient reference, we reproduce here the fluctuation-dissipation relation for the noise term from (2.77):

$$\begin{aligned}\overline{\boldsymbol{\theta}(\boldsymbol{r},t)} &= 0\,,\\ \overline{\theta_i(\boldsymbol{r}',t')\theta_j(\boldsymbol{r}'',t'')} &= 2Dk_BT\delta_{ij}\delta(\boldsymbol{r}'-\boldsymbol{r}'')\delta(t'-t'')\,.\end{aligned} \tag{5.22}$$

Now, the order parameter rapidly relaxes to its equilibrium value at the surface, and is not a conserved quantity. Therefore, one assumes a relaxational (Model A) kinetics for the order parameter at the surface. This yields (ignoring thermal fluctuations)

$$\lambda^{-1}\frac{\partial}{\partial t}\psi(\boldsymbol{\rho},0,t) = [(q-2)J_s+J-k_BT]\psi+\frac{J}{2}\frac{\partial\psi}{\partial z}\bigg|_{z=0}+H_1\,, \tag{5.23}$$

where λ^{-1} sets the time-scale. The lateral-diffusion term has been neglected as the order parameter rapidly saturates to its equilibrium value at the surface. Finally, we have a no-flux condition at $z=0$, viz.,

$$0 = \left\{ D \frac{\partial}{\partial z} \left[k_B(T_c - T)\psi - \frac{k_B T}{3}\psi^3 + J\nabla^2\psi - V(z) \right] + \theta_z \right\}_{z=0} . \quad (5.24)$$

As in Sect. 3.3.1, it is convenient to rescale this model into dimensionless units. For the SDSD problem, it is appropriate to consider the case with $T < T_c$. Following (3.21), introduce the following rescaled quantities:

$$\begin{aligned}
\psi' &= \frac{\psi}{\psi_0}, \quad \psi_0 = \sqrt{3\left(\frac{T_c}{T} - 1\right)}, \\
\boldsymbol{r}' &= \frac{\boldsymbol{r}}{\xi_b}, \quad \xi_b = \left[\frac{q}{2}\left(1 - \frac{T}{T_c}\right)\right]^{-1/2}, \\
t' &= \frac{t}{t_0}, \quad t_0 = \frac{\xi_b^2}{D k_B (T_c - T)}, \\
\boldsymbol{\theta}' &= \frac{t_0}{\xi_b \psi_C}\boldsymbol{\theta}, \\
V'(z') &= \frac{1}{\psi_0 k_B (T_c - T)} V(\xi_b z') .
\end{aligned} \quad (5.25)$$

Here, ξ_b is the bulk correlation length. This yields the dimensionless CHC equation (dropping the primes):

$$\frac{\partial}{\partial t}\psi(\boldsymbol{r}, t) = \boldsymbol{\nabla} \cdot \left\{ \boldsymbol{\nabla} \left[-\psi + \psi^3 - \frac{1}{2}\nabla^2\psi + V(z) \right] + \boldsymbol{\theta} \right\}, \quad z > 0, \quad (5.26)$$

where

$$\begin{aligned}
\overline{\boldsymbol{\theta}(\boldsymbol{r}, t)} &= 0, \\
\overline{\theta_i(\boldsymbol{r}', t')\theta_j(\boldsymbol{r}'', t'')} &= 2\epsilon\delta_{ij}\delta(\boldsymbol{r}' - \boldsymbol{r}'')\delta(t' - t''),
\end{aligned} \quad (5.27)$$

with the noise amplitude

$$\epsilon = \frac{1}{3}\left(\frac{T_c}{T} - 1\right)^{-2} \xi_b^{-d} . \quad (5.28)$$

The corresponding dimensionless boundary conditions are as follows:

$$\tau_0 \frac{\partial}{\partial t}\psi(\boldsymbol{\rho}, 0, t) = h_1 + g\psi + \gamma \left.\frac{\partial \psi}{\partial z}\right|_{z=0}, \quad (5.29)$$

$$0 = \left\{ \frac{\partial}{\partial z} \left[-\psi + \psi^3 - \frac{1}{2}\nabla^2\psi + V(z) \right] + \theta_z \right\}_{z=0}, \quad (5.30)$$

where $\tau_0 = \lambda^{-1} D / \xi_b^2$. The other parameters in the boundary conditions are as follows:

$$h_1 = \frac{H_1}{\psi_0 k_B (T_c - T)} , \tag{5.31}$$

$$g = \frac{(q-2)J_s + J - k_B T}{k_B (T_c - T)} , \tag{5.32}$$

$$\gamma = \frac{J}{2\xi_b k_B (T_c - T)} . \tag{5.33}$$

Equation (5.29) rapidly relaxes the surface value of the order parameter to its equilibrium value, and will be replaced by its static version ($\tau_0 = 0$) subsequently. In general, the potential and the parameters ϵ, h_1, g, γ determine the equilibrium phase diagram of the surface [141,286,287].

Before proceeding, some further remarks are in order. Firstly, the dynamical model described above is appropriate for a semi-infinite geometry. The extension to a thin-film (or any other) geometry is straightforward – the boundary conditions in (5.23)-(5.24) have to be implemented on all surfaces with appropriate parameters and surface potentials. As discussed earlier, there has been a great deal of interest in binary mixtures confined in a thin-film geometry. For example, one could consider a physical situation where the walls are *symmetric* and have an attraction for the same component of the mixture. Alternatively, one could consider the *antisymmetric* case, where the walls attract different components. There now exists a good understanding of the equilibrium properties of mixtures confined in a thin film [214,288–293]. However, there have been fewer investigations of the dynamical aspects of such problems. The thin-film geometry gives rise to many novel physical features because of the interaction of surface-directed waves arising from different boundaries [215,294–296].

Secondly, the above modeling has been in the context of diffusive dynamics. However, many of the experiments on SDSD involve fluid mixtures (see Sect. 5.2), and one may expect hydrodynamic effects to play an important role, as they do for bulk phase separation (see Sect. 3.3.2). As in the diffusive case, one could consider microscopic-level models, e.g., MD simulations of mixtures near surfaces. There have been a number of such studies (which are discussed later), and they provide valuable information about the ongoing kinetic processes. Alternatively, one can study coarse-grained models like Model H (cf. (3.25)-(3.29)) at a surface. The boundary conditions on the order-parameter field are similar to those described earlier. However, these must now be supplemented with boundary conditions on the velocity field, e.g., the velocity vanishes at the surface. There have been a few studies of Model H at surfaces, and these will also be discussed shortly.

Finally, we should mention the work of Diehl and Janssen [297], who have studied the *critical dynamics* (rather than *far-from-equilibrium dynamics*) of Model B at a surface. They have used symmetry considerations to classify possible boundary terms in the order-parameter equation and the boundary conditions. Diehl and Janssen have also examined the physical implications of different boundary conditions for critical dynamics at a surface.

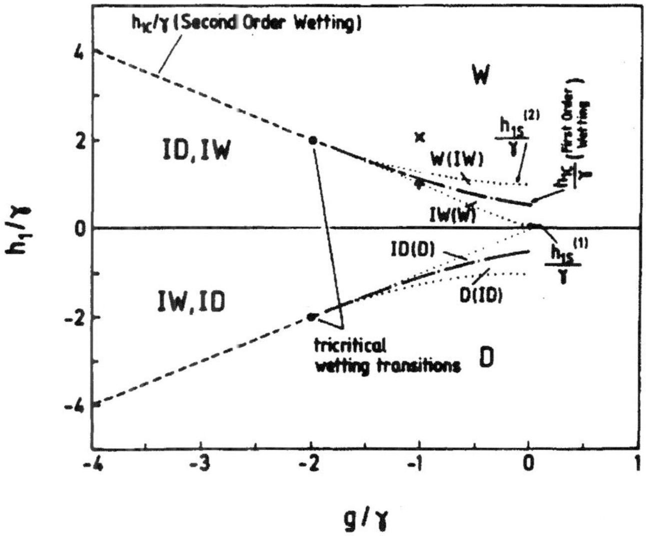

Fig. 5.4. Phase diagram in the $(h_1/\gamma, g/\gamma)$-plane for the case of a delta-function surface potential. The states are labeled as W (wet) or dry (D), and IW (incompletely wet) or ID (incompletely dry). The dashed line denotes a line of second-order transitions, and the dot-dashed line denotes a line of first-order transitions. These lines meet at a tricritical point. The dotted lines are the spinodals associated with the first-order phase transitions. Equations for the different lines are provided in [287]

5.3.4 Static Solutions for a Short-Ranged Surface Potential

It is useful to examine the static solutions of the deterministic version of the model in (5.26)-(5.30). The static solutions are obtained by setting $\partial\psi/\partial t = 0$. Ignoring the lateral fluctuations, the z-dependent static solution $\psi^s(z)$ is obtained as the solution of the Euler-Lagrange equation:

$$\psi^s - (\psi^s)^3 + \frac{1}{2}\frac{d^2\psi^s}{dz^2} - V(z) = 0\,, \tag{5.34}$$

with the boundary condition

$$\frac{h_1}{\gamma} + \frac{g}{\gamma}\psi^s(0) + \left.\frac{d\psi^s}{dz}\right|_{z=0} = 0\,. \tag{5.35}$$

Puri and Binder [287] have obtained solutions of these equations for the case of a short-ranged surface potential. The corresponding phase diagram in the $(h_1/\gamma, g/\gamma)$-plane is shown in Fig. 5.4. As expected, it is symmetric under $h_1 \rightarrow -h_1$. Furthermore, the phase diagram exhibits lines of both first-order and second-order transitions, meeting in a tricritical point.

Note that $\xi_b \to \infty$ and $\psi_0 \to 0$ as $T \to T_c^-$. In this limit, $h_1/\gamma \sim \xi_b^2 \to \infty$ for $h_1 > 0$. Furthermore, $g/\gamma \sim \xi_b \to \infty$ for $g > 0$, and $-\infty$ for $g < 0$. Thus, as the system approaches the critical point, it is deep in the wetting region of the phase diagram. In a typical experimental situation, one is interested in spinodal decomposition and wetting far from the bulk critical point. Therefore, we shall use the present model for parameter choices of $h_1/\gamma, g/\gamma$ of order unity as well.

5.3.5 Characterization of Surface-Directed Spinodal Decomposition

The characterization of SDSD involves understanding (a) the growth kinetics and morphology of the wetting layer, and (b) domain growth in the vicinity of the wetting layer. We already have a good understanding of phase-separation kinetics in the bulk, and this will serve as a useful reference point.

The morphology of the wetting layer is clarified by examining laterally averaged order-parameter profiles as in Figs. 5.1 and 5.2. The wetting-layer kinetics is characterized by the growth laws governing its thickness, e.g., the time-dependence of the zero-crossings of the laterally averaged profiles.

The characterization of domain growth is done via a generalization of the static formalism discussed in Sect. 5.3.2. In the non-equilibrium case, one again considers equal-time correlation functions and structure factors of the type defined in (5.16)-(5.17), which are now time-dependent since one focuses on far-from-equilibrium quenching experiments. The relevant quantities are as follows:
(i) the z-dependent order-parameter correlation function (cf. (3.6)):

$$C_{\parallel}(\boldsymbol{\rho}_1 - \boldsymbol{\rho}_2, z, t) = \langle \psi(\boldsymbol{\rho}_1, z, t)\psi(\boldsymbol{\rho}_2, z, t)\rangle - \langle \psi(\boldsymbol{\rho}_1, z, t)\rangle\langle \psi(\boldsymbol{\rho}_2, z, t)\rangle \,, \tag{5.36}$$

and
(ii) its Fourier transform (cf. (3.7)):

$$S_{\parallel}(\boldsymbol{k}_{\parallel}, z, t) = \int d\boldsymbol{\rho} e^{i\boldsymbol{k}_{\parallel}\cdot\boldsymbol{\rho}} C_{\parallel}(\boldsymbol{\rho}, z, t) \,. \tag{5.37}$$

Recall that, in (5.36), the angular brackets denote averages over both initial conditions and thermal noise.

We can also consider the more general correlation function:

$$C(\boldsymbol{\rho}_1 - \boldsymbol{\rho}_2, z_1, z_2, t) = \langle \psi(\boldsymbol{\rho}_1, z_1, t)\psi(\boldsymbol{\rho}_2, z_2, t)\rangle - \langle \psi(\boldsymbol{\rho}_1, z_1, t)\rangle\langle \psi(\boldsymbol{\rho}_2, z_2, t)\rangle \,, \tag{5.38}$$

where there is no translational invariance in the z-direction because of the surface located at $z = 0$. Its counterpart in Fourier space is defined as follows:

$$S(\boldsymbol{k}, z, t) = \int d\boldsymbol{\rho} \int_0^\infty dz' \exp\left[i\boldsymbol{k}_\| \cdot \boldsymbol{\rho} + k_\perp (z' - z)\right] C(\boldsymbol{\rho}, z', z, t) \,, \quad (5.39)$$

where the wave-vector $\boldsymbol{k} = (\boldsymbol{k}_\|, k_\perp)$, with $k_\perp$ denoting the component perpendicular to the surface. In both (5.37) and (5.39), we have used the translational invariance in the direction parallel to the surface. Note that, in this definition, $C_\|(\boldsymbol{\rho}_1 - \boldsymbol{\rho}_2, z, t) = C(\boldsymbol{\rho}_1 - \boldsymbol{\rho}_2, z, z, t)$. In analogy with (5.19), the surface excess of the time-dependent scattering intensity of the system is then

$$S_s(\boldsymbol{k}, t) = \int_0^\infty dz \left[S(\boldsymbol{k}, z, t) - S_b(\boldsymbol{k}, t)\right] \,, \quad (5.40)$$

where $S_b(\boldsymbol{k}, t)$ is the time-dependent bulk structure factor.

In studies of phase separation, length scales are defined from the decay of the correlation function or as inverse moments of the structure factor [96]. Here, one considers the characteristic length $L_\|(z, t)$, describing the growth of correlations in the parallel direction at a distance z from the surface. This is defined as the distance over which the appropriate correlation function decays to (say) half its maximum value, i.e.,

$$C_\|(L_\|, z, t) = \frac{C_\|(0, z, t)}{2} \,. \quad (5.41)$$

A perpendicular length scale $L_\perp(z, t)$ can be defined similarly. In the semi-infinite geometry, $L_\|(z, t)$ and $L_\perp(z, t)$ tend smoothly to the bulk length scale $L(t)$ as $z \to \infty$.

5.4 Analytical and Numerical Results

5.4.1 Analytical Approaches to Early-time (Linear) Behavior

The nonlinear model in (5.26)-(5.30) is not analytically tractable, in general. However, one can solve the linearized version of this model and thereby obtain useful information in certain relevant limits. In the bulk, solution of the linearized model is rather trivial. In the present case, however, one has to account for the two boundary conditions at the surface and this makes the problem considerably more difficult.

Binder and Frisch [284] first solved the linearized problem for the case of a delta-function surface potential, and a stable bulk with $T > T_c$. They studied the approach to equilibrium of a homogeneous initial condition, and showed that the resultant surface-enrichment profile has a transient minimum whose distance from the wall grows as $t^{1/2}$. A couple of remarks are in order here. Firstly, the kinetics of surface enrichment from a stable bulk is also an experimentally relevant problem [298,299]. This problem will be discussed at length in Sects. 5.5 and 5.6. Secondly, the mathematical solution of the linear problem is the same for both surface enrichment and SDSD. Of course, the physics is quite different. The solution of the surface-enrichment problem is reasonable for all times, and the only issue is whether strong surface fields can introduce sufficient nonlinearity to alter the physical properties of enrichment profiles. On the other hand, the solution of the linear SDSD problem is only valid for early times, as the growing fluctuations have to be saturated by nonlinearities in the system. Nevertheless, an early-time theory of SDSD is useful to clarify wavelengths of surface-directed profiles. Frisch et al. [300] studied the linear regime of SDSD, and found that the initial-stage behavior already replicated the damped concentration profiles seen in Fig. 5.1. Furthermore, the predicted wavelength of the concentration oscillations was also compatible with numerical results obtained from the fully nonlinear model [285].

In this context, another interesting study is due to Fischer et al. [301]. Recall that the boundary condition in (5.29) ignored surface diffusion, because the field homogenizes the order parameter at the surface. Fischer et al. have undertaken an analysis of linearized SDSD in the absence of a surface field but with surface diffusion. In this case, they demonstrated that novel surface instabilities can arise for certain parameter ranges.

5.4.2 Wetting for Critical Quenches ($\psi_0 = 0$)

Next, we discuss the kinetics of wetting-layer growth, and phase separation in the vicinity of the wetting layer. The discussion is based on (5.26)-(5.30), and follows Puri and Binder [302] and Puri et al. [303], who have studied SDSD for a wide range of mixture compositions and power-law potentials:

$$\begin{aligned} V(z) &= -V_0 \,, \quad z \le 1 \,, \\ &= -\frac{V_0}{z^n} \,, \quad z > 1 \,. \end{aligned} \tag{5.42}$$

Here, the lower cut-off is chosen to avoid the power-law singularity at $z = 0$. Such potentials are ubiquitous in the context of surface-molecule interactions, e.g., $n = \kappa - d$, with $\kappa = 6$ and 7 corresponds to cases with non-retarded and retarded van der Waals' interactions, respectively [304,305]. The short-ranged case is recovered in the limit $n \to \infty$. For simplicity, we will subsequently focus on the case with $E^{AA} = E^{BB}$ and $E_s^{AA} = E_s^{BB}$ in (5.9). In that case, the dimensionless surface field is $h_1 = -V(0) = V_0$.

Figure 5.5 shows the evolution of a critical binary mixture (i.e., average order parameter $\psi_0 = 0$) from a homogeneous initial condition. The surface (at $z = 0$) is completely wetted by the component A (marked in black). For these surface field strengths, the surface exhibits a multilayered morphology, i.e., wetting layer followed by depletion layer, etc. This morphology is time-dependent and propagates into the bulk.

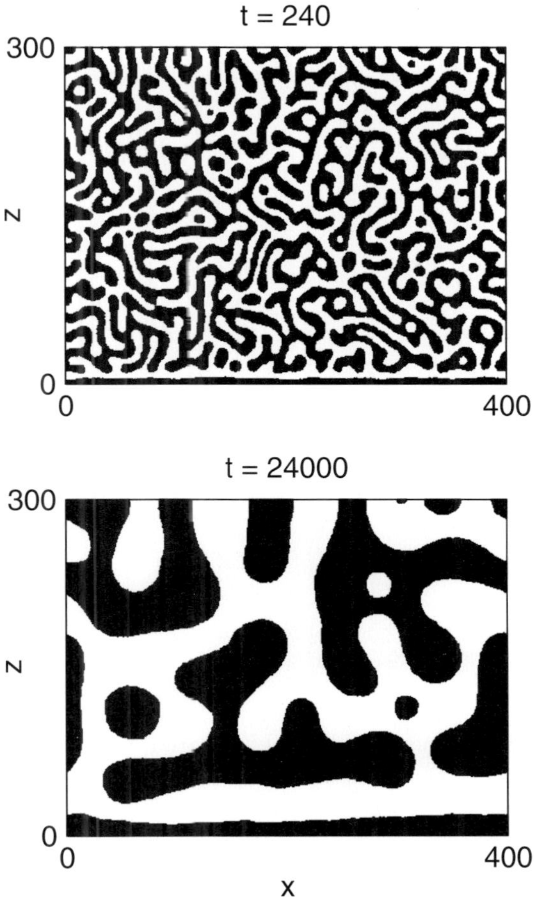

Fig. 5.5. Evolution of an unstable AB mixture from a homogeneous initial condition with a critical composition ($\psi_0 = 0$). The snapshots are obtained from an Euler-discretized implementation of (5.26)-(5.30) on a $d = 2$ square lattice of size $N_x \times N_z$ (with $N_x = 400$ and $N_z = 300$). The mesh sizes in space and time are $\Delta x = 1$ and $\Delta t = 0.03$. The surface is located at $z = 0$ and attracts A (with $\psi > 0$, marked in black) through a power-law potential with $n = 4$ and $V_0 = 0.8$. The other parameter values are $g = -0.4$ and $\gamma = 0.4$, corresponding to complete wetting in equilibrium. The noise amplitude is $\epsilon = 0.041$, which corresponds to a deep quench with $T \simeq 0.22T_c$ from (5.28). The upper and lower frames correspond to dimensionless times $t = 240$ and $t = 24000$, respectively

Figure 5.6 shows the laterally averaged profiles, $\psi_{av}(z, t)$ vs. z, corresponding to the evolution depicted in Fig. 5.5. These are obtained by averaging $\psi(x, z, t)$ along the x-direction for a typical evolution, and also averaging over 200 independent runs. This procedure is the numerical counterpart of the lateral-averaging which yielded the density-depth profiles in Figs. 5.1 and 5.2.

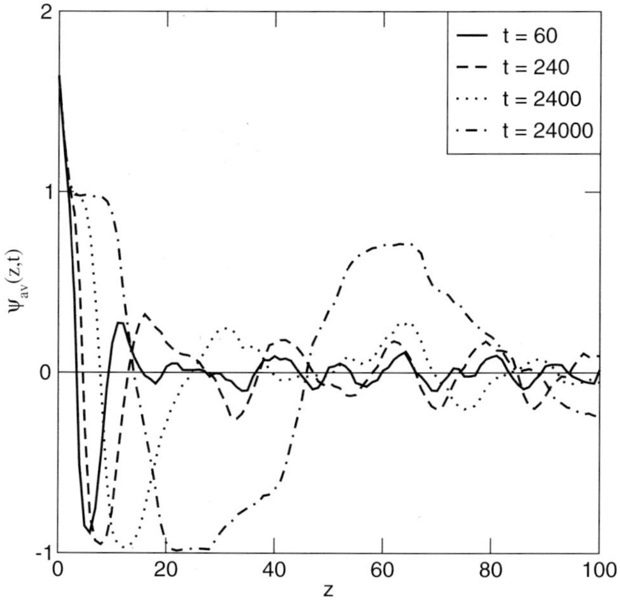

Fig. 5.6. Laterally averaged profiles for the evolution depicted in Fig. 5.5. The profiles are shown for dimensionless times $t = 60, 240, 2400$ and 2400, as indicated

Figure 5.6 clarifies the nature of the multilayered morphology. The averaging procedure gives $\psi_{\text{av}}(z,t) \simeq \psi_0 = 0$ in the bulk, where the phase-separation profiles are randomly oriented. However, a systematic behavior is seen at the surface. The wetting profiles are characterized by the zero-crossings of $\psi_{\text{av}}(z,t) - \psi_0$. The quantities $R_1(t)$ and $R_2(t)$ denote the first and second zeros, respectively. A phenomenological theory for these quantities will be formulated shortly.

It is also relevant to quantify the nature of domain growth in the vicinity of the wetting layer. Puri et al. [285,303] have studied z-dependent correlation functions and structure factors (defined in Sect. 5.3.5) for a critical quench and a range of surface potentials. Their studies suggest the following picture. Domains near the wetting layer are characterized by two length scales $L_{\parallel}(z,t), L_{\perp}(z,t) \sim t^{1/3}$, with $L_{\parallel} > L_{\perp}$ (see Fig. 5.5). The length scale parallel to the surface is larger because of the orientational effect of the multilayered morphology at the surface. However, the enhancement is only in the prefactor of the growth law. Furthermore, these length scales cross over to diffusive growth ($L_{\parallel}(z,t) \sim t^{1/2}$) as the domains are absorbed into the wetting layer which propagates into the bulk.

Next, consider the effect of asymmetric composition on the growth kinetics of the wetting layer. First, we consider the case where the wetting component is the minority component in the mixture.

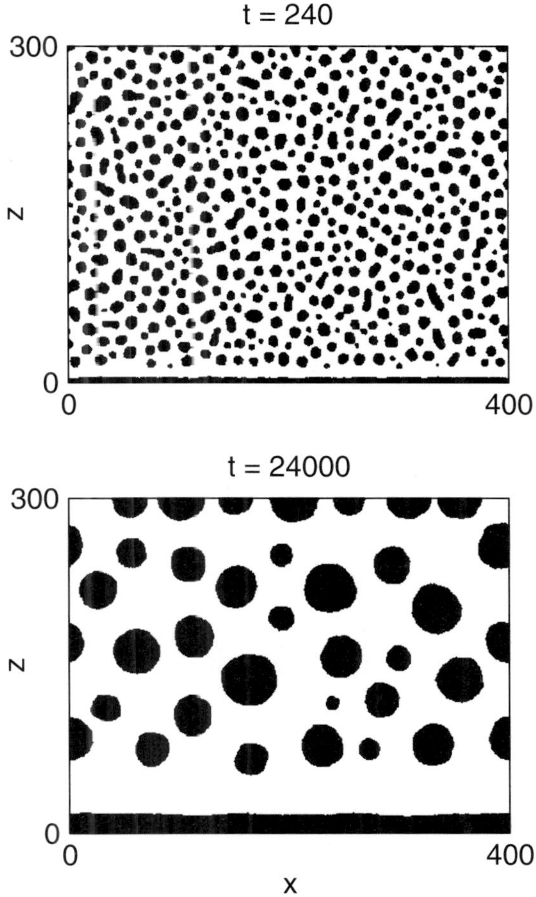

Fig. 5.7. Analogous to Fig. 5.5, but for the case where the wetting component is the minority component with $\psi_0 = -0.4$

5.4.3 Wetting by the Minority Component ($\psi_0 < 0$)

Figure 5.7 shows the evolution from a homogeneous initial condition for $\psi_0 = -0.4$, corresponding to a mixture with 30% A (the preferred component) and 70% B. The bulk (large z) is characterized by the usual droplet morphology for off-critical phase separation [96]. As in Fig. 5.5, there is a wetting layer of the preferred component at the surface, which is followed by a depletion layer. At comparable times, the thickness of the wetting and depletion layers is larger than in Fig. 5.5. Figure 5.8 shows the corresponding laterally averaged profiles as a function of depth from the surface.

Next, consider the evolution of an extremely off-critical case ($\psi_0 = -0.8$ or 10% A and 90% B) in Fig. 5.9. In this case, the thermal fluctuations are not sufficient to nucleate an A-rich droplet on the time-scale of the simulation.

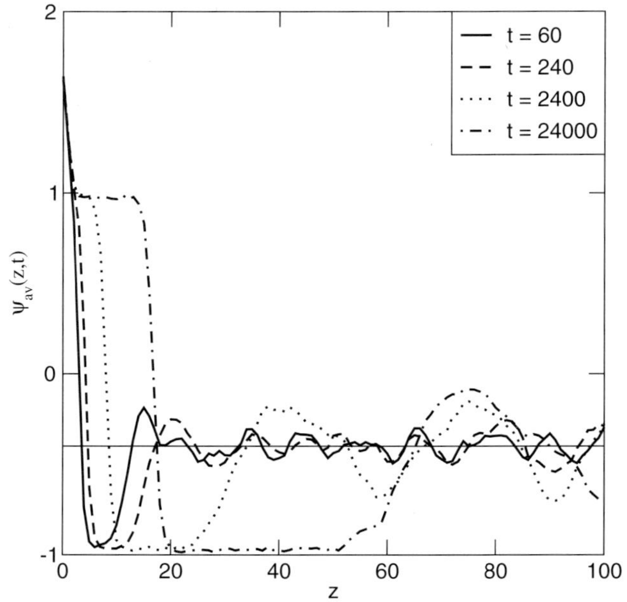

Fig. 5.8. Analogous to Fig. 5.6, but corresponding to the evolution in Fig. 5.7. The horizontal line denotes the average mixture composition, $\psi_0 = -0.4$

Thus, there is no phase separation in the bulk but there is a rapid growth of the wetting layer at the surface. Figure 5.10 shows the laterally averaged profiles corresponding to Fig. 5.9. The behavior is qualitatively different from that for $\psi_0 = 0.0, -0.4$, due to the absence of bulk phase separation. Notice that the A-rich wetting layer is followed by a layer which is moderately depleted in A, and extends deep into the bulk.

Focus first on the time-dependence of the wetting profiles seen in Figs. 5.6 and 5.8, where the bulk undergoes phase separation. We will consider the case of extremely off-critical quenches subsequently. The thickness of the depletion layer is denoted as $h(t) = R_2(t) - R_1(t)$. Consider the typical snapshots in Figs. 5.5 and 5.7. There are two contributions to the current which drives the growth of the wetting layer:

1. The surface-potential gradient drives A to the wetting layer with a current $-dV/dz|_{z=R_1}$.
2. The intrinsic chemical potential (due to local curvature) is higher on the curved surface of bulk A-rich droplets than on the flat wetting layer. This difference is estimated as σ/L, where $L(t)$ is the bulk domain size, and σ is the surface tension. (In these dimensionless units, $\sigma = 2/3$.) The corresponding current contribution at the wetting layer is $-\sigma/(Lh)$.

Thus the A-current in the z-direction is obtained by adding these two contributions as follows:

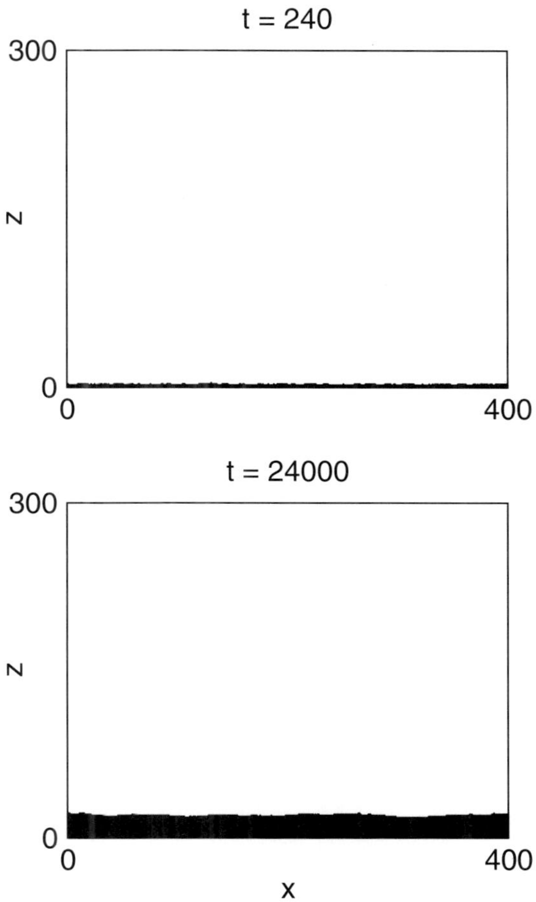

Fig. 5.9. Analogous to Fig. 5.5, but for $\psi_0 = -0.8$

$$J_z \simeq -\left.\frac{dV(z)}{dz}\right|_{z=R_1} - \frac{\sigma}{Lh} \,. \tag{5.43}$$

To estimate $h(t)$, assume that the wetting and depletion layers have an overall composition of ψ_0. This yields the relations:

$$\begin{aligned} R_2(t) &\simeq \frac{2}{1+\psi_0} R_1(t) \,, \\ h(t) &\simeq \frac{1-\psi_0}{1+\psi_0} R_1(t) \,. \end{aligned} \tag{5.44}$$

The validity of the scaling assumption in (5.44) has been confirmed by Puri and Binder [302].

Using the power-law form of the potential from (5.42), and $h(t)$ from (5.44), (5.43) yields [302]

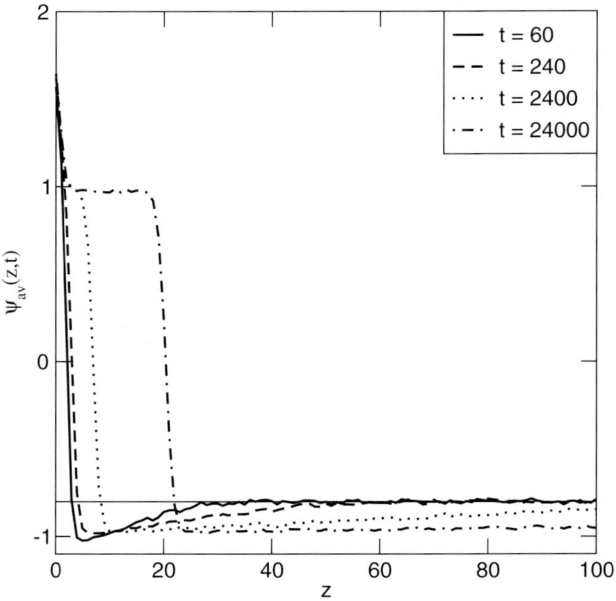

Fig. 5.10. Laterally averaged profiles for evolution in Fig. 5.9

$$\begin{aligned}\frac{dR_1}{dt} &= -J_z \\ &\simeq \frac{nh_1}{{R_1}^{n+1}} + \frac{\sigma}{LR_1}\left(\frac{1+\psi_0}{1-\psi_0}\right) .\end{aligned} \tag{5.45}$$

The bulk length scale obeys the LS growth law $L(t) = f(\psi_0)(\sigma t)^{1/3}$, where the function $f(\psi_0)$ is known analytically for $|\psi_0| \to 1$ [81,82], but only numerically for other values of ψ_0 [96]. Clearly, the first term on the RHS of (5.45) is dominant at early times (for $n > 1$) and the second term is dominant at late times. This yields the growth regimes as

$$\begin{aligned}R_1(t) &\sim (h_1 t)^{1/(n+2)} , \quad t \ll t_c , \\ &\sim \sqrt{\frac{(1+\psi_0)}{f(\psi_0)(1-\psi_0)}}\,(\sigma t)^{1/3} , \quad t \gg t_c .\end{aligned} \tag{5.46}$$

The scaling behavior of the crossover time-scale is then obtained by equating the early-time and late-time length scales as (for $n > 1$)

$$t_c \sim h_1^{3/(n-1)} \left[\frac{f(\psi_0)(1-\psi_0)}{(1+\psi_0)}\right]^{3(n+2)/[2(n-1)]} \sigma^{-(n+2)/(n-1)} . \tag{5.47}$$

The crossover between the potential-dependent growth regime and the universal regime ($R_1 \sim t^{1/3}$) can be extremely delayed, depending on the various system parameters and mixture composition. This clarifies the reason for

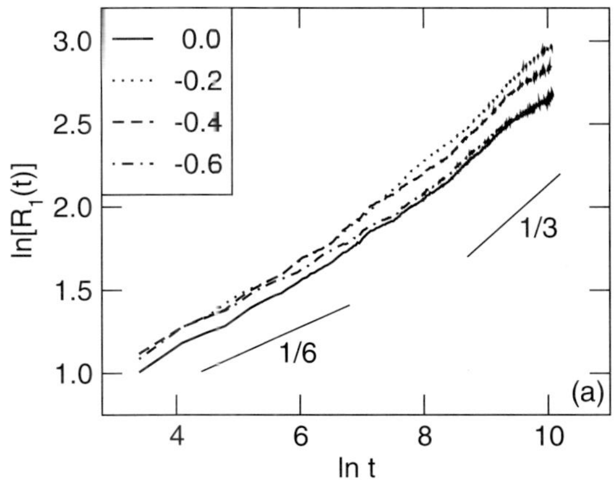

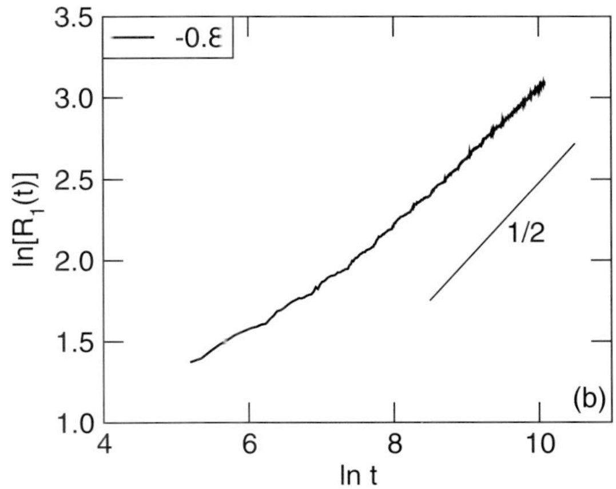

Fig. 5.11. (**a**) Time-dependence of wetting-layer thickness, $\ln(R_1)$ vs. $\ln t$, for mixtures with different composition: $\psi_0 = 0.0, -0.2, -0.4, -0.6$, as indicated. The straight lines have slopes 1/6 and 1/3, respectively. The exponent $\phi = 1/6$ corresponds to potential-dependent growth for the case of nonretarded van der Waals' interactions with $n = 4$ in $d = 2$. (**b**) Plot of $\ln(R_1)$ vs. $\ln t$ for an extremely off-critical composition with $\psi_0 = -0.8$. The straight line has a slope of 1/2

diverse growth exponents reported by various experiments and numerical simulations. Figure 5.11(a) plots $\ln[R_1(t)]$ vs. $\ln t$ for $\psi_0 = 0.0, -0.2, -0.4, -0.6$ and illustrates this crossover behavior. Figure 5.11(b) shows the corresponding data for $\psi_0 = -0.8$, which will be discussed shortly.

Before proceeding, it is relevant to discuss the applicability of the above arguments in the limit of a critical quench, $\psi_0 \to 0$. As both components are present in equal proportions, the bulk is bicontinuous and has surfaces with both positive and negative curvatures. Then, it is appropriate to replace the second term on the RHS of (5.45) by its average value, which changes sign as ψ_0 goes through zero. This suggests a divergence of the crossover time in (5.47) as $\psi_0 \to 0$. However, Fig. 5.11 does not show this because the above arguments do not account for the fact that, even the $\psi_0 = 0$ evolution morphology is characterized by A-rich droplets in the region subsequent to the depletion layer (see the upper frame of Fig. 5.5). These droplets are a result of the flow of A to the wetting layer through the depletion layer.

It is relevant to separately consider the cases of the power-law potential with $n = 1$; and the short-ranged potential $V(z) = -V_0 \exp(-z/z_0)$, where z_0 is the characteristic decay length. For $V(z) \sim -z^{-1}$, both terms on the RHS of (5.45) are comparable for all times and the resultant growth law is the LS law, $R_1(t) \sim t^{1/3}$. On the other hand, the short-ranged potential yields a logarithmic early-time growth, $R_1(t) \sim z_0 \ln(h_1 t/z_0^2)$, which rapidly crosses over to the universal LS growth law. However, thermal fluctuations may interfere with the observation of the logarithmic growth regime [306].

Finally, consider the case of extremely off-critical quenches ($\psi_0 \ll 0$), where there is no bulk phase separation as in Fig. 5.9. Here, there are no droplets in the bulk to feed the wetting layer. Thus, the intrinsic chemical potential in the bulk is the uniform value $\mu_0 = {\psi_0}^3 - \psi_0$. The corresponding current to the wetting layer is $-\mu_0/h$, where $h(t)$ is now the characteristic scale on which the order parameter exponentially saturates to its bulk value (see Fig. 5.10). We neglect lateral fluctuations and assume a simple form (motivated by Fig. 5.10) for $\psi(z,t)$ as follows:

$$\begin{aligned}\psi(z,t) &\simeq 1\,, \quad z < R_1(t)\,,\\ &\simeq \psi_0 - B_0 e^{-(z-R_1)/h}\,, \quad z > R_1(t)\,,\end{aligned} \tag{5.48}$$

where B_0 is a parameter. This functional form will be justified when we discuss the kinetics of surface enrichment in Sect. 5.5. The composition constraint then yields

$$h(t) \simeq \frac{(1-\psi_0)}{B_0} R_1(t)\,. \tag{5.49}$$

Thus, (5.45) is modified as

$$\begin{aligned}\frac{dR_1}{dt} &\simeq \frac{nh_1}{{R_1}^{n+1}} + \frac{\mu_0 B_0}{1-\psi_0}\frac{1}{R_1}\\ &= \frac{nh_1}{{R_1}^{n+1}} + \frac{|\psi_0|(1+\psi_0)B_0}{R_1}\,.\end{aligned} \tag{5.50}$$

The corresponding growth regimes in this case are (for any value of n)

$$\begin{aligned}R_1(t) &\sim (h_1 t)^{1/(n+2)}\,, \quad t \ll t_c\,,\\ &\sim \left[|\psi_0|(1+\psi_0)B_0\right]^{1/2} t^{1/2}\,, \quad t \gg t_c\,.\end{aligned} \tag{5.51}$$

In this case, there is a crossover from potential-dependent growth to a universal diffusive growth law at the crossover time

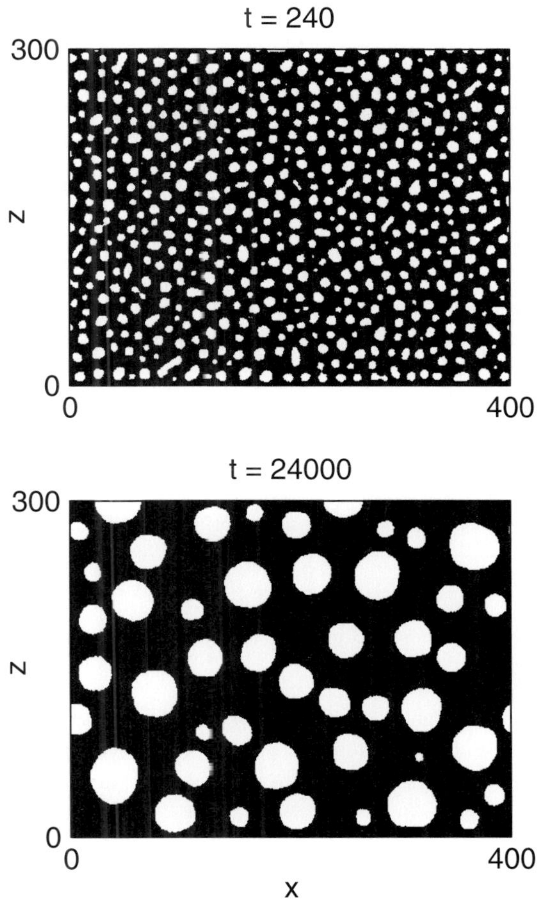

Fig. 5.12. Analogous to Fig. 5.5, but for the case where the wetting component is the majority component, with $\psi_0 = 0.4$

$$t_c \sim h_1^{2/n} \left[|\psi_0|(1+\psi_0)B_0\right]^{-(n+2)/n} . \tag{5.52}$$

Figure 5.11(b) shows data for $\ln[R_1(t)]$ vs. $\ln t$ for $\psi_0 = -0.8$, illustrating the asymptotically diffusive growth of the wetting layer. For a short-ranged surface potential, the initial growth regime is logarithmic, as before.

5.4.4 Wetting by the Majority Component ($\psi_0 > 0$)

Next, consider the case where $\psi_0 > 0$, so that the majority component wets the surface. Figure 5.12 shows the evolution of a homogeneous initial condition with $\psi_0 = 0.4$. Now, the droplets are of the non-wetting component. A thin wetting layer is formed and grows very slowly. The depletion layer

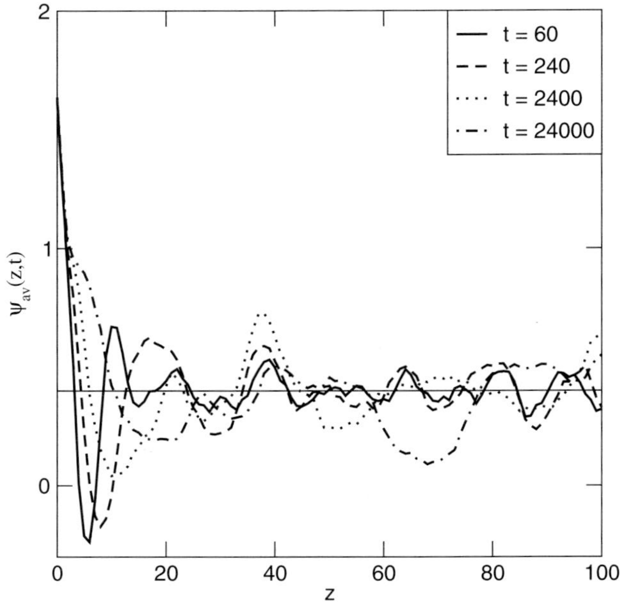

Fig. 5.13. Laterally averaged profiles for the evolution shown in Fig. 5.12

that forms adjacent to the wetting layer is hardly apparent in this case. Figure 5.13 shows the laterally averaged profiles corresponding to the evolution depicted in Fig. 5.12, and confirms the extremely slow growth of the wetting layer.

It is somewhat counter-intuitive that the wetting layer grows so slowly in the case where the majority component wets the surface. Essentially, the bulk droplets now compete with (rather than feed) the wetting layer for the preferred component A, as the intrinsic chemical potential for A is lower on the surface of the drops. Thus, only the first term on the RHS of (5.45) is operational, and accretion on the wetting layer is driven only by the surface potential with the growth law $R_1(t) \sim t^{1/(n+2)}$. As a matter of fact, the wetting-layer growth should be even slower than that, due to the competing bulk droplets – the intrinsic chemical-potential gradient actually drives A into the bulk [302]. Similar considerations apply for other values of $\psi_0 > 0$, when the bulk undergoes phase separation.

The extremely off-critical case (with $\psi_0 \gg 0$) is analogous to enrichment kinetics seen for $T > T_c$. Of course, this assumes that there is no nucleation of droplets over extended time-scales. If droplets are nucleated, the scenario described for $\psi_0 = 0.4$ is applicable again. Sections 5.5 and 5.6 discuss surface enrichment at length, and we do not duplicate the discussion here.

5.4.5 Other Relevant Studies for Case of Diffusive Transport

The discussion so far has followed the work of Puri et al. [302,303]. Now, we discuss other important theoretical studies of this problem. First, consider studies involving a diffusive mechanism for phase separation. In early work, Brown and Chakrabarti [307] presented results from a Langevin simulation of SDSD in $d = 2$ with both short-ranged and long-ranged surface fields. Their model was similar to that described in Sect. 5.3.3, but their parameter values corresponded to a PW surface morphology. Brown and Chakrabarti found that the thickness of the enriched surface layer obeyed the LS growth law $R_1(t) \sim t^{1/3}$. Furthermore, they showed that length scales perpendicular and parallel to the surface are also consistent with a LS growth law, but the typical domain sizes parallel to the surface are larger than those perpendicular to the surface. This is in conformity with the results described in Sect. 5.4.2. Subsequently, Bhattacharya et al. [295] used a similar model to study spinodal decomposition in a strip geometry. Also, Brown et al. [308] reported an interesting study of *surface-induced nucleation*, where they consider a metastable binary mixture in contact with a surface which can nucleate droplets of the preferred phase with much greater facility than in the bulk. This corresponds to the case discussed in Sect. 5.4.4, with $\psi_0 > 1/\sqrt{3}$, the mean-field spinodal. Brown et al. found that the thickness of the wetting layer exhibits very slow growth, which is consistent with the discussion in Sect. 5.4.4.

Brown and Chakrabarti also studied SDSD in block copolymer melts in both semi-infinite and strip geometries [309]. Following Oono and Shiwa [310], they modeled the block copolymer system by the usual ψ^4-free-energy functional in conjunction with a long-ranged interaction term. The surfaces were introduced through boundary conditions on the order-parameter field. For bulk phase separation, it is known that the length scale of the coarsening block copolymer system saturates at an equilibrium value, which depends inversely on the strength of the long-ranged interaction potential [310,311]. Analogous effects occur in the context of SDSD, and some of these are discussed by Brown and Chakrabarti [309].

A comprehensive cell dynamical system (CDS) [96] study of this problem was reported by Marko [312], who also studied SDSD with both short-ranged and long-ranged surface potentials. Marko undertook simulations in $d = 2, 3$, and also investigated both the PW and CW limits. Marko carefully clarified the morphologies of domain growth and the laterally averaged profiles in both limits, though his primary focus was the case with a PW morphology. In this case, he found results consistent with those of Brown and Chakrabarti [307] for the growth of wetting layers. On the other hand, in the CW case, he found a drastic slowing down of the growth of the wetting layer. This is consistent with the potential-dependent growth regime discussed in Sect. 5.4.3.

There have also been studies of the corresponding microscopic model. For example, Sagui et al. [313] reported a $d = 3$ MC study of surface-affected phase separation in an Ising model with a short-ranged surface field. They studied situations with both zero and non-zero surface fields. As discussed earlier, phase separation can occur in the surface layer, even if the bulk is stable, by an appropriate adjustment of the interactions. For the case of a stable bulk and unstable surface, Sagui et al. found that lateral domain growth in the surface layer was consistent with the LAC growth law, viz., $L_{\|}(t) \sim t^{1/2}$ (see Sect. 3.2). For the case where both the bulk and surface are unstable, Sagui et al. found that the growth exponent in the surface layer ranged from $\phi = 1/2 \to 1/3$, depending on the quench depth. The lower exponents correspond to deeper quenches. Nevertheless, in both cases above, Sagui et al. found that the scaled structure factor in the surface layer is consistent with the OJK form [88] in (3.12).

Finally, there have also been Langevin simulations of SDSD on patterned substrates by Karim et al. [263]. These were done in conjunction with experiments on polymer blends by the same authors, and suggested ways of controlling the phase-separation morphology near the surface. Lee et al. [314] have also adapted the models described here to study the effect of immobile obstacles with non-rectangular shapes (e.g., spheres or cylinders) on spinodal decomposition.

Next, we consider the role of hydrodynamic effects in SDSD. In Sect. 3.3.2, we have stressed that domain connectivity is crucial in enabling hydrodynamic transport. For off-critical mixtures, where one has isolated droplets of the minority component, SDSD in binary fluids is driven by diffusive transport as above. For near-critical mixtures, we expect novel features due to the presence of hydrodynamic effects.

5.4.6 Role of Hydrodynamic Effects

In Sect. 3.3.2, we saw that bulk binary fluids coarsen considerably faster than binary solids due to the presence of hydrodynamic modes. In the case of bicontinuous morphologies, the primary coarsening mechanism was identified by Siggia [117] as advective transport of the order parameter by the hydrodynamic velocity field. This results in the bulk growth law $L(t) \sim (\sigma/\eta)t$ in the *viscous hydrodynamic* regime, where σ is the surface tension and η is the fluid viscosity (cf. (3.33)).

In this subsection, we derive the growth exponents characterizing SDSD in binary fluids. We will also discuss some numerical studies of this problem. In many experiments on SDSD, the surface is initially coated by the preferred component, regardless of the equilibrium morphology. The dynamics of coating has been discussed by Tanaka [275,276]. The bicontinuous tubes in the bulk are in contact with the surface, and the pressure gradient between the bulk and surface portions results in a flux $\sim (\sigma/\eta)L^2$, where L is the characteristic tube size. Under strong wetting conditions, this flux results mainly in the lateral spreading of the layer droplet of size $L_s(t)$, whose growth (in $d = 3$) is then determined by $dL_s^2/dt \sim (\sigma/\eta)L^2$. Using the Siggia growth law, $L(t) \sim (\sigma/\eta)t$, one obtains the surface droplet growth law

$L_s(t) \sim (\sigma/\eta)^{3/2} t^{3/2}$. This is in accordance with the experimental observations of a fast mode by Wiltzius, Cumming and others [267,268]. The above arguments also yield the crossover time to complete coating as $t_c \sim \eta/\sigma$.

However, the experiments of Harrison et al. [270] found a fast mode (with exponent $\phi \simeq 1.1 - 1.5$) even when the bulk growth was of the slower LS form $L(t) \sim (\sigma t)^{1/3}$. For diffusion-driven growth, recall that the chemical potential on the bulk tube can be estimated as $\mu \sim \sigma/L$, whereas the chemical potential at the flat surface layer is $\mu \sim 0$. The corresponding current is $J_z \sim \sigma/L^2$ and the flux per tube is estimated as σ. Thus, the surface coating droplet grows as $L_s(t) \sim (\sigma t)^{1/2}$, which is considerably slower than the experimental results of Harrison et al. [270]. Troian [273] has demonstrated how domain coalescence near the surface may speed this slow growth to the experimentally-observed levels.

Next, we consider the dynamics of the wetting layer which grows into the system. If one neglects the force due to attraction from the surface which leads to potential-dependent growth, we can quantify the growth as follows. The wetting layer grows due to the flux of material from the tube into the surface layer. For the hydrodynamic problem, this yields $S(dR_1/dt) \sim (\sigma/\eta)L^2 \times (S/L^2)$, where S is the surface area and S/L^2 is the number of tubes. Thus, $R_1(t) \sim (\sigma/\eta)t$ for the wetting-layer dynamics also. However, this result is at variance with the experiments of Guenoun et al. [274], who observed much slower growth of the wetting layer. This discrepancy could be the result of long-lived transient growth laws dependent upon the form of the surface potential, as discussed in Sect. 5.4.3.

One of the earliest numerical studies of this problem is due to Ma et al. [271,272], who performed MD simulations of AB fluids confined in a thin film. They considered two different cases: one in which both A and B were equally repulsed by the walls, and the other in which one of the walls had a preferential attraction for A. In the first case, Ma et al. examined density-correlation functions in the parallel direction, and found that these exhibited reasonable dynamical scaling, except when the capillary was too thin, i.e., comparable to the molecular diameter. The associated length scale was $L_{\parallel} \sim t^{\phi}$ with $\phi \simeq 0.54$, which they claimed to be consistent with the experiments of Guenoun et al. [274]. However, the two physical situations are considerably different so the comparison may not be appropriate. The second case of Ma et al. is appropriate to the SDSD problem. In this case, they observe a *fast mode* in the surface layer, but with an exponential growth rather than power-law growth.

Another early MD study is due to Zhang and Chakrabarti [315], who considered a fluid mixture confined in a $d = 1$ capillary. These authors focused on static rather than dynamical issues. In particular, they investigated the conditions under which tube (or CW) and plug (or PW) morphologies are obtained for the phase-separated mixture in the capillary.

As stated earlier, a reasonable model for SDSD in a binary fluid is Model H, supplemented with appropriate boundary conditions. Chen and Chakrabarti [316] have studied SDSD (in $d = 2$) through numerical solutions of Model H in a semi-infinite geometry. They consider a surface with a long-ranged potential and impose *no-slip* conditions on the velocity field at the surface. Their primary result is that the wetting-layer growth crosses over from $R_1(t) \sim t^{1/3}$ (characteristic of bulk diffusive growth in any dimension $d > 1$) to $R_1(t) \sim t^{2/3}$ (characteristic of bulk hydrodynamic growth in $d = 2$ [317]). This crossover occurs because domains of the preferred component establish contact with the surface layer and drain rapidly.

Another study of Model H (in $d = 3$) in a semi-infinite geometry is due to Tanaka and Araki [276,318]. They find that the wetting-layer thickness grows initially as $R_1(t) \sim t^{1/3}$, and then crosses over to the hydrodynamic regime with $R_1(t) \sim t$. Tanaka and Araki also find that the oscillatory profile of SDSD waves originating at the surface is destroyed more rapidly in the hydrodynamic case than in the diffusion-driven case. These authors also studied characteristic length scales in the layers parallel to the surface. In the vicinity of the surface, they found a faster growth than $L_{\parallel} \sim t$, the usual bulk law. However, it is difficult to assign an unambiguous exponent to this faster growth. Furthermore, the time-regime of the fast mode is considerably later than the formation of the CW layer. This suggests that the fast growth observed by Tanaka and Araki should be identified with the anisotropic growth (with $L_{\perp} < L_{\parallel}$) of domains due to orientational effects of the wetting layer, rather than the fast mode of [267–270]. As we have discussed earlier, this fast mode is associated with the coating dynamics which results in a CW layer.

A recent MD study by Toxvaerd [319] investigated a critical AB mixture with particles interacting through Lennard-Jones potentials. This mixture was placed in contact with two parallel walls, which also interacted with the fluid particles through a Lennard-Jones potential. Toxvaerd focused on the morphologies which arise for different wall-types, e.g., one wall attracts A whereas the other wall attracts B; both walls attract A and B equally, etc. In all cases, he finds that the system evolves into a layered morphology, with the layer being parallel to the surface walls. These are probably metastable configurations which evolve exceedingly slowly due to the low effective dimensionality of the system.

Moreover, Bastea et al. [320] studied SDSD in $d = 3$ binary fluids by a direct integration of the Vlasov-Boltzmann equations. Their modeling is at an intermediate level between the microscopic and coarse-grained models we have discussed so far. Earlier, Bastea and Lebowitz [126] had used such mesoscale models to study bulk segregation in fluids. The simulations of Bastea et al. clarified the wetting-layer kinetics in both the PW and CW cases. In Figs. 5.14 and 5.15, we show their evolution pictures for the PW and CW cases, respectively.

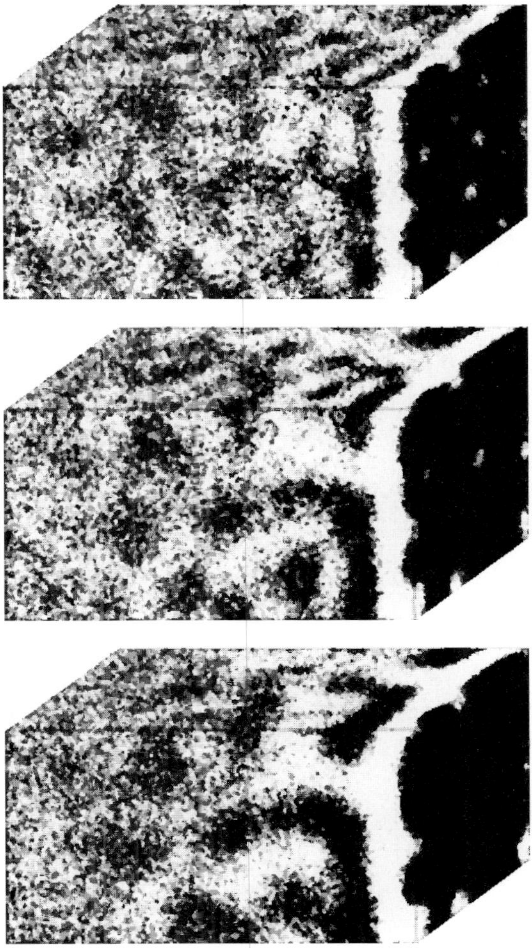

Fig. 5.14. Evolution pictures for SDSD in an unstable binary fluid with critical composition [320]. The surface (located at the right) is partially wetted by the preferred component, which is marked in black. From top to bottom, the frames correspond to times $t = 60, 120, 180$

Finally, Kuksenok et al. [321,322] have used lattice-Boltzmann and CDS models to study AB fluids flowing on a patterned substrate. In their simulations, the substrate is decorated with a checkerboard pattern, with different patches preferring A or B. Kuksenok et al. studied the interplay between the intrinsic drive for phase separation and the competing drive for alignment with the surface pattern.

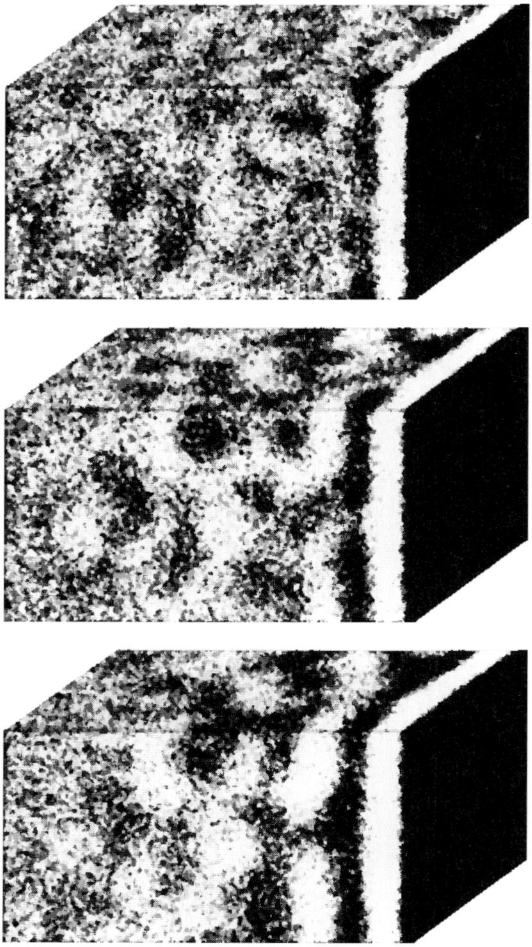

Fig. 5.15. Analogous to Fig. 5.14, but for the case of a CW surface morphology

5.5 Kinetics of Surface Enrichment for Stable Binary Mixtures

Next, consider the problem of surface enrichment, which has already been referred to on a few occasions. This occurs when a stable binary mixture with $T > T_c$ is placed in contact with a surface (at $z = 0$) which prefers one of the components. For large z, the stable mixture continues to be homogeneous. However, the surface again becomes enriched in the preferred component, resulting in a time-dependent surface-enrichment profile, which propagates into the bulk. Jones et al. [298,299] have studied the kinetics of surface enrichment for polymer mixtures, and found that the enrichment profiles are characterized by diffusive length-scales. There have also been some numerical studies of this problem (with both short-ranged and long-ranged surface fields) via MC [278] and Langevin [323] simulations. Furthermore, Binder and Frisch [284] have studied this problem for a short-ranged surface potential in the framework of a linear theory. Following Frisch et al. [324,325], we present an analytical solution of the linear problem for a broad class of surface potentials. The regime of validity and the characteristic properties of this solution are also discussed.

Recall the nonlinear model in (5.26)-(5.30). Consider its deterministic version with $T = 0$. Let us examine the evolution of an initial condition consisting of fluctuations about a uniform background, viz., $\psi(\boldsymbol{r},t) = \psi_0 + \psi'(\boldsymbol{r},t)$. If we linearize (5.26), (5.29) and (5.30) in the fluctuation field ψ', we obtain (dropping the primes)

$$\frac{\partial}{\partial t}\psi(\boldsymbol{r},t) = -\nabla^2\left[\left(\mathrm{sgn}(T_c - T) - 3\psi_0^2\right)\psi + \frac{1}{2}\nabla^2\psi - V(z)\right], \quad z > 0, \tag{5.53}$$

in conjunction with the boundary conditions:

$$\tau_0\frac{\partial}{\partial t}\psi(\boldsymbol{\rho},0,t) = h_1 + g\psi_0 + g\psi + \gamma\left.\frac{\partial\psi}{\partial z}\right|_{z=0}, \tag{5.54}$$

$$0 = \frac{\partial}{\partial z}\left[\left(\mathrm{sgn}(T_c - T) - 3\psi_0^2\right)\psi + \frac{1}{2}\nabla^2\psi - V(z)\right]\Bigg|_{z=0}. \tag{5.55}$$

Here, we have explicitly incorporated the possibility that T may be above or below T_c by using the sgn-function, $\mathrm{sgn}(x) = x/|x|$. The earlier discussion of SDSD had used the static version of the boundary condition in (5.29). Here, we retain the dynamical version, as we are also interested in the time-dependent behavior of the order parameter at the surface. Finally, we focus on potentials which are flat at the origin (cf. (5.42)), so $V'(0)$ is subsequently set to zero in (5.55).

For $T < T_c$, the fluctuations grow in time (when $3\psi_0^2 < 1$) and the linearized equations are not valid after the early-time regime. For $T > T_c$, the fluctuations do not grow and (5.53)-(5.55) are valid for all time – provided that the surface field is sufficiently weak that the value of ψ stays small even in the vicinity of the surface. Thus, the subsequent analysis of the linearized model will be valid at early times for arbitrary temperatures; and for late times when $T > T_c$ and the surface field is weak. Furthermore, the solution of the linearized model is also valid for $T < T_c$ and highly off-critical quenches ($3\psi_0^2 - 1 \gg 0$) provided that there is no nucleation of droplets. This case was briefly discussed in Sects. 5.4.3 and 5.4.4.

It is convenient to Fourier transform (5.53)-(5.55) in the direction parallel to the wall, resulting in the following equations:

$$\frac{\partial}{\partial t}\psi(\boldsymbol{k}_\parallel, z, t) = \left[\left(\mathrm{sgn}(T_c - T) - 3\psi_0^2\right) - \frac{1}{2}\boldsymbol{k}_\parallel^2\right]\boldsymbol{k}_\parallel^2\psi - \left[\left(\mathrm{sgn}(T_c - T) - 3\psi_0^2\right) - \boldsymbol{k}_\parallel^2\right]\frac{\partial^2\psi}{\partial z^2} - \frac{1}{2}\frac{\partial^4\psi}{\partial z^4} + V''(z)\delta(\boldsymbol{k}_\parallel)\,, \tag{5.56}$$

$$\tau_0\frac{\partial}{\partial t}\psi(\boldsymbol{k}_\parallel, 0, t) = (h_1 + g\psi_0)\delta(\boldsymbol{k}_\parallel) + g\psi + \gamma\left.\frac{\partial\psi}{\partial z}\right|_{z=0}, \tag{5.57}$$

$$0 = \frac{\partial}{\partial z}\left[\left(\mathrm{sgn}(T_c - T) - 3\psi_0^2 - \frac{1}{2}\boldsymbol{k}_\parallel^2\right)\psi + \frac{1}{2}\frac{\partial^2\psi}{\partial z^2}\right]\Bigg|_{z=0}, \tag{5.58}$$

where $\boldsymbol{k}_\parallel$ denotes the wave-vector parallel to the surface. Inclusion of a surface diffusion term modifies the parameter g in the second term on the RHS of (5.57) [301].

To be specific, we consider critical binary mixtures (viz., $\psi_0 = 0$) and focus on the case with $T > T_c$. The solution trivially generalizes to cases with arbitrary ψ_0 and $T < T_c$. Of course, it should be kept in mind that the solution for $T < T_c$ and $3\psi_0^2 - 1 < 0$ is only valid for early times before the fluctuations become too large.

Furthermore, we consider the case where the order-parameter field is homogeneous parallel to the surface (viz., $\psi(\boldsymbol{\rho}, z, t) \equiv \psi(z, t)$). This is reasonable because the bulk is strongly driven towards a uniform homogeneous state, and the surface field promotes homogeneity at the surface. In any case, the method of solution outlined below is also applicable to the laterally inhomogeneous case, with an appropriate modification of parameters. This is apparent from a comparison of (5.56)-(5.58) and the corresponding laterally-homogeneous version.

Thus, the relevant model to be solved is (5.56)-(5.58) with $\psi_0 = 0, T > T_c$, and $\boldsymbol{k}_\parallel = 0$:

$$\frac{\partial}{\partial t}\psi(z, t) = \frac{\partial^2\psi}{\partial z^2} - \frac{1}{2}\frac{\partial^4\psi}{\partial z^4} + V''(z)\,, \quad z > 0\,, \tag{5.59}$$

$$\tau_0\frac{\partial}{\partial t}\psi(0, t) = h_1 + g\psi + \gamma\left.\frac{\partial\psi}{\partial z}\right|_{z=0}, \tag{5.60}$$

$$0 = \left.\frac{\partial\psi}{\partial z}\right|_{z=0} - \frac{1}{2}\left.\frac{\partial^3\psi}{\partial z^3}\right|_{z=0}. \tag{5.61}$$

We Laplace-transform (5.59)-(5.61) in time to obtain

$$s\tilde{\psi}(z, s) = \frac{\partial^2\tilde{\psi}}{\partial z^2} - \frac{1}{2}\frac{\partial^4\tilde{\psi}}{\partial z^4} + \frac{V''(z)}{s}, \quad z > 0\,, \tag{5.62}$$

$$s\tilde{\psi}(0, s) = \frac{h_1}{s} + g\tilde{\psi} + \gamma\left.\frac{\partial\tilde{\psi}}{\partial z}\right|_{z=0}, \tag{5.63}$$

$$0 = \left.\frac{\partial \tilde{\psi}}{\partial z}\right|_{z=0} - \frac{1}{2}\left.\frac{\partial^3 \tilde{\psi}}{\partial z^3}\right|_{z=0}, \tag{5.64}$$

where (as usual)

$$\tilde{\psi}(z,s) = \int_0^\infty dt e^{-st}\psi(z,t), \tag{5.65}$$

and we have used the homogeneous initial condition $\psi(z, t = 0) = 0$. Notice that we have absorbed τ_0 into the definition of the parameters h_1, g, γ.

Equation (5.62) is an inhomogeneous fourth-order differential equation with two boundary conditions. The solution of (5.62)-(5.64) can be obtained using the solutions of the homogeneous part of (5.62) [326]. Some algebra yields the final result as follows [324]:

$$\begin{aligned}
\tilde{\psi}(z,s) = {} & \frac{h_1\xi_- + 2s(s-g)(\xi_+\alpha - \xi_-\beta)\xi_-}{D(s)} e^{-z/\xi_+} - \xi_+\alpha e^{-z/\xi_+} \\
& - \frac{h_1\xi_+ + 2s(s-g)(\xi_+\alpha - \xi_-\beta)\xi_+}{D(s)} e^{-z/\xi_-} + \xi_-\beta e^{-z/\xi_-} \\
& - \frac{\xi_+}{s\Delta}\left[e^{-z/\xi_+}\int_0^z dz' e^{z'/\xi_+}V''(z') + e^{z/\xi_+}\int_z^\infty dz' e^{-z'/\xi_+}V''(z')\right] \\
& + \frac{\xi_-}{s\Delta}\left[e^{-z/\xi_-}\int_0^z dz' e^{z'/\xi_-}V''(z') + e^{z/\xi_-}\int_z^\infty dz' e^{-z'/\xi_-}V''(z')\right],
\end{aligned} \tag{5.66}$$

where

$$\begin{aligned}
\alpha &= \frac{1}{s\Delta}\int_0^\infty dz' e^{-z'/\xi_+}V''(z'), \\
\beta &= \frac{1}{s\Delta}\int_0^\infty dz' e^{-z'/\xi_-}V''(z'), \\
\Delta &= \xi_+^{-2} - \xi_-^{-2}, \quad \xi_\pm^{-2} = 1 \pm \sqrt{1-2s}, \\
D(s) &= \frac{\sqrt{s}(s-g)}{\sqrt{2}}\left(\xi_+^{-1} - \xi_-^{-1}\right) + \gamma\sqrt{2s}\sqrt{1-2s}.
\end{aligned} \tag{5.67}$$

The first four terms on the RHS of (5.66) are recognized as arising from the homogeneous part of (5.62), with ξ_+ and ξ_- being appropriate length scales. The terms involving integrals over $V''(z)$ in (5.66) correspond to the particular solution of the inhomogeneous (5.62).

The expression in (5.66) is unwieldy and has to be inverse-transformed numerically to obtain the temporal evolution of enrichment profiles. However, it is relatively straightforward to obtain the asymptotic time-dependence of important profile characteristics. Recall that we are primarily interested in potentials which exhibit power-law decays for distances larger than a cut-off, i.e., $V(z) \sim -z^{-n}$.

5.6 Asymptotic Properties of Enrichment Profiles

An important characteristic of the profile is the time-dependence of $\psi(0,t)$, the value of the order parameter at the surface. To evaluate this, we put $z=0$ in (5.66) and obtain

$$\tilde{\psi}(0,s) = \frac{h_1(\xi_- - \xi_+) - 2(\xi_+\alpha - \xi_-\beta)\gamma\sqrt{2s}\sqrt{1-2s}}{s(s-g)(\xi_- - \xi_+) + \gamma\sqrt{2s}\sqrt{1-2s}} \,. \tag{5.68}$$

The asymptotic ($t \to \infty$) behavior of $\psi(0,t)$ is determined by the $s \to 0$ behavior of $\tilde{\psi}(0,s)$. To ascertain this, the leading s-dependence (as $s \to 0$) of various quantities on the RHS of (5.68) is required. These are computed as follows:

$$\begin{aligned}
\xi_- &= \frac{1}{\sqrt{s}}(1+O(s))\,, \\
\xi_+ &= \frac{1}{\sqrt{2}}(1+O(s))\,, \\
\alpha &= -\frac{A}{2s} + O(1)\,, \quad A = -\int_0^\infty dz e^{-z\sqrt{2}} V''(z)\,, \\
\beta &= \frac{h_1}{2\sqrt{s}} - \frac{B}{2} + O(\sqrt{s})\,, \quad B = -\int_0^\infty dz V(z)\,.
\end{aligned} \tag{5.69}$$

Replacing these in (5.68), we obtain

$$\begin{aligned}
\tilde{\psi}(0,s) = \frac{1}{s}\left(\frac{h_1 + A\gamma + h_1\gamma\sqrt{2}}{\gamma\sqrt{2}-g}\right) \\
- \frac{1}{\sqrt{s}}\frac{1}{(\gamma\sqrt{2}-g)}\left[\frac{g(h_1 + A\gamma + h_1\gamma\sqrt{2})}{\sqrt{2}(\gamma\sqrt{2}-g)} - \frac{h_1}{\sqrt{2}} - B\gamma\sqrt{2}\right] + O(1)\,.
\end{aligned} \tag{5.70}$$

The asymptotic time-dependence of $\psi(0,t)$ is obtained by inverse-transforming $\tilde{\psi}(0,s)$ as follows:

$$\psi(0,t) = \left(\frac{h_1 + A\gamma + h_1\gamma\sqrt{2}}{\gamma\sqrt{2} - g}\right) - \frac{1}{\sqrt{\pi}(\gamma\sqrt{2} - g)} \left[\frac{g(h_1 + A\gamma + h_1\gamma\sqrt{2})}{\sqrt{2}(\gamma\sqrt{2} - g)} - \frac{h_1}{\sqrt{2}} - B\gamma\sqrt{2}\right] t^{-1/2}, \tag{5.71}$$

where we have retained only the leading-order correction to the constant value of $\psi(0,t)$ as $t \to \infty$. Thus, $\psi(0,t)$ saturates diffusively to its equilibrium value.

Next, consider the s-dependence of the profile moments, which are defined as

$$\langle z^m \rangle = \int_0^\infty dz z^m \tilde{\psi}(z,s)\,. \tag{5.72}$$

Replacing the functional form for $\tilde{\psi}(z,s)$ from (5.66) in (5.72), some algebra leads to the following result:

$$\begin{aligned}
\langle z^m \rangle &= \Gamma(m+1)\xi_+^{m+1} \left[\frac{h_1\xi_- + 2s(s-g)(\xi_+\alpha - \xi_-\beta)\xi_-}{D(s)} - \xi_+\alpha\right] \\
&\quad - \Gamma(m+1)\xi_-^{m+1} \left[\frac{h_1\xi_+ + 2s(s-g)(\xi_+\alpha - \xi_-\beta)\xi_+}{D(s)} - \xi_-\beta\right] \\
&\quad - \frac{\xi_+}{s\Delta} \int_0^\infty dz V''(z) \left[\int_0^\infty dx e^{-x/\xi_+}(z+x)^m + \int_0^z dx e^{-x/\xi_+}(z-x)^m\right] \\
&\quad + \frac{\xi_-}{s\Delta} \int_0^\infty dz V''(z) \left[\int_0^\infty dx e^{-x/\xi_-}(z+x)^m + \int_0^z dx e^{-x/\xi_-}(z-x)^m\right].
\end{aligned} \tag{5.73}$$

Clearly, moments are not defined to all orders for power-law potentials. Thus, for a potential $V(z) \sim -z^{-n}$, moments are only defined upto $\langle z^n \rangle$. The asymptotic behavior of $\langle z^m \rangle$ as $t \to \infty$ is determined by the leading terms in (5.73) as $s \to 0$. The dominant contributions arise from the second, fifth and sixth terms on the RHS of (5.73), which are of order $s^{-(m+3)/2}$ and $s^{-(m+2)/2}$. However, there is a delicate cancellation of contributions of order $s^{-(m+3)/2}$ arising from the different terms. The resultant answer, to leading order, is

$$\langle z^m \rangle = \frac{1}{s^{\frac{m}{2}+1}} \Gamma(m+1) \left[\frac{h_1\sqrt{2} + g(A + h_1\sqrt{2})}{2(\gamma\sqrt{2} - g)} + B\right]. \tag{5.74}$$

Finally, the expression in (5.74) is inverse-transformed to obtain the leading asymptotic time-dependence as

$$\langle z^m \rangle = -\frac{\Gamma(m+1)}{\Gamma(\frac{m}{2}+1)} \left[\frac{h_1\sqrt{2} + g(A + h_1\sqrt{2})}{2(\gamma\sqrt{2} - g)} + B \right] t^{m/2}, \tag{5.75}$$

which also corresponds to diffusive behavior.

The results presented here are universal as regards the time-dependence of $\psi(0,t)$ and $\langle z^m \rangle$ for a wide range of physical potentials. Of course, these results have been obtained in the context of a linear theory. There is a regime of *weak* surface fields in which there is no appreciable difference between the results obtained from the nonlinear and linear models. (The results for the nonlinear model are obtained by numerical solution of the corresponding equations [324,325].) In this regime, the profiles in (5.66) constitute a complete solution of the surface enrichment problem. As the field strength is increased, the validity of linear theory should break down, at least in the vicinity of the surface. Though this is true for the time-dependent profiles, Frisch et al. [324] demonstrated that the diffusive behavior of various profile characteristics is unaffected, even in the strongly nonlinear regime.

6 Phase Ordering Dynamics in the Complex Ginzburg-Landau Equation

6.1 Introduction

There has been intense interest in problems of pattern formation in biological systems and chemical reactions [327–331]. In general, many physical systems exhibit fascinating spatio-temporal dynamics, resulting from the emergence and interaction of spatially-extended structures, e.g., vortices, spirals, etc. In this context [327–333], much attention has focused on the complex Ginzburg-Landau (CGL) equation, which has the general form:

$$\frac{\partial}{\partial t}\psi(\boldsymbol{r},t) = \psi + (1+i\alpha)\nabla^2\psi - (1+i\beta)|\psi|^2\psi\,. \tag{6.1}$$

In (6.1), $\psi(\boldsymbol{r},t)$ $(= \psi_1(\boldsymbol{r},t) + i\psi_2(\boldsymbol{r},t))$ is a complex order-parameter field; and α, β are real parameters. The CGL equation is a generalization of the deterministic version of the XY model ((3.13) with $n = 2$) introduced in Chap. 3. The real and imaginary parts of ψ in (6.1) are identified as the two components of the vector $\boldsymbol{\psi} = (\psi_1, \psi_2)$ in (3.13). However, the complex notation for ψ is particularly convenient in the present context, and we will use it throughout this chapter.

The importance of the CGL equation arises from the fact that it provides a generic description of the slow modulation of oscillations in a spatially-extended system near a Hopf bifurcation [327–331]. Apart from its extensive applications in biophysics and chemical physics, the CGL equation arises in diverse physical systems, e.g., thermal convection in binary fluids [334,335], multi-mode lasers [336], etc.

More generally, the CGL equation is of great interest as a paradigm for studies of pattern formation. It exhibits a rich range of dynamical behavior with variation of the parameters α and β, and the *phase diagram* in $d = 1, 2$ has been investigated by various authors [337–341]. In a large range of parameter space, spiral (and antispiral) defects play an important role in determining the morphology. In this chapter, we will use the concepts introduced in Chap. 3 to investigate nonequilibrium dynamics in the CGL equation [342–344]. In analogy with the phase ordering problem, we consider the evolution dynamics resulting from a small-amplitude random initial condition. As in the case of phase ordering systems, we find that defect structures (*spirals*, in this case) enable a quantitative characterization of the evolution morphology.

This chapter is organized as follows. In Sect. 6.2, we review some analytical results which are relevant to this problem. In Sect. 6.3, we present analytical results for the correlation function of a single-spiral morphology. In Sects. 6.4 and 6.5, we discuss numerical results obtained from simulations of the CGL equation in $d = 2$ and $d = 3$, respectively. Finally, Sect. 6.6 concludes this chapter with a summary and discussion.

6.2 Overview of Relevant Analytical Results

In Sect. 3.2, we showed that the evolution of a ferromagnet from an unstable disordered state is a strongly nonlinear process. In appropriate dimensionless units, this evolution is described by the TDGL equation (3.4), which (for $H = T = 0$) is analogous to (6.1) with $\psi(\boldsymbol{r}, t)$ real, and $\alpha = \beta = 0$. The ferromagnet evolves by the formation and growth of domains which are enriched in either up or down spins, and are characterized by a time-dependent length scale, $L(t)$. In the case of a pure and isotropic ferromagnet, the characteristic domain size obeys the LAC growth law, $L(t) \sim t^{1/2}$ [83]. The primary mechanism for domain growth is the curvature-driven motion and annihilation of interfaces (or defects).

Next, let us consider the dynamical XY model, which is (6.1) with $\psi(\boldsymbol{r}, t)$ complex, but $\alpha = \beta = 0$. In this case, the relevant defect structures (for dimensionality $d \geq 2$) are vortices, vortex lines, etc., and domain growth is driven by the motion and annihilation of vortices and anti-vortices. Puri [345] obtained the time-dependent correlation function for the XY model, using singular-perturbation methods [346–348]. This result was subsequently generalized to the n-component TDGL equation (with $n \leq d$) by Bray and Puri (BP) [90] and Toyoki (T) [91]. The BPT correlation function is given in (3.17).

There is also a good understanding of the CGL equation in various other limits. For example, when $\alpha = \beta$, the CGL equation can be transformed into the XY model (with complex time) by a change of variables as $\psi' = \psi \exp(i\beta t)$. Clearly, the appropriate defects which characterize the evolution are again vortices. Furthermore, when $\alpha, \beta \to \infty$, the CGL equation reduces to the nonlinear Schrödinger equation (NLSE). The $d = 1$ NLSE is an example of an integrable infinite-dimensional Hamiltonian system, and is of great interest in the context of *soliton equations* [349,350]. The higher-dimensional NLSE with a potential term is usually referred to as the Gross-Pitaevskii (GP) equation [351], and is of topical interest in the context of *Bose-Einstein condensation* [352].

The present chapter focuses on ordering dynamics in the CGL equation with $\alpha = 0$, $\beta \neq 0$:

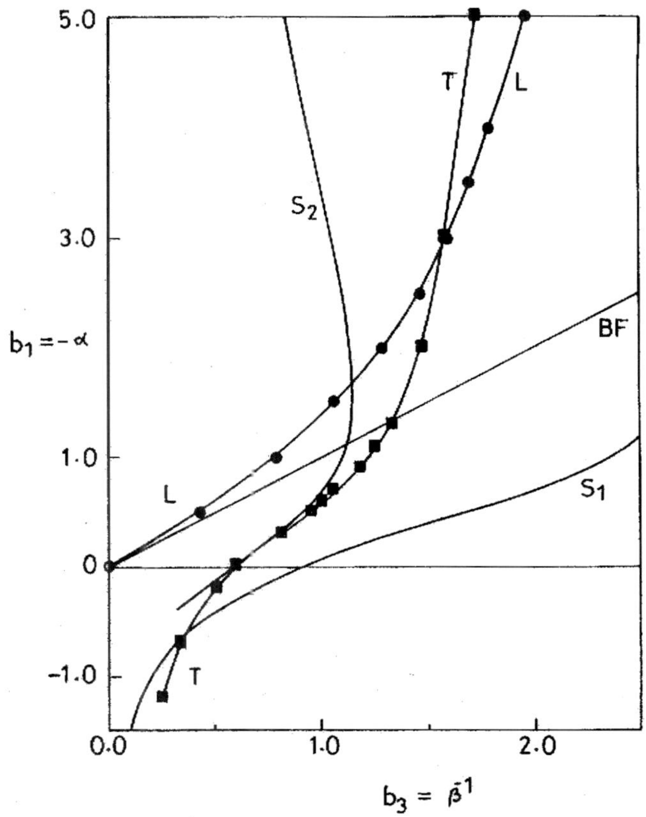

Fig. 6.1. Phase diagram of the $d = 2$ CGL equation in $(b_1 = -\alpha, b_3 = \beta^{-1})$-space [341]. All plane-wave solutions are unstable above the Benjamin-Feir (BF) line, $b_1 = b_3$. Spirals are linearly stable to the right of the line S_1. Between the lines S_1 and S_2, spirals are linearly unstable but fluctuations are advected away. To the left of the line S_2, spirals are globally unstable. The observed morphologies are as follows: *phase turbulence* (i.e., no defects), between lines L and BF; *defect turbulence*, between lines T and S_2; and metastable *frozen states*, to the right of T. For details of the dynamics in overlapping regions, we refer the reader to [341]

$$\frac{\partial}{\partial t}\psi(\boldsymbol{r}, t) = \psi + \nabla^2\psi - (1 + i\beta)|\psi|^2\psi\,. \tag{6.2}$$

The analytical results presented here correspond to (6.2) in $d = 2$, where spirals are point defects. (The appropriate defects in $d = 3$ are spiral lines.) However, these results can be easily extended to the case with $\alpha \neq 0$ and $d \geq 2$, as the underlying paradigm remains the same, i.e., spiral defects still determine the morphology in large regions of parameter space and for higher dimensionality.

Following Hagan [337], Aranson et al. [338], and Chate and Manneville [341], we discuss the phase diagram of the $d = 2$ CGL equation. Figure 6.1 shows the phase diagram in ($b_1 = -\alpha, b_3 = \beta^{-1}$)-space, obtained by Chate and Manneville [341]. We consider the case with $\alpha = 0, \beta \geq 0$. For $0 \leq \beta \leq \beta_1$ ($\beta_1 \simeq 1.397$ [337]), spirals (which are asymptotically plane-waves) are *linearly stable* to fluctuations. For $\beta_1 < \beta \leq \beta_2$ ($\beta_2 \simeq 1.82$ [338,341]), spirals are linearly unstable, but the growing fluctuations are advected away, i.e., the spiral structure is *globally stable.* Finally, for $\beta_2 < \beta$, the spirals are globally unstable structures and cannot exist for extended times [338]. The results in this chapter correspond to the parameter regime $\beta \leq \beta_2$. Here, the CGL morphology is characterized by spirals and the system evolves into metastable *frozen* states, where the $|\psi|$-field is almost stationary in time.

Hagan [337] also studied analytical solutions of the $d = 2$ CGL equation for general α, β. He found a family of spiral solutions with the following functional form (for $\alpha = 0$):

$$\psi(\boldsymbol{r}, t) = \rho(r) \exp\left[-i\beta(1 - q^2)t + im\theta - i\phi(r)\right] , \tag{6.3}$$

where $\boldsymbol{r} \equiv (r, \theta)$; $q \geq 0$ is a constant which is determined uniquely as a function of β; and m is the number of arms in the spiral. The cases with $m > 0$ and $m < 0$ correspond to spirals and antispirals, respectively. The limiting forms of the functions $\rho(r)$ and $\phi(r)$ are

$$\begin{aligned} &\rho(r) \to (1 - q^2)^{1/2}, \quad \phi'(r) \to q, \quad \text{as} \quad r \to \infty, \\ &\rho(r) \to ar^m, \quad \phi'(r) \to r, \quad \text{as} \quad r \to 0, \end{aligned} \tag{6.4}$$

where the constant a is determined by finiteness conditions. Hagan has presented explicit solutions for $q(\beta)$ in the cases with $m = 1, 2$. We will focus on the case with $m = \pm 1$, as only the 1-armed spirals are expected to be stable in the evolution [337]. Figure 6.2 plots Hagan's solution for $q(\beta)$ (with $\beta \leq 1.5$) in the case with $m = \pm 1$. In the simple limit $\beta = 0$, we have $q = 0$, and the spiral solution simplifies to the vortex solution. For the $m = \pm 1$ vortex, the lines of constant phase correspond to radial lines with constant θ. The stability properties of spiral solutions for the general case with $\alpha, \beta \neq 0$ have been discussed by Aranson et al. [338].

6.3 Correlation Function for a Single-Spiral Morphology

Figure 6.3 shows the typical evolution of the $d = 2$ CGL equation from a small-amplitude random initial condition. It is clear that the evolving morphology is characterized by spirals and their interactions. (We use the term *spiral* for both spirals and antispirals, unless specifically stated otherwise.) This should be compared with the evolution shown in Fig. 3.2, where the corresponding defects were vortices. The morphology in Fig. 6.3 has a typical

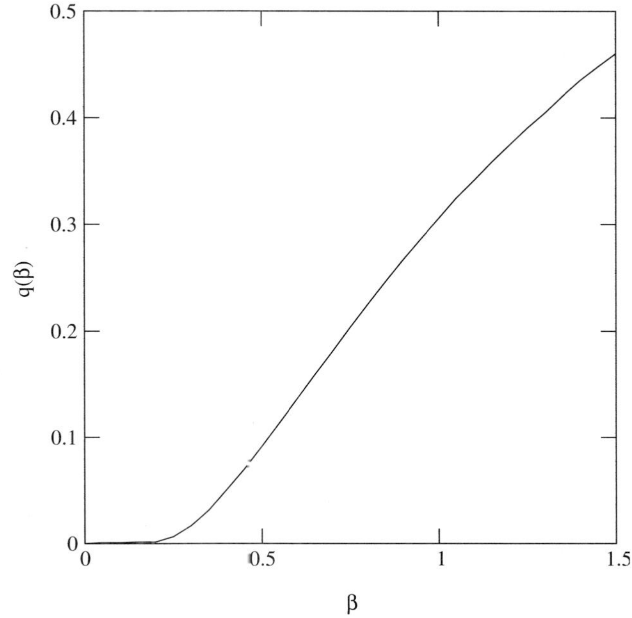

Fig. 6.2. Plot of the Hagan parameter $q(\beta)$ vs. β for the 1-armed spiral ($m = +1$) or antispiral ($m = -1$) solution of the CGL equation with $\alpha = 0$. Notice that $q(\beta) \to 0$ as $\beta \to 0$, corresponding to the vortex limit. (cf. Fig. 5 of [337])

length scale, viz., inter-spiral spacing or the square root of the inverse defect density, which is denoted as L.

We would like to quantitatively describe the evolution morphology shown in Fig. 6.3. The standard tools for this are the correlation function and structure factor of the order-parameter field, which were defined in (3.15). At the simplest level of approximation, the morphology in the snapshots of Fig. 6.3 can be interpreted as consisting of disjoint spirals, each of size L. (The modulations of the order-parameter field at spiral-spiral boundaries will be discussed later.) Therefore, it is useful to compute the correlation function for a single-spiral solution [342].

We are interested in the correlation function for a 1-armed spiral at large length scales, so (6.3) can be simplified as

$$\psi(\boldsymbol{r}, t) \simeq \sqrt{1-q^2} \exp\left[-i\beta(1-q^2)t + i(\theta - qr)\right] . \tag{6.5}$$

The correlation between points $\boldsymbol{r}_1$ and $\boldsymbol{r}_2$ for a single spiral is then determined as

$$\begin{aligned} C(\boldsymbol{r}_1, \boldsymbol{r}_2, t) &= \text{Re}\left\{\psi(\boldsymbol{r}_1, t)\psi^*(\boldsymbol{r}_2, t)\right\} \\ &\simeq (1-q^2)\text{Re}\left\{\exp[i(\theta_1 - qr_1) - i(\theta_2 - qr_2)]\right\} \\ &\equiv C(\boldsymbol{r}_1, \boldsymbol{r}_2) . \end{aligned} \tag{6.6}$$

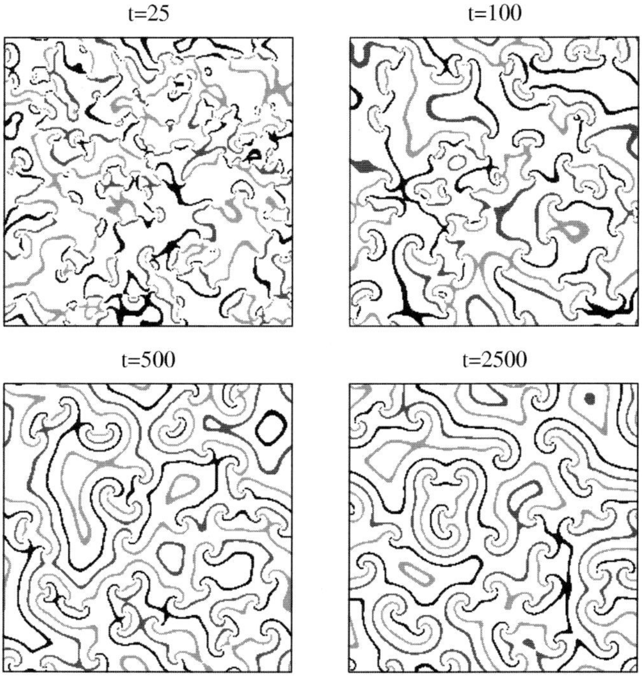

Fig. 6.3. Evolution of the CGL equation (with $\alpha = 0, \beta = 0.75$) from a random initial condition. These pictures were obtained from an Euler-discretized version of (6.1). The discretization mesh sizes were $\Delta t = 0.01$ and $\Delta x = 1.0$, and the lattice size was 256^2. The snapshots show regions of constant phase $\theta_\psi = \tan^{-1}(\text{Im}\psi/\text{Re}\psi)$, measured in radians. The color coding is the same as that in Fig. 3.2, i.e., $\theta_\psi \in [1.85, 2.15]$ (black); $\theta_\psi \in [3.85, 4.15]$ (dark grey); and $\theta_\psi \in [5.85, 6.15]$ (light grey)

The average correlation function is obtained by integrating over the point $\boldsymbol{r}_1$, setting $\boldsymbol{r}_2 = \boldsymbol{r}_1 + \boldsymbol{r}$ (cf. (3.15)):

$$\begin{aligned} C(r) &= \frac{1}{V} \int d\boldsymbol{r}_1 C(\boldsymbol{r}_1, \boldsymbol{r}_1 + \boldsymbol{r}) h(L - |\boldsymbol{r}_1 + \boldsymbol{r}|) \\ &= \frac{(1-q^2)}{V} \text{Re} \int d\boldsymbol{r}_1 \exp[i(\theta_1 - \theta_2 - q r_1 + q|\boldsymbol{r}_1 + \boldsymbol{r}|)] \\ &\quad \times h(L - |\boldsymbol{r}_1 + \boldsymbol{r}|) , \end{aligned} \tag{6.7}$$

where V is the spiral volume. In (6.7), we introduce the step function, $h(x) = 1$ (0) if $x \geq 0$ ($x < 0$), which ensures that we do not include points which lie outside the defect of size L.

At this stage, two remarks are in order. First, recall that the definition of the correlation function in (3.15) involved subtractions of the term $\sum_i \langle \psi_i(\boldsymbol{r}_1, t) \rangle \langle \psi_i(\boldsymbol{r}_1 + \boldsymbol{r}, t) \rangle$. In the present case, this term is identically zero.

Second, the time-dependence of the correlation function of the ordering system enters through the time-dependence of the length scale L.

For $d = 2$, the vector notation $\boldsymbol{r}_2 = \boldsymbol{r}_1 + \boldsymbol{r}$ is equivalent to $r_2 e^{i\theta_2} = r_1 e^{i\theta_1} + r e^{i\theta}$. Thus,

$$e^{i\theta_2} = \frac{r_1 e^{i\theta_1} + r e^{i\theta}}{\left[r_1^2 + r^2 + 2r_1 r\cos(\theta_1 - \theta)\right]^{1/2}}, \tag{6.8}$$

and

$$C(r) = \frac{(1-q^2)}{V}\mathrm{Re}\int_0^L dr_1 r_1 \int_0^{2\pi} d\theta_1 \frac{r_1 + re^{i(\theta_1-\theta)}}{\left[r_1^2 + r^2 + 2r_1 r\cos(\theta_1 - \theta)\right]^{1/2}}$$
$$\times \exp\left(-iq\left\{r_1 - \left[r_1^2 + r^2 + 2r_1 r\cos(\theta_1 - \theta)\right]^{1/2}\right\}\right) h(L - |\boldsymbol{r}_1 + \boldsymbol{r}|)\,. \tag{6.9}$$

Introduce the variables $x = r/L$, $y = r_1/L$, $\theta_1 - \theta = \theta'$, to obtain

$$C(r) = \frac{(1-q^2)}{\pi}\mathrm{Re}\int_0^1 dy y \int_0^{2\pi} d\theta \frac{y + xe^{i\theta}}{(y^2 + x^2 + 2yx\cos\theta)^{1/2}}$$
$$\times \exp\left[-iqL\left\{y - (y^2 + x^2 + 2yx\cos\theta)^{1/2}\right\}\right]$$
$$\times h\left[1 - (y^2 + x^2 + 2yx\cos\theta)^{1/2}\right], \tag{6.10}$$

where we have dropped the prime on θ', and used $V = \pi L^2$ in $d = 2$. Thus, the scaling form of the single-spiral correlation function is

$$\frac{C(r)}{C(0)} = g\left(\frac{r}{L}, q^2 L^2\right). \tag{6.11}$$

In general, there is no scaling with the spiral size because of the additional factor qL. We recover scaling only in the limit $q = 0$ ($\beta = 0$), which corresponds to the case of a vortex. Spirals of different sizes are not morphologically equivalent because there is more rotation in the phase as one goes out further from the core.

In Fig. 6.4, we plot $C(r)/C(0)$ vs. r/L for the case with $\beta = 0.75$ ($q \simeq 0.203$ from Fig. 6.2). The functional form in Fig. 6.4 decays almost monotonically for small values of L (i.e., in the vortex limit). On the other hand, it exhibits oscillatory behavior for larger values of L, as is expected from the integral expression.

Before proceeding, we should point out that the imaginary part of the integral in (6.10) is non-zero, in general. This corresponds to a weak correlation between the real and imaginary parts of the order-parameter field, which is absent in the XY case. The imaginary part can also be obtained with relative ease. However, we will confine our discussion to the conventional definition of the correlation function in (6.7).

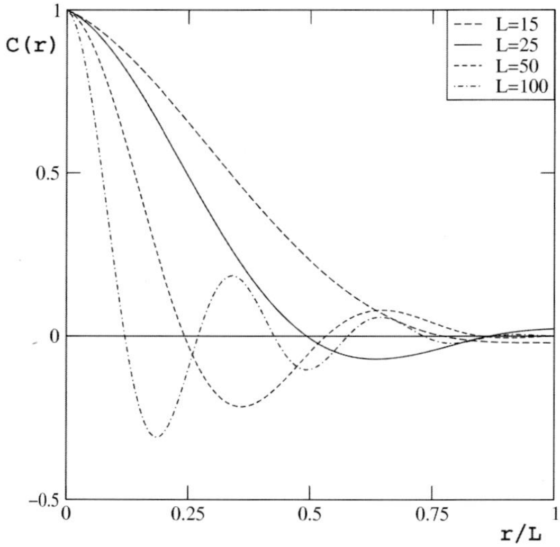

Fig. 6.4. Plot of $C(r)/C(0)$ vs. r/L for the 1-armed spiral solution when $\beta = 0.75$ ($q \simeq 0.203$). These results are obtained by a direct numerical integration of (6.10) for different spiral sizes, $L = 15, 25, 50, 100$, as denoted

Following the discussion in Sect. 3.2, let us next consider the asymptotic behavior of the correlation function in the limit $r/L \to 0$. Of course, r remains larger than the size of the defect core ξ, i.e., $r/\xi > 1$. The approximation in (6.5) is not valid if the latter condition is not satisfied.

6.3.1 Case with $\beta = 0$

For simplicity, we first focus on the case with $\beta = 0$, corresponding to vortex defects. The asymptotic analysis in this case will already illustrate the important features to be understood.

In the case with $\beta = 0$, we have $q = 0$ and the integral expression in (6.10) simplifies as

$$C(r) = \frac{1}{\pi}\mathrm{Re}\int_0^1 dyy \int_0^{2\pi} d\theta \frac{y + xe^{i\theta}}{(y^2 + x^2 + 2yx\cos\theta)^{1/2}} \times h\left[1 - (y^2 + x^2 + 2yx\cos\theta)^{1/2}\right] . \qquad (6.12)$$

As discussed in Chap. 3, the behavior in the $x \to 0$ limit is of considerable interest as it determines the large-wave-vector ($kL \to \infty$) behavior of the structure factor. In that case, we can neglect the step function on the RHS of (6.12) as it only provides corrections at the vortex boundary. We do not provide details of the subsequent analysis here as it is a limiting case of the calculation for $\beta \neq 0$, which is detailed in Sect. 6.3.2. Rather, we present the final result, which is as follows [342]:

$$C(r) = \frac{1}{\pi} \sum_{n=0}^{\infty} \frac{\Gamma(n+\frac{1}{2})^2}{n!^2} \left[A_n(x) - B_n(x)\right], \tag{6.13}$$

where

$$A_n(x) = \frac{1}{(n-1)} \left(\frac{4n+1}{2n+3} x^2 - x^{2n}\right)(1-\delta_{n1}) + \left(\frac{2}{5}x^2 - 2x^2 \ln x\right)\delta_{n1}, \tag{6.14}$$

and

$$B_n(x) = \frac{(2n+1)}{2n(n+1)} \left(\frac{4n+3}{2n+3} x^2 - x^{2n+2}\right)(1-\delta_{n0}) + \left(\frac{1}{3}x^2 - x^2 \ln x\right)\delta_{n0}, \tag{6.15}$$

where we have used the Kronecker delta, $\delta_{ij} = 1$ if $i = j$, 0 if $i \neq j$.

In the limit $x \to 0$, the singular terms in $C(r)$ arise from $A_1(x)$ and $B_0(x)$, and can be computed as

$$C_{\text{sing}}(r) = \frac{1}{2} x^2 \ln x . \tag{6.16}$$

Recall that (3.16) contains the corresponding result for arbitrary n and d $(n \leq d)$ [86]. Thus, it should yield the same result for $n = d = 2$. To obtain the correct result from (3.16), we set $n = 2 + \epsilon$ and take the limit $\epsilon \to 0$. Then, we have

$$g(x) = 1 + \pi^{\epsilon/2} \frac{\Gamma\left(\frac{3+\epsilon}{2}\right)^2 \Gamma\left(-1-\frac{\epsilon}{2}\right)}{\Gamma\left(2+\frac{\epsilon}{2}\right)\Gamma\left(1+\frac{\epsilon}{2}\right)} x^{2+\epsilon} + \text{higher-order terms}. \tag{6.17}$$

Notice that

$$\Gamma\left(-1-\frac{\epsilon}{2}\right) = \left(-\frac{\epsilon}{2}\right)^{-1}\left(-1-\frac{\epsilon}{2}\right)^{-1}\Gamma\left(1-\frac{\epsilon}{2}\right), \quad \text{and}$$
$$x^{2+\epsilon} = x^2\left[1 + \epsilon \ln x + O(\epsilon^2)\right]. \tag{6.18}$$

Thus, in the limit $\epsilon \to 0$, we can extract the leading singularity of $g(x)$ as

$$g_{\text{sing}}(x) \equiv C_{\text{sing}}(r) = \frac{\pi}{2} x^2 \ln x . \tag{6.19}$$

This result is identical to (6.16) if one accounts for the fact that Bray and Humayun [86] defined the length scale as the square root of the inverse defect density, viz., $\sqrt{\pi L^2}$ in the present case. As discussed in Sect. 3.2.2, this non-analytic behavior of the correlation function as $x \to 0$ corresponds to a power-law tail in the structure factor, $S(k) \simeq 4\pi L^2 (kL)^{-4}$. This result is referred to as the *generalized Porod law* [85,90].

6.3.2 Case with $\beta \neq 0$

We would like to undertake a similar asymptotic analysis in the general case with $\beta \neq 0$. As we are only interested in the limit $x \to 0$, we again discard the step function on the RHS of (6.10) to obtain

$$C(r) = \frac{(1-q^2)}{\pi} \mathrm{Re} \sum_{n=0}^{\infty} \frac{(iqL)^n}{n!} \int_0^1 dyye^{-iqLy} \int_0^{2\pi} d\theta (y + x\cos\theta) \times \left(y^2 + x^2 + 2yx\cos\theta\right)^{(n-1)/2} . \tag{6.20}$$

It is convenient to separately consider the cases with n odd and n even in the above series:

(i) n odd:
In this case, $n = 2p+1$, and the angular integral in (6.20) is

$$\begin{aligned} \tilde{I}_{2p+1}(y,x) &= \int_0^{2\pi} d\theta (y + x\cos\theta)(y^2 + x^2 + 2yx\cos\theta)^p \\ &= 2yx^{2p} \int_0^{\pi} d\theta \left(1 + \frac{y^2}{x^2} + \frac{2y}{x}\cos\theta\right)^p \\ &\quad +2x^{2p+1} \int_0^{\pi} d\theta \cos\theta \left(1 + \frac{y^2}{x^2} + \frac{2y}{x}\cos\theta\right)^p \\ &\equiv 2yx^{2p} I_1 + 2x^{2p+1} I_2 . \end{aligned} \tag{6.21}$$

The integrals I_1 and I_2 are obtained from Gradshteyn and Ryzhik [65], and the overall result is

$$\begin{aligned} \tilde{I}_{2p+1}(y,x) = 2\pi \Bigg[&\sum_{k=0}^{p} \binom{p}{k}^2 y^{2k+1} x^{2(p-k)} \\ &+ x^2 \sum_{k=0}^{\left[\frac{p-1}{2}\right]} \binom{p}{k} \binom{p+k}{k+1} y^{2k+1} x^{2k} (y^2 + x^2)^{p-2k-1} \Bigg] , \end{aligned} \tag{6.22}$$

where

$$\binom{p}{k} \equiv {}^p C_k \equiv \frac{p!}{k!(p-k)!} , \tag{6.23}$$

and $[*]$ denotes the integer part of $*$. The corresponding contribution to $C(r)$ is

$$C_1(r) = \frac{1-q^2}{\pi} \sum_{n=1,3,5,..}^{\infty} (-1)^{\frac{n-1}{2}} \frac{(qL)^n}{n!} \int_0^1 dyy \sin(qLy) \tilde{I}_n(y,x) . \tag{6.24}$$

The relevant feature here is that the expression for $C_1(r)$ only contains powers of x^2. Therefore, the overall contribution to $C(r)$ from this set of terms is analytic as $x \to 0$. In the limiting case $q = 0$ ($\beta = 0$), the above contribution is identically 0.

(ii) n even:

In this case, $n = 2p$, and the angular integral in (6.20) is

$$\tilde{I}_{2p}(y,x) = \int_0^{2\pi} d\theta (y + x\cos\theta)\left(y^2 + x^2 + 2yx\cos\theta\right)^{p-1/2} . \tag{6.25}$$

Introduce $\rho_< = \min(y,x)$ and $\rho_> = \max(y,x)$ to obtain

$$\begin{aligned}\tilde{I}_{2p}(y,x) &= 2\rho_>^{2p-1}\int_0^{\pi} d\theta (y + x\cos\theta)\left(1 + \frac{\rho_<^2}{\rho_>^2} + \frac{2\rho_<}{\rho_>}\cos\theta\right)^{p-1/2} \\ &\equiv 2\rho_>^{2p-1}(yI_3 + xI_4) .\end{aligned} \tag{6.26}$$

Here, the integrals I_3 and I_4 are related to hypergeometric functions as follows [65]:

$$I_3 = \pi F\left(\frac{1}{2} - p, \frac{1}{2} - p; 1; \frac{\rho_<^2}{\rho_>^2}\right) , \tag{6.27}$$

and

$$\begin{aligned}I_4 &= \pi\frac{\rho_<}{\rho_>}\left[\left(\frac{1}{2} + p\right) F\left(\frac{1}{2} - p, \frac{1}{2} - p; 2; \frac{\rho_<^2}{\rho_>^2}\right) - F\left(\frac{1}{2} - p, \frac{1}{2} - p; 1; \frac{\rho_<^2}{\rho_>^2}\right)\right] \\ &= \pi\left(p - \frac{1}{2}\right)\frac{\rho_<}{\rho_>} F\left(\frac{3}{2} - p, \frac{1}{2} - p; 2; \frac{\rho_<^2}{\rho_>^2}\right) .\end{aligned} \tag{6.28}$$

The first line of (6.28) has been simplified by using the identity [65]

$$(c - a - 1)F(a,b;c;z) + aF(a+1,b;c;z) = (c-1)F(a,b;c-1;z) , \tag{6.29}$$

with $a = 1/2 - p$, $b = 1/2 - p$, $c = 2$.

Using the results for I_3 and I_4 in (6.26),

$$\begin{aligned}\tilde{I}_{2p}(y,x) = 2\pi\rho_>^{2p-1}&\left[yF\left(\frac{1}{2} - p, \frac{1}{2} - p; 1; \frac{\rho_<^2}{\rho_>^2}\right)\right. \\ &\left.+x\frac{\rho_<}{\rho_>}\left(p - \frac{1}{2}\right) F\left(\frac{3}{2} - p, \frac{1}{2} - p; 2; \frac{\rho_<^2}{\rho_>^2}\right)\right] .\end{aligned} \tag{6.30}$$

The corresponding terms in the correlation function are

$$\begin{aligned}
C_2(r) = (1-q^2) & \sum_{n=0,2,4,..}^{\infty} (-1)^{n/2} \frac{(qL)^n}{n!} \int_0^1 dyy \cos(qLy) \\
& \times \left[2y\rho_>^{n-1} F\left(\frac{1-n}{2}, \frac{1-n}{2}; 1; \frac{\rho_<^2}{\rho_>^2}\right) \right. \\
& \left. + x\rho_>^{n-2} \rho_< (n-1) F\left(\frac{3-n}{2}, \frac{1-n}{2}; 2; \frac{\rho_<^2}{\rho_>^2}\right) \right] \\
\equiv (1-q^2)(T_1+T_2) \,. &
\end{aligned} \tag{6.31}$$

The non-analytic contributions to $C(r)$ as $x \to 0$ arise only from $C_2(r)$, as $C_1(r)$ is analytic in x. The singular terms in T_1 and T_2 are obtained as follows:

$$\begin{aligned}
T_1 = 2 & \sum_{p=0}^{\infty} (-1)^p \frac{(qL)^{2p}}{(2p)! \Gamma\left(\frac{1}{2}-p\right)^2} \sum_{m=0}^{\infty} \frac{\Gamma\left(\frac{1}{2}-p+m\right)^2}{m!^2} \\
& \times \int_0^1 dy \cos(qLy) y^2 \frac{\rho_<^{2m}}{\rho_>^{2(m-p)+1}} \,,
\end{aligned} \tag{6.32}$$

where we have expanded the hypergeometric function. The integral on the RHS of (6.32) can be split up as

$$I_5 = \frac{1}{x^{2(m-p)+1}} \int_0^x dy \cos(qLy) y^{2m+2} + x^{2m} \int_x^1 dy \cos(qLy) y^{-2(m-p)+1} \,. \tag{6.33}$$

The first term on the RHS of (6.33) is analytic as $x \to 0$. This is seen by Taylor-expanding the cosine-function, and observing that the resultant terms in the integrand contain only even powers of the integration variable y. The second term contributes singular terms only if $m \geq p+1$. Thus, we have the result

$$I_5 = (-1)^{m-p} \frac{(qL)^{2(m-p-1)}}{[2(m-p-1)]!} x^{2m} \ln x + \text{analytic terms} \,. \tag{6.34}$$

Replacing this in the expression for T_1, we obtain

$$\begin{aligned}
T_1 = & \sum_{p=0}^{\infty} \sum_{m=0}^{\infty} (-1)^{p+m+1} \frac{(qL)^{2(p+m)}}{(2p)!(2m)!} \frac{\Gamma\left(\frac{3}{2}+m\right)^2}{\Gamma\left(\frac{1}{2}-p\right)^2 (m+p+1)!^2} x^{2(m+p+1)} \ln x \\
& + \text{analytic terms} \,.
\end{aligned} \tag{6.35}$$

A similar analysis for T_2 yields

$$\begin{aligned}
T_2 = & \sum_{p=0}^{\infty} \sum_{m=0}^{\infty} (-1)^{p+m} \frac{(qL)^{2(p+m)}}{(2p)!(2m)!} \frac{\Gamma\left(\frac{1}{2}+m\right)^2}{\Gamma\left(\frac{1}{2}-p\right)^2 (m+p)!^2} \frac{(2m+1)}{(m+p+1)} \\
& \times x^{2(m+p+1)} \ln x + \text{ analytic terms} \,.
\end{aligned} \tag{6.36}$$

Finally, we can combine the non-analytic contributions from T_1 and T_2 to obtain the singular part of the correlation function as follows [342]:

$$
\begin{aligned}
C_{\text{sing}}(r) &= \frac{1}{2}\sum_{p=0}^{\infty}\sum_{m=0}^{\infty}(-1)^{p+m}\frac{(qL)^{2(p+m)}}{(2p)!(2m)!}\frac{\Gamma\left(\frac{1}{2}+m\right)^2}{\Gamma\left(\frac{1}{2}-p\right)^2(m+p+1)!^2} \\
&\quad\times(2m+1)(2p+1)x^{2(m+p+1)}\ln x \\
&= \frac{x^2\ln x}{2}\sum_{k=0}^{\infty}\frac{(qL)^{2k}}{2^{4k}(k+1)!^2}(-x^2)^k\left(\sum_{m=0}^{k}\frac{(2m+1)!(2k-2m+1)!}{m!^2(k-m)!^2}\right).
\end{aligned}
\tag{6.37}
$$

The leading singularity in (6.37) arises for $k = 0$, viz., $C_{\text{sing}}(r) \simeq \frac{1}{2}x^2 \ln x$, as in the case with $\beta = 0$ (cf (6.16)). This is in accordance with the intuitive expectation that there is no difference between a spiral and a vortex, sufficiently close to the core. In addition to the vortex-like singularity, there is now a sequence of sub-dominant singularities proportional to $(qL)^2 x^4 \ln x$, $(qL)^4 x^6 \ln x$, etc., and these become increasingly important as the length scale L increases. Their relative strengths can be computed by taking ratios of successive terms in the series expansion of (6.37).

These sub-dominant terms in $C_{\text{sing}}(r)$ are reminiscent of the leading-order singularities for defects in the n-component TDGL equation, where n is even [81,86]. Of course, in the context of the TDGL equation, these singularities only arise for $n \leq d$ as there are no topological defects unless this condition is satisfied. In the present context, all these terms are already present for $d = 2$. The corresponding structure-factor tail exhibits a sequence of power-law decays with $S(k) \sim (qL)^{2(m-1)} L^d/(kL)^{d+2m}$, where $m = 1, 2$, etc. Thus, though the asymptotic behavior in $d = 2$ is still the generalized Porod tail, $S(k) \sim L^2(kL)^{-4}$, it may be difficult to distinguish this from other power-law decays.

In principle, the results presented here only determine the short-distance behavior of the correlation function, or the large-wave-vector behavior of the structure factor. This is because small length-scales only probe individual defects. Nevertheless, as our subsequent discussion will demonstrate, the single-spiral correlation function agrees with the correlation function for multi-spiral morphologies (obtained numerically) over a considerable range of distances. This is somewhat fortuitous, as will be discussed later. A more complete theory must explicitly account for the order-parameter modulation resulting from spiral-spiral interactions [353].

6.4 Numerical Results for $d = 2$

There have been many numerical studies of the $d = 2$ CGL equation by various authors. In the context of phase ordering dynamics, it is relevant to study the evolution from a small-amplitude random initial condition, corresponding to the *high-temperature* disordered phase of the system. The evolution morphology and the asymptotic ($t \to \infty$) state of the system depend upon the parameter values, as summarized in the phase diagram of Fig. 6.1. As stated earlier, we will focus upon parameter values where the emergent morphology is characterized by spirals, i.e., to the right of the line T in Fig. 6.1.

An early study of this problem is due to Huber et al. [354], who demonstrated the emergence of a multi-spiral morphology from a random initial condition. This is analogous to the evolution depicted in Fig. 6.3. The asymptotic state (e.g., $t = 2500$ snapshot in Fig. 6.3) is statistically frozen in a metastable state due to a repulsive defect-defect interaction [353,355,356]. In this context, the term *frozen* refers to a state in which the $|\psi|$-field is almost independent of time. Huber et al. referred to this state as a *vortex glass*, because the slowing-down of kinetic processes is reminiscent of *structural glasses*. A subsequent study of the vortex glass state is due to Chate and Manneville [341]. More generally, these authors mapped out the phase diagram of the $d = 2$ CGL equation in (α, β)-space, and this was shown in Fig. 6.1. The vortex glass morphology was also studied by Bohr et al. [357], who focused upon the structure of interfaces between spirals. The above studies have mainly focused on the properties of the frozen state. A more recent study by Kevrekidis et al. [358] has investigated the time-dependent evolution resulting due to *quenches* from the vortex glass state to the turbulent state.

These and related studies [331] have been useful in clarifying the qualitative features of the emergent morphology in the frozen state. Our goal in this chapter is to demonstrate how the concepts of phase ordering dynamics (discussed in Chap. 3) can be used to obtain a quantitative understanding of (a) the time-dependent evolution, and (b) the statistical properties of the multi-spiral morphology.

The results presented here correspond to the case with $\alpha = 0$ and β arbitrary. Figure 6.3 showed the typical system evolution for $\beta = 0.75$. Figure 6.5 compares the evolution for $\beta = 0.75$ (upper frames) and $\beta = 1.25$ (lower frames). The spiral cores are marked as filled circles in the snapshots of Fig. 6.5, and their density is seen to diminish with time. Notice that the spiral sizes at $t = 1000$ are smaller for larger values of β. The final frames in Fig. 6.5 (i.e., at $t = 1000$) depict the frozen morphology of the CGL equation for the appropriate value of β. This should be contrasted with the evolution of the XY model, where the annihilation of vortices and anti-vortices continues for all time, resulting in an ever-increasing length scale. We will further clarify this issue subsequently.

Next, we present detailed results for the growth laws, correlation functions and structure factors for typical evolution morphologies, as shown in Figs. 6.3 and 6.5.

6.4.1 Spiral Growth Laws

The ordering of the CGL system depicted in Figs. 6.3 and 6.5 proceeds by the emergence and annihilation of spirals. At early times, the defect scales are

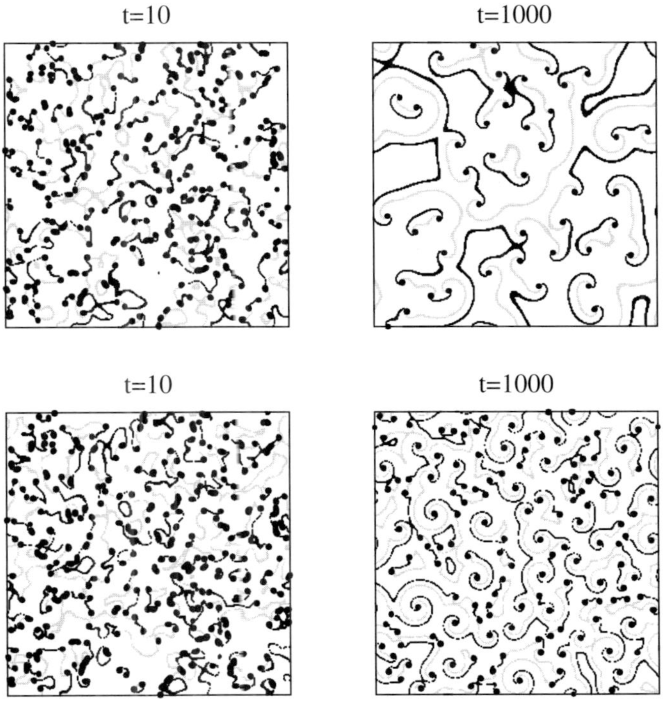

Fig. 6.5. Comparison of evolution for the $d = 2$ CGL equation with $\alpha = 0$ and $\beta = 0.75$ (upper frames), $\beta = 1.25$ (lower frames). The numerical details are the same as for Fig. 6.3. The snapshots show regions of constant phase with the coding: $\theta_\psi \in [1.85, 2.15]$ (black); and $\theta_\psi \in [3.85, 4.15]$ (grey). The filled circles denote spiral cores, which are defined as regions where the order parameter amplitude $|\psi| < 0.5$

small and spirals are equivalent to vortices. As discussed in Sect. 3.2.2, the growth law for vortex annihilation in the $d = 2$ XY model is $L(t) \sim (t/\ln t)^{1/2}$ [92]. Therefore, in Fig. 6.6, we plot the square of the characteristic length scale, $L(t)^2$ vs. $t/\ln t$ for different values of β. The XY growth law would correspond to a straight line in Fig. 6.6. The data for $\beta = 0.25$ is comparable to that for the XY model over simulation time-scales.

From (6.5), we see that spiral effects become significant on length scales $qr > 1$, where q is the Hagan parameter. Thus, the crossover length-scale from vortex-mediated growth to spiral-mediated growth scales as $L_c \sim q^{-1}$. Figure 6.7(a) plots L_c vs. q^{-1} for $\beta \in [0.5, 1.25]$, and demonstrates the validity of this scaling law. As $\beta \to 0$ (or $q \to 0$), we expect the crossover scale $L_c \to \infty$, as there is no freezing of domain growth in the XY model. As we have mentioned earlier, the evolving morphology freezes due to a repulsive spiral-antispiral interaction [353,355,356]. It is also relevant to study the scaling behavior of the saturation length scale, L_s. On the simulation time-scales, we see saturation clearly for $\beta > 0.75$. The simplest scaling scenario suggests

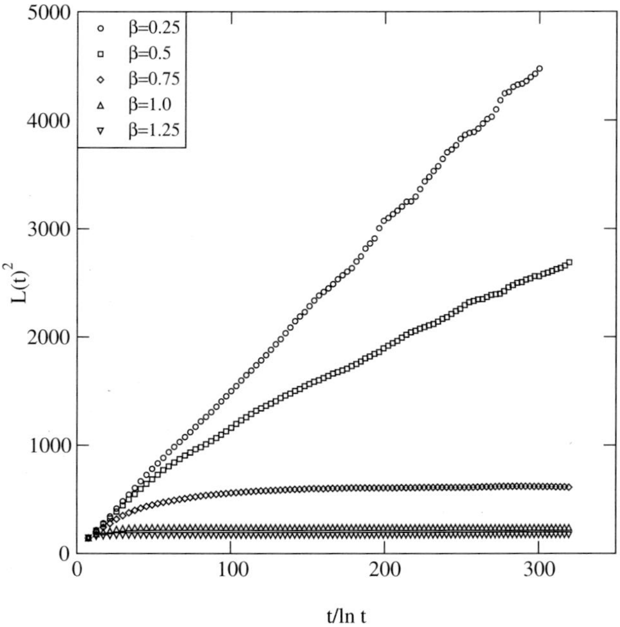

Fig. 6.6. Plot of $L(t)^2$ vs. $t/\ln t$ for the $d = 2$ CGL equation. The typical defect size is obtained as $\pi L(t)^2 = (N\Delta x)^2/N_d(t)$, where $(N\Delta x)^2$ is the system area and $N_d(t)$ is the total number of defects at time t. All statistical data presented here for the $d = 2$ CGL equation is obtained as an average over 5 independent runs for system sizes $N = 1024$. The parameter values are $\alpha = 0$ and $\beta = 0.25, 0.5, 0.75, 1.0, 1.25$, as indicated

that $L_s \sim L_c \sim q^{-1}$. Figure 6.7(b) plots L_s vs. q^{-1} for $\beta \in [0.75, 1.25]$, and demonstrates the validity of this scaling law.

6.4.2 Equal-Time Correlation Functions

The correlation function has been defined in (3.15). Figure 6.8 shows the time-dependence of the spherically-averaged correlation function for $\beta = 0.75, 1.25$, corresponding to the evolution depicted in Fig. 6.5. The data shows a prominent dip as time increases, and the spiral nature of defects become more apparent (cf. Fig. 6.4). Thus, there is no scaling of the correlation function with the defect size, as discussed earlier. The early-time ($t = 5$) correlation function in Figs. 6.8(a)-(b) decays monotonically, and is described well by the $n = 2$ BPT function in (3.17). The correlation function at later times ($t = 1000$) is described well by the single-spiral correlation function in (6.10) upto the first minimum. The appropriate q-value was obtained from Hagan's solution for $q(\beta)$ in the case of 1-armed spirals (cf. Fig. 6.2).

Recall that the single-spiral correlation function is expected to be applicable only for length scales $L \gg r > \xi$, as the expression in (6.7) neglects

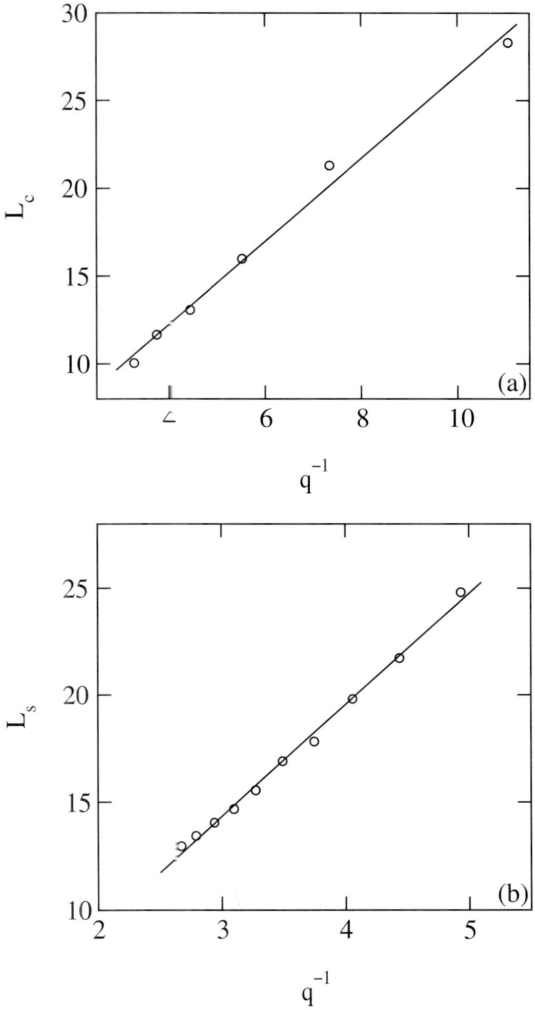

Fig. 6.7. (**a**) Plot of crossover length-scale, L_c vs. $q(\beta)^{-1}$, for the $d = 2$ CGL equation with $\alpha = 0$ and $\beta \in [0.5, 1.25]$. The solid line denotes the best linear fit to the data. We determine $L_c(\beta)$ from the length-scale data in Fig. 6.6 as follows. The first 3 points of each data set are fitted by a straight line, and the crossover point is defined as the first point which deviates from this straight line. (**b**) Analogous to (a), but for the saturation length scale, L_s. The parameter values are $\alpha = 0$ and $\beta \in [0.75, 1.25]$

order-parameter modulations at spiral-spiral boundaries. However, the fits in Figs. 6.8(a)-(b) suggest that the single-spiral correlation function describes the multi-spiral results somewhat better than expected. This is because the spirals are effectively isolated from each other by a second class of defects

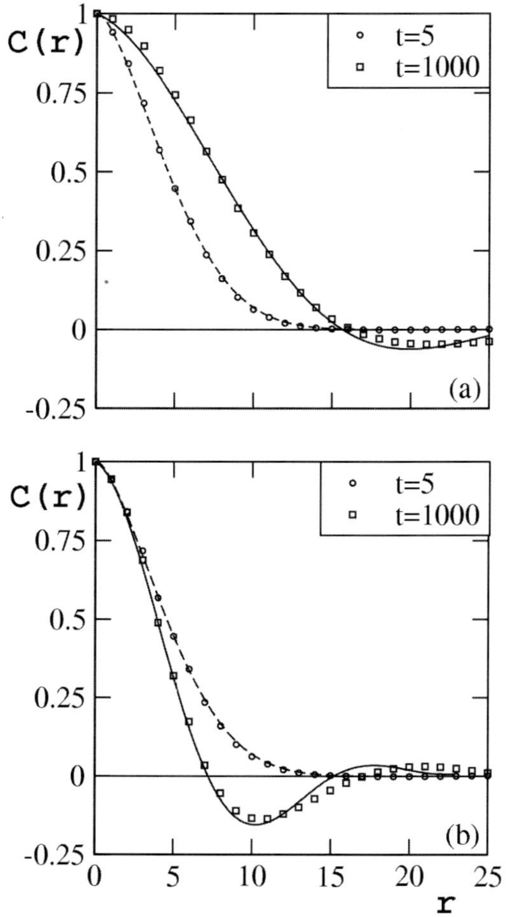

Fig. 6.8. (**a**) Plot of correlation function, $C(r,t)$ vs. r, for the $d = 2$ CGL equation from $t = 5, 1000$. The parameter values are $\alpha = 0$ and $\beta = 0.75$, as depicted in the upper frames of Fig. 6.5. The early-time data ($t = 5$) is fitted to the BPT function for the XY model (denoted as a dashed line), with the length-scale L as a fitting parameter. The best-fit value is $L \simeq 4.40$. The late-time data ($t = 1000$) is fitted to (6.10) (denoted as a solid line) with L as a fitting parameter. The best-fit value is $L \simeq 24.0$. (**b**) Analogous to (a), but for $\beta = 1.25$, as depicted in the lower frames of Fig. 6.5. The best-fit values are $L \simeq 4.40$ ($t = 5$); and $L \simeq 19.25$ ($t = 1000$)

which we have not discussed so far, i.e., *shock-line* defects [331,341]. These shock lines correspond to local maxima of the $|\psi|$-field, and result from the interference of waves emitted by spirals [357,359,360]. The order-parameter amplitude decays exponentially through the shock-lines – therefore, the correlation function is dominated by contributions from single spirals.

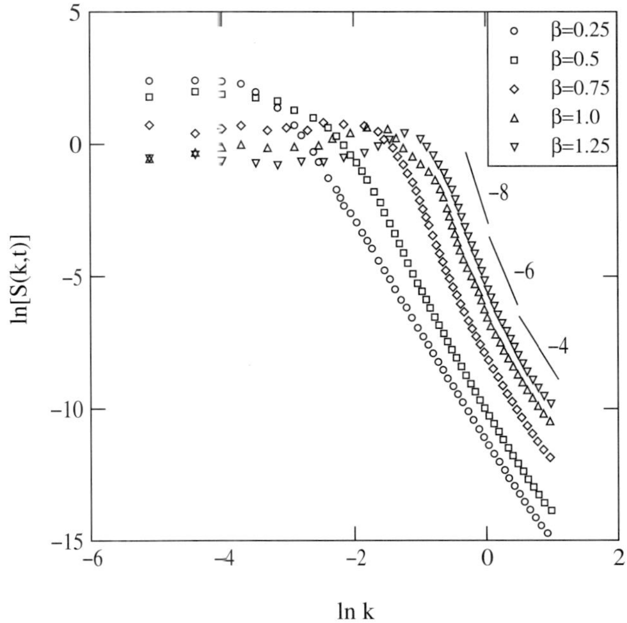

Fig. 6.9. Plot of structure factor, $\ln[S(k,t)]$ vs. $\ln k$, for the $d = 2$ CGL equation with $\alpha = 0$ and $\beta = 0.25, 0.5, 0.75, 1.0, 1.25$. All data sets correspond to $t = 1000$. The spherically-averaged data is normalized as $\int_0^\infty dk S(k,t) = 1$. The solid lines have slopes -4, -6 and -8, as indicated

6.4.3 Equal-Time Structure Factors

As discussed in Sect. 6.3, the consequence of the singularities in the correlation function as $x \to 0$ is a sequence of power-law decays in the tail of the structure factor, $S(k) \sim (qL)^{2(m-1)} L^d/(kL)^{d+2m}$, where $m = 1, 2$, etc. In $d = 2$, the appropriate power-law decays are $S(k) \sim q^{2(m-1)} L^{-2} k^{-(2m+2)}$. Clearly, the asymptotic behavior is still $S(k) \sim k^{-4}$, as in the XY model, but there are also intermediate decays of the form k^{-6}, k^{-8}, etc.

Figure 6.9 shows data for $\ln[S(k,t)]$ vs. $\ln k$ for different values of β at $t = 1000$. The solid lines in Fig. 6.9 have slopes -4, -6 and -8, corresponding to $m = 1, 2$ and 3, respectively. For smaller values of β (e.g., $\beta = 0.25$), the contributions from the higher-order decays are small and the tail is dominated by $S(k,t) \sim k^{-4}$. However, for higher values of β, the higher-order contributions are not negligible and one sees intermediate regions which are consistent with the behavior $S(k,t) \sim k^{-6}$, $\sim k^{-8}$, etc. for $k < 1$. Nevertheless, the tail is always asymptotically parallel to the line with slope -4.

Before concluding this section, we would like to highlight two recent studies which are relevant to the present discussion. Firstly, in recent work, Mazenko [361] has also studied phase ordering dynamics in the $d = 2$ CGL equation. Mazenko focused upon the velocity-distribution function for spiral defects, and found that this was the same as the corresponding function for the XY model [362]:

$$P(\boldsymbol{v},t)d\boldsymbol{v} = \frac{1}{\pi\bar{v}^2}\left(1+\frac{\boldsymbol{v}^2}{\bar{v}^2}\right)^{-2} d\boldsymbol{v}\,, \tag{6.38}$$

where $\bar{v} \sim dL/dt$ is the characteristic defect velocity. Notice that $\bar{v} \simeq 0$ on finite time-scales for the vortex glass state due to the freezing of the morphology. Clearly, none of the spatial features described above manifest themselves in the defect velocities, which are associated with the defect core. Recall that spirals and vortices are indistinguishable in the core region.

Secondly, Brito et al. [363] have undertaken large-scale numerical studies of the vortex glass state. They focus on the case with $\alpha = 0$ and $\beta_1 \geq \beta > 0$, which is also the parameter regime for most of the results presented in this chapter. Brito et al. confirm that the *frozen* multi-spiral morphologies actually evolve, though on extremely slow time-scales. Furthermore, they subclassify the multi-spiral states as *vortex liquid* and *vortex glass* states. The vortex liquid arises for $\beta^* > \beta > 0$ ($\beta^* \simeq 0.845$), where the repulsive spiral-spiral interaction decays monotonically. In the vortex liquid state, the system evolution occurs due to spiral diffusion. On the other hand, the vortex glass arises for $\beta_1 > \beta > \beta^*$, where the spiral-spiral interaction is oscillatory, promoting the formation of long-lived metastable structures. The evolution in the vortex glass state is intermittent and chaotic. Of course, the results presented in this chapter are independent of whether or not the system is actually relaxing towards *equilibrium*. Rather, they depend upon the existence of well-defined spirals with a characteristic length scale.

6.5 Numerical Results for $d = 3$

Next, let us study the evolution of the $d = 3$ CGL equation from a random initial condition. In this case, the defect dimensionality is 1 ($= d - n$, where n is the number of components of the order parameter). Thus, the defect cores meet and form lines, which we will refer to as *spiral lines*. The annealing of these spiral lines gives rise to ordering dynamics in the $d = 3$ CGL equation. These structures are also referred to as *vortex filaments* or *scroll waves* [331], and have been realized in a variety of experiments on chemical [364] and biological [365,366] systems.

There have also been some analytical studies of spiral line dynamics in the CGL equation. For example, there is a good understanding of vortex line dynamics in the NLSE, which arises for $\alpha, \beta \to \infty$. This is obtained by invoking analogies with vortex kinetics in hydrodynamics [330,367]. For the general CGL equation, Gabbay et al. [368] studied the collapse of spiral rings and the evolution of twist in a straight spiral line. These authors also investigated the kinetics of spiral reconnection in the CGL equation.

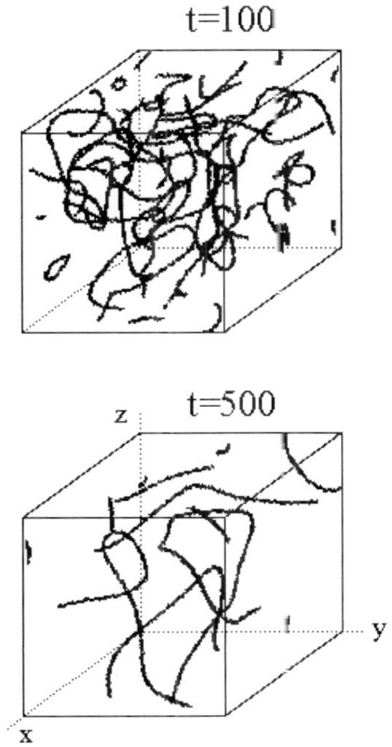

Fig. 6.10. Evolution of the CGL equation (with $\alpha = 0, \beta = 1.25$) from a random initial condition. The discretization mesh sizes were $\Delta t = 0.01$, $\Delta x = 1.0$, and the lattice size was 128^3. The solid lines denote spiral cores, defined as regions where $|\psi| < 0.5$

Finally, we discuss some relevant numerical studies. There have been a number of simulations of spiral line dynamics in the CGL equation and related reaction-diffusion systems [369,370]. These have partially categorized the morphologies which arise for different parameter values. In particular, there have been detailed studies of multi-vortex morphologies in the NLSE limit [371].

Inspite of the above studies, our understanding of the $d = 3$ CGL equation is considerably less than that of the $d = 2$ case. For example, there is no systematic classification of the morphologies which arise for different (α, β)-values, as shown in Fig. 6.1 for the $d = 2$ case. In this section, we present numerical results for statistical properties of the $d = 3$ CGL equation. As in Sect. 6.4, we will focus on parameter values where the system evolution is characterized by spiral line defects. The evolution will be quantitatively analyzed using tools from phase ordering dynamics.

Figure 6.10 shows the evolution of the CGL equation with $\beta = 1.25$ from a disordered initial condition. Initially, there is a large density of spiral lines with various topological entanglements. These include spiral rings, which are comprised of closed-loop structures. At early times and small defect sizes, spiral lines are indistinguishable from vortex lines. Therefore, we expect the early-time dynamics to be identical to the dynamics of the XY model in $d = 3$. With the passage of time, the spirals annihilate and the spiral length scale grows. Then, the evolution morphology consists of an assembly of spirals and antispirals. At comparable values of β, saturation of the length scale occurs considerably later in the $d = 3$ case than in the $d = 2$ case. This is because the spiral defects can relax in the perpendicular direction without experiencing a strong repulsive potential.

Next, we present results for the growth laws, correlation functions and structure factors for the $d = 3$ CGL equation.

6.5.1 Spiral Growth Laws

A spiral line defect consists of a spiral translated in the perpendicular direction. The spiral length scale is measured by counting the number of spiral cores in all possible planar cross-sections along various directions. The defect scale is then obtained as $\pi L(t)^2 = (N\Delta x)^2/N_d(t)$, where $(N\Delta x)^2$ is the planar area and $N_d(t)$ is the average number of defects in the plane. In general, spiral lines are not oriented along the x-, y- or z-axis (see Fig. 6.10). Therefore, this definition underestimates the actual length scale by a constant factor, assuming that the spiral lines are randomly oriented with respect to any arbitrary plane.

The domain growth law is $L(t) \sim t^{1/2}$ for the $d = 3$ XY model. In Fig. 6.11, we plot $L(t)^2$ vs. t for different values of β. The data for $\beta = 0.25$ is consistent with XY-like growth over the duration of the simulation. The data for $\beta = 1.0, 1.25, 1.5$ shows signs of freezing, but at considerably later times than the $d = 2$ case (see Fig. 6.6). In analogy with the $d = 2$ case, the crossover length scale from vortex-mediated growth to spiral-mediated growth scales as $L_c \sim q^{-1}$ [344].

6.5.2 Equal-Time Correlation Functions

As in the $d = 2$ case, we spherically average the correlation function $C(\boldsymbol{r}, t)$ (which depends on vector $\boldsymbol{r}$) to obtain $C(r, t)$. The evolving system (see Fig. 6.10) consists of an isotropic mixture of spiral line defects with random orientation, so the spherical-averaging procedure is appropriate. Of course, an individual line defect is highly anisotropic with strong correlation along the line, and weak correlation in the plane of the spiral. Therefore, in the disjoint-spiral approximation, it is reasonable to compare the $d = 3$ correlation function with the single-spiral function in (6.10). However, the quantity L should now be treated as the lateral defect size.

Figure 6.12(a) shows the time-dependent correlation function for $\beta = 1.25$ at different times. The early-time data ($t = 100$) decays monotonically and is well-fitted by the BPT function for the XY model. The late-time

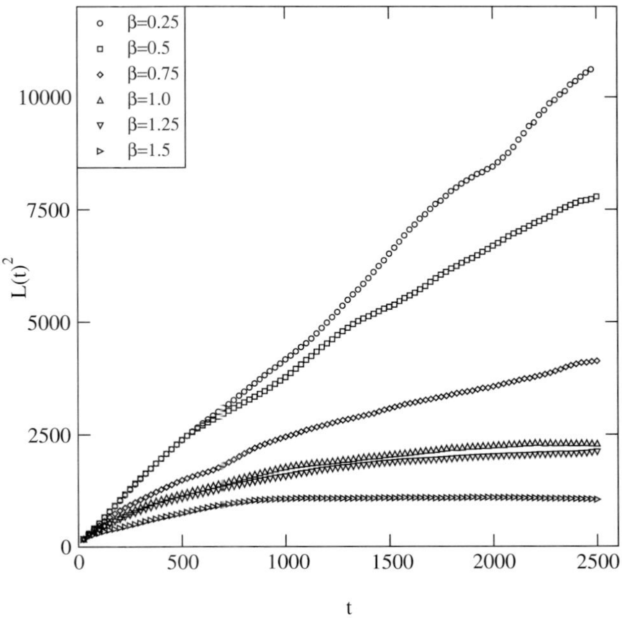

Fig. 6.11. Plot of $L(t)^2$ vs. t for the $d = 3$ CGL equation with $\alpha = 0$ and $\beta = 0.25, 0.5, 0.75, 1.0, 1.25, 1.5$. The length-scale data is obtained as an average over 5 independent runs for lattice sizes 180^3

data ($t = 500, 1000$) exhibits oscillations characteristic of spiral defects. The solid lines denote best-fits to the integral expression in (6.10). Again, the fits are surprisingly good upto the first minimum. Figure 6.12(b) shows the corresponding results for $\beta = 1.0$.

6.5.3 Equal-Time Structure Factors

In $d = 3$, the tail of the structure factor decays as $S(k) \sim q^{2(m-1)} L^{-2} k^{-(2m+3)}$, where $m = 1, 2$, etc. Figure 6.13 plots the structure factor, $\ln[S(k,t)]$ vs. $\ln k$, for different values of β at $t = 1000$. The solid lines have slopes -5, -7, and -9, corresponding to $m = 1, 2$ and 3, respectively. As in the $d = 2$ case, the contributions from higher-order decays are small for smaller values of β (e.g., $\beta = 0.25$, 0.5), and the tail is dominated by the generalized Porod law $S(k,t) \sim k^{-5}$. The higher-order power-law decays become more prominent at higher values of β.

6.6 Summary and Discussion

We conclude with a summary and discussion of the results presented here. Before getting into specific details, it is useful to discuss the contents of this

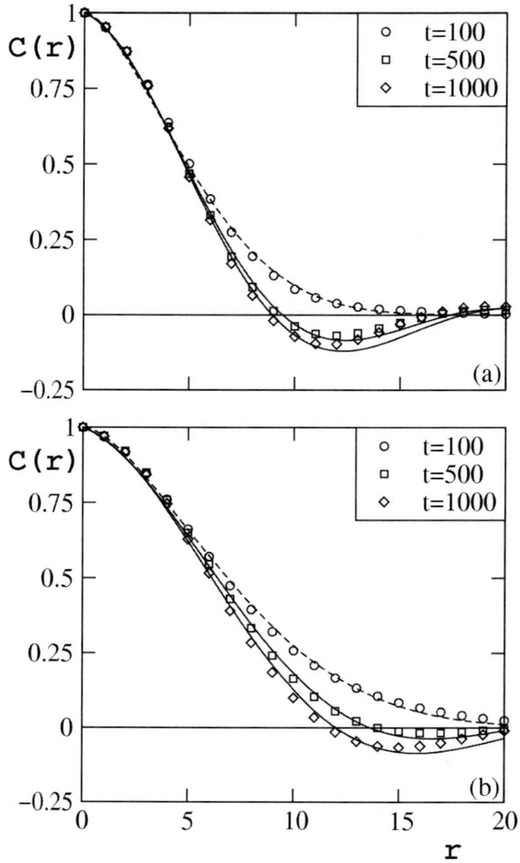

Fig. 6.12. Plot of correlation function, $C(r,t)$ vs. r, for the $d = 3$ CGL equation from times $t = 100, 500$ and 1000. The correlation-function data is obtained from 5 independent runs for lattice sizes 128^3. (**a**) Numerical results for $\alpha = 0$ and $\beta = 1.25$. The $t = 100$ data is fitted to the BPT function (dashed line) and the best-fit length scale is $L \simeq 4.9$. The data for $t = 500, 1000$ is fitted to (6.10) (solid lines). The best-fit length scales are $L \simeq 13.6$ ($t = 500$), and $L \simeq 16.2$ ($t = 1000$). (**b**) Analogous to (a), but for $\alpha = 0$ and $\beta = 1.0$. The best-fit length scales are $L \simeq 6.85$ ($t = 100$); $L \simeq 13.75$ ($t = 500$); $L \simeq 18.0$ ($t = 1000$)

chapter in the context of earlier chapters. Recall that Chaps. 1-3 provided a general framework for the discussion of dissipative phenomena in condensed matter physics. In particular, Chap. 3 focused on elementary aspects of phase ordering dynamics. This research area has been of great topical interest in diverse physical problems. In Chaps. 4 and 5, we saw how experimentally relevant effects can drastically modify the nature of domain growth kinetics.

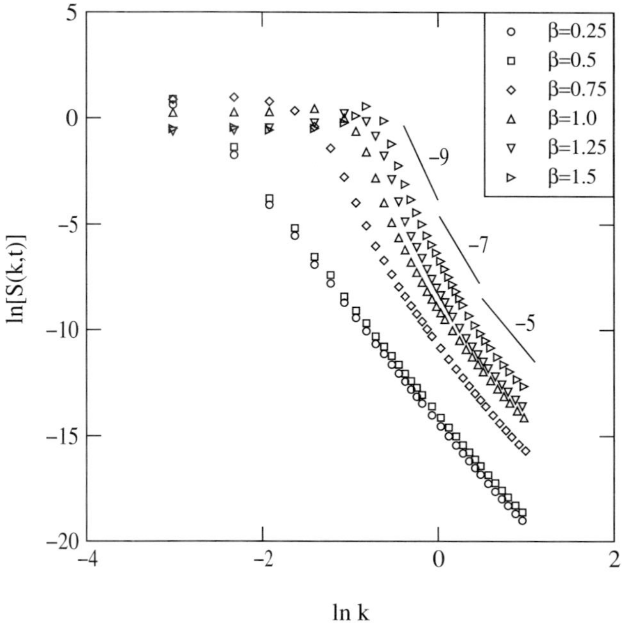

Fig. 6.13. Plot of $\ln[S(k,t)]$ vs. $\ln k$ for the $d = 3$ CGL equation with $\alpha = 0$ and $\beta = 0.25, 0.5, 0.75, 1.0, 1.25, 1.5$. All data sets correspond to $t = 1000$. The solid lines have slopes -5, -7 and -9, as indicated

The present chapter takes a somewhat different direction. It demonstrates that the general principles of phase ordering dynamics can be of great utility in understanding pattern formation in a different context. Thus, spiral defects provide a good basis for quantifying pattern dynamics in the CGL equation in a large region of parameter space. The universal nature of the CGL equation has made it one of the most widely-studied systems in the areas of *nonlinear dynamics* and *pattern formation*. Nevertheless, the statistical approach discussed in this chapter provides a novel framework for understanding the quantitative properties of pattern formation in the CGL equation.

Let us now return to a discussion of what we have learnt in this chapter. As stated earlier, we have used spirals and antispirals to characterize the evolution morphology of the CGL equation from a random (disordered) initial condition. We focused on the case with $\alpha = 0$ and $\beta \neq 0$, but the results are generic to regions of parameter space where spirals determine the CGL morphology.

The presence of spirals gives rise to many novel physical features:

1. Firstly, there is a repulsive defect-defect potential which *freezes* the evolving system into a metastable state consisting of an assembly of spirals and antispirals of characteristic length $L_s(\beta) \to \infty$ as $\beta \to 0$. We should stress that the term *freezing* is used in a statistical sense. The local order param-

eter is time-dependent in this frozen state [363], and the spirals continue to rearrange for all times. We may expect that the metastable state finally decays into an ordered homogeneous state with $\psi(\boldsymbol{r}, t) = \exp(-i\beta t)$. However, the time-scales for this are far beyond present computational facilities, though the decay may be facilitated by the presence of thermal fluctuations. Of course, only a finite system would order into a homogeneous state. An infinite system would always be characterized by the presence of spirals.
2. Secondly, an asymptotic analysis of the short-distance behavior of the single-spiral correlation function exhibits a sequence of singularities. The dominant singularity is equivalent to that for the case characterized by vortex defects. The sub-dominant singularities are analogous to those which arise for the n-component TDGL model with n even. Of course, the n-component TDGL model exhibits topological defect structures only in the case $n \leq d$. However, in the present case, all singularities are already present for $d = 2$. The sequence of singularities in the short-distance correlation function gives rise to a sequence of power-laws in the structure-factor tail.

We have also presented detailed numerical results in this chapter. In the context of the correlation function, we find that the agreement between the numerical results (obtained for a *multi-spiral morphology*) and the analytical results (obtained for a *single-spiral morphology*) is surprisingly good. This is because spirals are effectively isolated from each other by shock-line defects [331,341], and the independent-spiral approximation is physically reasonable. A more sophisticated theory should account for the order-parameter modulation due to the formation of spiral-spiral bound states [353]. There have been initial efforts in this direction, but they have not yet resulted in a substantially better agreement between theory and numerical results [372]. Furthermore, a complete calculation of the correlation function or structure factor should include contributions from the shock-line defects. However, it should be stressed that the dimensionality of the shock-lines is lower than that of the spirals, and they have a correspondingly weaker contribution to the correlation function.

Finally, we should emphasize that appropriate experiments, e.g., on chemically-reacting systems, would assist in clarifying the quantitative features of systems characterized by spirals. Such morphologies are ubiquitous, and arise in diverse systems ranging from chemical and biological physics to astrophysics. There have already been many numerical simulations of the CGL equation and related reaction-diffusion systems. Clearly, it is relatively easy to obtain quantitative results from these simulations. Unfortunately, experiments in this context have primarily focused on classifying emergent morphologies. Given the quantitative nature of analytical results now available, it would be desirable for these to be subjected to experimental confirmation.

7 Quantum Dissipative Systems

7.1 Introduction

Our discussion so far has been restricted to *classical* dissipative systems. We now turn our attention to systems which are *quantum mechanical*, and yet are in contact with a heat bath which injects dissipation. Quantum mechanics introduces additional frequency scales in the problem through *deterministic* evolution. The interplay of these frequencies with the heat-bath-induced frequencies (such as the coefficient of friction, relaxation rate, etc.) leads to fascinating new physics. The latter has been the focus of intense activity in condensed matter physics. It is our objective here to give the reader a flavor of this excitement by choosing a few illustrative examples. Our treatment will be based on concepts already developed in Chaps. 1 and 2.

In accordance with our stated objective, we adopt once again the system-plus-bath approach (SBA), in which the total Hamiltonian is split up as (cf. (1.73))

$$\mathcal{H} = \mathcal{H}_s + V + \mathcal{H}_B \,. \tag{7.1}$$

Recall that $\mathcal{H}_s$ is the Hamiltonian for the system of interest, $\mathcal{H}_B$ is the Hamiltonian for the bath, and V is the system-bath coupling. In our earlier discussion in Chap. 1, $\mathcal{H}_s$ was taken to be the Hamiltonian of an Ising (or Ising-like) model and hence, was classical We would now like to consider examples in which $\mathcal{H}_s$ has to be tackled quantum mechanically. Further, in a major departure from the usual cases, in which the heat-bath-induced effects are modeled as classical stochastic processes, we now consider the influence of the bath Hamiltonian explicitly.

There is one issue which we would like to dispense with before proceeding. We are interested in calculating either the expectation value of an observable A or its correlation function in equilibrium:

$$Q(t) = \langle A(t) \rangle \,, \tag{7.2}$$

$$C_{AA}(t) = \langle A(0)A(t) \rangle_0 \,, \tag{7.3}$$

where the angular brackets with the subscript 0 imply that the correlation function is the average over an *equilibrium* ensemble of the system. We should reiterate that the quantity A occurring inside the angular brackets is a quantum operator which does not commute with itself at different times. Now, the time-varying quantities can be treated either in the Schrödinger picture by considering the time-evolution of the density operator ρ (as in our discussion in Appendix 1.A), or in the Heisenberg picture by considering the time-evolution of the observable itself. Thus, $Q(t)$ can be written in either of two alternative forms:

$$Q(t) = \mathrm{Tr}\left[\rho(t)A(0)\right] , \tag{7.4}$$

or

$$Q(t) = \mathrm{Tr}\left[\rho(0)A(t)\right] , \tag{7.5}$$

the equivalence of which can be easily established by using the cyclic property of the trace operation and noting that (cf. (1.A.2))

$$\rho(t) = e^{-i\mathcal{H}t/\hbar}\rho(0)e^{i\mathcal{H}t/\hbar} , \tag{7.6}$$

$$A(t) = e^{i\mathcal{H}t/\hbar}A(0)e^{-i\mathcal{H}t/\hbar} . \tag{7.7}$$

Earlier, in our derivation of the quantum master equation, we had worked in the Schrödinger picture (see Appendix 1.A). We would now like to adopt the Heisenberg picture, which is convenient for calculating correlation functions in equilibrium, in order to provide a direct link with experiments. Thus, we can rewrite (7.2) and (7.3) as

$$Q(t) = \mathrm{Tr}\left[\rho(0)A(t)\right] , \tag{7.8}$$

and

$$C_{AA}(t) = \mathrm{Tr}\left[\rho_{\mathrm{eq}}A(0)A(t)\right] , \tag{7.9}$$

where the density operator will be assumed to be of the factorized form (cf. (1.A.14)):

$$\rho(0) = \rho_B \otimes \rho_s(0) , \tag{7.10}$$

$$\rho_{\mathrm{eq}} = \rho_B \otimes \rho_s^{\mathrm{eq}} . \tag{7.11}$$

In both cases, ρ_B represents the density operator of the heat bath in equilibrium. Further, because A is a subsystem operator that commutes with ρ_B, both (7.8) and (7.9) can be expressed in terms of the *averaged time-development operator*:

$$\left(U(t)\right)_{\mathrm{av}} = \mathrm{Tr}_B\left[\rho_B \exp\left(iLt\right)\right] . \tag{7.12}$$

Thus

$$Q(t) = \mathrm{Tr}_s\left[\rho_s(0)\left(U(t)\right)_{\mathrm{av}} A(0)\right] , \tag{7.13}$$

and

$$C_{AA}(t) = \mathrm{Tr}_s\left[\rho_s^{\mathrm{eq}} A(0) \left(U(t)\right)_{\mathrm{av}} A(0)\right] . \tag{7.14}$$

Here, L denotes the Liouvillian associated with the total Hamiltonian $\mathcal{H}$, and $\mathrm{Tr}_s\{...\}$ denotes a trace over the Hilbert space of $\mathcal{H}_s$ alone, following the notation of Appendix 1.A. Thus, in both cases, we need to evaluate the single quantity in (7.12), with the difference that in (7.13), the subsystem is viewed to be in an arbitrary initial state, whereas in (7.14), the subsystem is in thermal equilibrium.

The averaged time-development operator can be computed either in the time-domain or in the frequency-domain (see Appendix 1.A). For analyzing spectroscopy experiments, however, it is appropriate to work in the frequency-domain. Following a treatment which is equivalent to that in Appendix 1.A, the resolvent expansion of (7.12) in Laplace-transform notation can be written as follows:

$$\left(\tilde{U}(s)\right)_{\mathrm{av}} = \frac{1}{s - iL_s - \left(\tilde{M}^c(s)\right)_{\mathrm{av}}} , \tag{7.15}$$

where

$$\begin{aligned}\left(\tilde{M}^c(s)\right)_{\mathrm{av}} &= P(iL_I)(1-P)\left(\frac{1}{s - iL_s - iL_B - (1-P)(iL_I)(1-P)}\right) \\ &\quad \times (1-P)(iL_I)P . \end{aligned} \tag{7.16}$$

Here L_I is the Liouvillian associated with V, and P is a projection operator defined by

$$PA = \mathrm{Tr}_B(\rho_B A) . \tag{7.17}$$

In particular, upto second order in the interaction Hamiltonian V, we may write

$$-\left(\tilde{M}^c(s)\right)_{\mathrm{av}} = \tilde{\Sigma}(s) = \left(L_I \frac{1}{s - iL_s - iL_B} L_I\right)_{\mathrm{av}} , \tag{7.18}$$

and

$$\left(\tilde{U}(s)\right)_{\mathrm{av}} = \frac{1}{s - iL_s + \tilde{\Sigma}(s)} , \tag{7.19}$$

where $\tilde{\Sigma}(s)$ is referred to as the *self-energy*. In the Markovian limit,

$$\tilde{\Sigma}(s) \simeq \tilde{\Sigma}(s=0) = \int_0^\infty dt \left(L_I e^{i(L_s + L_B)t} L_I\right)_{\mathrm{av}} . \tag{7.20}$$

Equation (7.20), which we have seen before as (1.A.51), will form the basis of our discussions in the present chapter, as well as Chap. 8.

With this general background, we would now like to make a distinction between two classes of quantum dissipative systems. In the first class, which will be discussed in the present chapter, the heat bath represented by the Hamiltonian $\mathcal{H}_B$ will be treated *classically*. In the second class, which is discussed in Chap. 8, $\mathcal{H}_B$ will be treated *quantum mechanically*. Of course, in both cases, the subsystem represented by the Hamiltonian $\mathcal{H}_s$ is purely quantum mechanical. We will have occasion to remark on the crucial differences between these two classes of problems in that the quantum nature of the heat bath will be shown to give rise to a richer structure in the underlying physics.

While the second-order perturbation result embodied in (7.19)-(7.20) implies a *weak* coupling, there are situations in which the coupling should be retained to all orders. Thus, we shall need the full form of $(\tilde{M}^c(s))_{\rm av}$ as given in (7.16). Such *strong-coupling* cases, which we refer to as *quantum Brownian motion*, will be dealt with in Chap. 9. Thus, Chaps. 7-9 all involve quantum dissipative systems, but are presented in order of increasing degree of mathematical complexity. This chapter discusses systems whose Hamiltonian is treated partly quantum mechanically and partly classically. Chap. 8 discusses systems whose Hamiltonian is treated entirely quantum mechanically, but the interaction between the subsystem and the bath is weak in some sense. The term *weak coupling* requires a qualifying remark: it turns out that, even within the second-order approximation in (7.19), certain cases with *strong coupling* can be dealt with by a suitable redefinition of the interaction Hamiltonian (see Chap. 8). Finally, Chap. 9 deals with systems whose Hamiltonian needs to be treated quantum mechanically and *exactly*. However, and perhaps inevitably, the applications discussed in these chapters are in *decreasing* levels of physical complexity. Accordingly, in this chapter, we will take the subsystem to be a mutually and strongly-interacting entity. However, in Chaps. 8 and 9, we will be concerned with subsystems that are composed of independent, non-interacting entities.

7.2 Transverse Ising Model and Applications

The Ising model, discussed earlier in Chap. 2, is entirely classical as it can be diagonalized by choosing the z-direction as the quantization axis. Consequently, there is no dynamics of the spin variable σ_z in this model because σ_z commutes with the Hamiltonian. The Ising model has a special status in statistical mechanics because of its wide applications in condensed matter physics. It is therefore tempting to add a quantum component to the Ising model but retain most of its original features. The simplest such model is the so-called *transverse Ising model* (TIM), represented by the Hamiltonian

$$\mathcal{H} = -\sum_{i>j=1}^{N} J_{ij}\sigma_{zi}\sigma_{zj} - \Gamma\sum_{i=1}^{N}\sigma_{xi}\,. \tag{7.21}$$

As the transverse field Γ couples to the x-component of the spin in (7.21), there is now dynamics in all three components of the magnetization. We have earlier considered several applications of the Ising model, especially in the context of dissipative dynamics. It is interesting, therefore, to examine the additional effects which arise due to the presence of the transverse field Γ.

Although the model described by (7.21) is simple, it finds important applications in condensed matter physics, e.g., in the context of hydrogen-bonded ferroelectric and antiferroelectric crystals such as KH_2PO_4 [373]. The *left* and *right* positions of the H-atom in an O-H...O bond are mapped onto an Ising pseudospin variable σ_z, which assumes eigenvalues $+1$ and -1. The interaction Hamiltonian for two hydrogen atoms separated by distances much larger than the bond length is governed by the usual exchange coupling J_{ij}. The additional term proportional to Γ accounts for the fact that the hydrogen atom can tunnel between its left and right positions. As this motion is governed by quantum laws and not thermal activation, it persists down to zero temperature.

The application of the TIM to ferroelectric and antiferroelectric crystals has been extensively reviewed in the literature. Moreover, in recent work, Chakrabarti et al. [374] have devoted an entire monograph to the statistical mechanics of (and hysteretic behavior in) the TIM. We shall therefore avoid duplication and instead focus on dissipative effects in the TIM and (subsequently) in the disordered TIM (DTIM). For the present, we consider a recent application of the TIM, which will later constitute the basis for a discussion of the DTIM. This involves a dipolar-coupled ferromagnet $LiHoF_4$, which is so well described by the Hamiltonian in (7.21) that this system is often referred to as a *model quantum magnet* [375].

The system $LiHoF_4$ belongs to a class of compounds $LiRF_4$, where R is a rare-earth element. The rare-earth moments of Ho^{3+} are dipolar-coupled but, because of the presence of a large crystal field, only the lowest doublet of the 17 crystal field-split states is appreciably populated at low temperatures ($T < 2$ K). Further, the sign of the crystal field is negative, hence the lowest doublet corresponds to the highest spin substates. Consequently, from angular-momentum matrix relations, it follows that the off-diagonal terms of the dipolar interaction are effectively *quenched*, yielding the so-called *truncated dipolar Hamiltonian* [376]. The latter is of the Ising form with J_{ij} given by

$$J_{ij} = J\left(\frac{1 - 3\cos^2\theta_{ij}}{r_{ij}^3}\right), \tag{7.22}$$

where θ_{ij} is the angle between the Ising axis (which turns out to be the c-axis of the crystal) and the vector connecting the spins. The dependence of J_{ij} on the inter-spin distance r_{ij} is proportional to r_{ij}^{-3} and is considerably long-ranged. This suggests that a mean-field (MF) treatment would be appropriate for such a system. On the other hand, the angular dependence allows for J_{ij} to change sign, implying that the interaction switches from ferromagnetic ($J_{ij} > 0$) to antiferromagnetic ($J_{ij} < 0$) for appropriate angles. While this occurrence of competing interactions may raise doubts about the uniqueness of the ground state, a classical analysis by Luttinger and Tisza [377] showed that the ground state is indeed that of a dipolar ferromagnet for the b.c.c. structure of $LiHoF_4$. The experimentally measured critical temperature for the paramagnetic-ferromagnetic transition turns out to be $T_c = 1.53$ K. Whereas this is as far as the Ising interaction between the spin-1/2 eigenstates (representing the lowest doublet) is concerned, what about the transverse term proportional to Γ? This is realized by the application of a magnetic field H_t perpendicular to the c-axis, which can cause an admixture of the crystal field-split states. This yields (in perturbation theory) a spin Hamiltonian that is proportional to σ_x, with a prefactor $\Gamma \propto H_t^2$ [378].

The transverse field Γ acts as a disordering field, like the temperature T. This is because, as Γ increases, a stronger exchange interaction is required for the spins to order along the z-axis. Thus, even at zero temperature, there is a critical value Γ_c such that the system is in a paramagnetic phase for $\Gamma > \Gamma_c$. This phenomenon is called a *quantum phase transition*, because it occurs due to purely quantal fluctuations [379–381]. Quantum phase transitions, which differ fundamentally from classical phase transitions, are of interest in a variety of systems involving, e.g., transition-metal oxides [382,383], the apparent *non-Fermi liquid* behavior of highly correlated f-electron compounds [384,385], and the unusual normal-state properties of high-T_c superconducting cuprates [386–388]. The fascinating properties of each of these systems have been attributed to the presence of a $T = 0$ quantum critical point. While unambiguous experimental study of these systems is quite difficult, $LiHoF_4$ provides a clean model system in which high-precision, controlled investigation of quantum critical phenomena can be performed.

In one other respect, $LiHoF_4$ is unusual in that the observed magnetic behavior is *quantitatively* described by MF theory, in *both* the quantum and thermal regimes. The fact that the quantum critical behavior in $d = 3$ is MF-like is not surprising as it is known that the d-dimensional quantum phase transition (at $T = 0$) for the TIM is equivalent to the $(d+1)$-dimensional thermal phase transition for the ordinary Ising model [389,390]. The MF nature of the thermal phase transition observed in $LiHoF_4$ is also not surprising as renormalization-group (RG) theory shows that, for a dipolar-coupled Ising system, the marginal dimensionality is $d_c = 3$. (Recall that we expect MF behavior for $d > d_c$ [391].)

Let us next study the static behavior of the TIM in the MF approximation. In subsequent sections, we will turn our attention to dissipative dynamics in the TIM.

7.3 Static Behavior of the Transverse Ising Model in Mean-Field Theory

In the MF approximation, the terms involving σ_i in the Hamiltonian of (7.21) can be rewritten as

$$\mathcal{H}_i = -\left(H_i\sigma_{zi} + \Gamma\sigma_{xi}\right) , \tag{7.23}$$

where the *local field* is approximated as

$$H_i = \sum_{j=1}^{N} J_{ij}\langle\sigma_{zj}\rangle . \tag{7.24}$$

Furthermore, in the *uniform* case, $\langle\sigma_{zj}\rangle$ is independent of the site index j. Therefore

$$H_i \equiv H = J(0)\langle\sigma_z\rangle , \quad \text{where}$$
$$J(0) = \sum_{j=1}^{N} J_{ij} . \tag{7.25}$$

Thus, from (7.23),

$$\mathcal{H}_i = -\left(H\sigma_{zi} + \Gamma\sigma_{xi}\right) . \tag{7.26}$$

The partition function is given by

$$\begin{aligned} Z_i &= \mathrm{Tr}\left(e^{-\beta\mathcal{H}_i}\right) \\ &= \mathrm{Tr}\left[e^{\beta(H\sigma_z+\Gamma\sigma_x)}\right] . \end{aligned} \tag{7.27}$$

Using the properties of Pauli matrices [22], we obtain

$$e^{\beta(H\sigma_z+\Gamma\sigma_x)} = \cosh(\beta h) + \sinh(\beta h)\left(\frac{H\sigma_z + \Gamma\sigma_x}{h}\right) , \tag{7.28}$$

where

$$h = \sqrt{H^2 + \Gamma^2} . \tag{7.29}$$

Therefore,

$$Z_i = 2\cosh(\beta h) . \tag{7.30}$$

On the other hand,

$$\begin{aligned} \langle\sigma_z\rangle &= \frac{1}{2\cosh(\beta h)}\mathrm{Tr}\left[\sigma_z e^{\beta(H\sigma_z+\Gamma\sigma_x)}\right] \\ &= \frac{H}{h}\tanh(\beta h) . \end{aligned} \tag{7.31}$$

Similarly,

$$\langle \sigma_x \rangle = \frac{\Gamma}{h} \tanh(\beta h) . \tag{7.32}$$

Denoting the LHS of (7.31) and (7.32) as the components of the magnetization m_z and m_x respectively, we obtain the transcendental equations:

$$m_z = \frac{J(0)m_z}{\sqrt{[J(0)m_z]^2 + \Gamma^2}} \tanh\left(\beta\sqrt{[J(0)m_z]^2 + \Gamma^2}\right) , \tag{7.33}$$

$$m_x = \frac{\Gamma}{\sqrt{[J(0)m_z]^2 + \Gamma^2}} \tanh\left(\beta\sqrt{[J(0)m_z]^2 + \Gamma^2}\right) . \tag{7.34}$$

Equations (7.33)-(7.34) constitute self-consistent relations for the two relevant components of the magnetization. Note that m_z goes to zero in the limits of either Γ or T approaching infinity. This confirms our earlier physical arguments that the transverse field Γ acts like a disordering field.

In order to deduce the phase diagram predicted by (7.33), we designate m_z as the order parameter ψ and introduce the critical temperature for the pure (i.e., $\Gamma = 0$) Ising model as (cf. (2.12))

$$T_c^0 = \frac{J(0)}{k_B} . \tag{7.35}$$

Equation (7.33) then yields (for the solution corresponding to $\psi \neq 0$)

$$1 = \frac{T_c^0/T}{\sqrt{(T_c^0\psi/T)^2 + \overline{\Gamma}^2}} \tanh\left(\sqrt{\left(\frac{T_c^0}{T}\psi\right)^2 + \overline{\Gamma}^2}\right) , \tag{7.36}$$

where $\overline{\Gamma} = \beta\Gamma$.

The phase boundary can be obtained by setting $\psi = 0$ in (7.36), yielding

$$T_c = \frac{T_c^0}{\overline{\Gamma}_c} \tanh\overline{\Gamma}_c , \tag{7.37}$$

which is equivalent to

$$\frac{\Gamma}{J(0)} = \tanh\left(\frac{\Gamma}{k_B T_c}\right) . \tag{7.38}$$

The resultant phase diagram is shown in Fig. 7.1. The following three regions can be distinguished:

1. *Region I*, where Γ is *large*, and T is *small* – quantum effects are dominant in this regime.
2. *Region II*, where Γ is *small* – classical effects are dominant in this regime, and the TIM is analogous to the usual Ising model.
3. *Region III*, where both Γ and T are *moderate*.

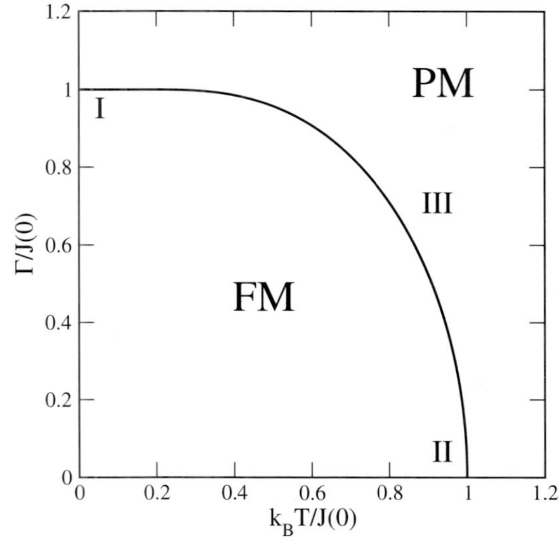

Fig. 7.1. Phase diagram for the transverse Ising model (TIM) in the MF approximation. The transverse field is Γ, and the temperature is T. The terms PM and FM denote the paramagnetic and ferromagnetic phases, respectively. Other details are provided in the text

In order to analyze the static susceptibility in region III, we add a small aligning field ΔH along the z-axis. In that case, (7.33) is modified as

$$\psi = \frac{\beta \Delta H + T_c^0 \psi / T}{\sqrt{(\beta \Delta H + T_c^0 \psi / T)^2 + \overline{\Gamma}^2}} \tanh \left[\sqrt{\left(\beta \Delta H + \frac{T_c^0}{T} \psi \right)^2 + \overline{\Gamma}^2} \right] . \tag{7.39}$$

Since both $\beta \Delta H$ and ψ are small, we have (in the linearized regime)

$$\psi \left(1 - \frac{T_c^0}{T} \frac{\tanh \overline{\Gamma}}{\overline{\Gamma}} \right) = \beta \Delta H \frac{\tanh \overline{\Gamma}}{\overline{\Gamma}} , \tag{7.40}$$

which yields the susceptibility

$$\chi = \frac{\psi}{\Delta H} = \frac{\tanh \overline{\Gamma}}{k_B \overline{\Gamma}} \left(T - T_c^0 \frac{\tanh \overline{\Gamma}}{\overline{\Gamma}} \right)^{-1} . \tag{7.41}$$

Thus, the susceptibility diverges at the critical temperature T_c given by (7.37) with a critical exponent $\gamma = 1$. Alternatively, we may rewrite χ as

$$\chi = \tanh \overline{\Gamma} \left(\Gamma - k_B T_c^0 \tanh \overline{\Gamma} \right)^{-1} , \tag{7.42}$$

which again exhibits a divergence at $\Gamma = \Gamma_c$ (given by (7.37)) with a critical exponent $\gamma = 1$ [392,393].

7.4 Relaxation Kinetics in Mean-Field Theory

We have seen in Chap. 2 that the MF behavior of the Ising model with Glauber kinetics, in terms of the magnetization and the spin-spin correlation function, can be described by the kinetics of a *single* spin in an effective field. In the context of the TIM, a stochastic approach using Glauber kinetics is not possible because of the inherently quantum nature of the subsystem Hamiltonian. Thus, one has to revert to the master equation for the full density operator of the system (see Appendix 1.A). However, this route is quite complicated. As our goal is to provide an approximate MF description, we adopt the same strategy as in the $\Gamma = 0$ case. Thus, we consider a single spin in an effective field H along the z-axis and a transverse field Γ along the x-axis, and embed this spin in a heat bath with which it interacts, as in the SBA of Sects. 1.6 and 1.7. The resultant rate equations have the same validity as the MF rate equations for the ordinary Ising model [394].

We proceed as in Sect. 1.6. The subsystem Hamiltonian in the MF approximation can be written as

$$\mathcal{H}_s = -\left(H\sigma_z + \Gamma\sigma_x\right) , \quad \text{where} \quad H = J(0)\langle\sigma_z\rangle . \tag{7.43}$$

To obtain the coupling term V between the spin and the heat bath, it is useful to first diagonalize $\mathcal{H}_s$ by rotating the coordinate system about the y-axis in the clockwise direction by an angle θ, where

$$\tan\theta = \frac{\Gamma}{H} , \quad \cos\theta = \frac{H}{\sqrt{H^2+\Gamma^2}} . \tag{7.44}$$

The corresponding rotation operator is

$$S = \exp\left(-i\theta\sigma_y\right) . \tag{7.45}$$

Note that in the rotated frame,

$$\begin{aligned} \tilde{H}_s &\equiv S^{-1}\mathcal{H}_s S = -h\sigma_z , \\ h &= \sqrt{H^2+\Gamma^2} , \end{aligned} \tag{7.46}$$

which is of the same form as the subsystem Hamiltonian in the $\Gamma = 0$ case (cf. (1.55)). Recall that the coupling term was taken to be $V = g\hat{b}\sigma_x$ in the $\Gamma = 0$ case (see (1.74)). This suggests that we should choose the coupling term $\tilde{V}$ in the rotated frame to be of the same form as earlier, which implies that in the laboratory frame,

$$\begin{aligned} V \equiv S\tilde{V}S^{-1} &= S\left(g\hat{b}\sigma_x\right)S^{-1} \\ &= \frac{g}{h}\hat{b}\left(H\sigma_x - \Gamma\sigma_z\right) . \end{aligned} \tag{7.47}$$

Introducing the density operator in the rotated frame as

$$\tilde{\rho}(t) = S^{-1}\rho(t)S\,, \tag{7.48}$$

we find that $\tilde{\rho}(t)$ satisfies an identical equation as the density operator $\rho(t)$ in the laboratory frame for $\Gamma = 0$:

$$\begin{aligned} i\hbar\frac{\partial}{\partial t}\tilde{\rho}(t) &= S^{-1}\left(i\hbar\frac{\partial\rho(t)}{\partial t}\right)S \\ &= S^{-1}[\mathcal{H},\rho(t)]S \\ &= [\tilde{\mathcal{H}},\tilde{\rho}(t)]\,. \end{aligned} \tag{7.49}$$

In (7.49), we have the transformed Hamiltonian

$$\begin{aligned} \tilde{\mathcal{H}} &= \tilde{\mathcal{H}}_s + \tilde{V} + \tilde{\mathcal{H}}_B \\ &= -h\sigma_z + g\hat{b}\sigma_x + \mathcal{H}_B\,. \end{aligned} \tag{7.50}$$

Therefore, the spin dynamics (and the corresponding rate equations) in the rotated frame should be identical to that in the laboratory frame, *but* for the case $\Gamma = 0$. Note that

$$m_\mu(t) = \mathrm{Tr}\,[\rho(t)\sigma_\mu] = \mathrm{Tr}\left[\tilde{\rho}(t)S^{-1}\sigma_\mu S\right]\,, \quad \mu = x, y, z\,. \tag{7.51}$$

Further,

$$\begin{aligned} S^{-1}\sigma_z S &= \frac{1}{h}\left(H\sigma_z - \Gamma\sigma_x\right)\,, \\ S^{-1}\sigma_x S &= \frac{1}{h}\left(H\sigma_x + \Gamma\sigma_z\right)\,, \\ S^{-1}\sigma_y S &= \sigma_y\,. \end{aligned} \tag{7.52}$$

Therefore, we obtain

$$\begin{aligned} m_z(t) &= \frac{H}{h}m_z^R(t) - \frac{\Gamma}{h}m_x^R(t)\,, \\ m_x(t) &= \frac{H}{h}m_x^R(t) + \frac{\Gamma}{h}m_z^R(t)\,, \\ m_y(t) &= m_y^R(t)\,, \end{aligned} \tag{7.53}$$

where the superscript R on the magnetization denotes the corresponding component in the rotated frame. In the rotated frame, all equations are identical to those in the $\Gamma = 0$ case, and the equations of motion are obtained by simply substituting the corresponding equations derived earlier for the $\Gamma = 0$ case (see (1.64), (1.115), (1.116)):

$$\begin{aligned} \frac{d}{dt}m_z(t) &= -\frac{\lambda H}{h}\left[m_z^R(t) - \tanh(\beta h)\right] - \frac{\Gamma}{h}\left[\frac{2h}{\hbar}m_y^R(t) - \lambda m_x^R(t)\right]\,, \\ \frac{d}{dt}m_x(t) &= \frac{H}{h}\left[\frac{2h}{\hbar}m_y^R(t) - \lambda m_x^R(t)\right] - \frac{\lambda\Gamma}{h}\left[m_z^R(t) - \tanh(\beta h)\right]\,, \\ \frac{d}{dt}m_y(t) &= -\frac{2h}{\hbar}m_x^R(t) - \lambda m_y^R(t)\,. \end{aligned} \tag{7.54}$$

Further, eliminating the components of magnetization in the rotated frame in favor of those in the laboratory frame, we obtain

$$\begin{aligned} m_z^R(t) &= \frac{1}{h}\left[Hm_z(t) + \Gamma m_x(t)\right] , \\ m_x^R(t) &= \frac{1}{h}\left[Hm_x(t) - \Gamma m_z(t)\right] , \\ m_y^R(t) &= m_y(t) . \end{aligned} \tag{7.55}$$

Substituting in (7.54), we finally derive

$$\begin{aligned} \frac{d}{dt}m_z(t) &= -\frac{2\Gamma}{\hbar}m_y(t) - \lambda\left[m_z(t) - \frac{H}{h}\tanh(\beta h)\right] , \\ \frac{d}{dt}m_x(t) &= \frac{2H}{\hbar}m_y(t) - \lambda\left[m_x(t) - \frac{\Gamma}{h}\tanh(\beta h)\right] , \\ \frac{d}{dt}m_y(t) &= \frac{2}{\hbar}\left[\Gamma m_z(t) - Hm_x(t)\right] - \lambda m_y(t) . \end{aligned} \tag{7.56}$$

As expected, for $\Gamma = 0$, (7.56) reduce to the corresponding rate equations for the Ising model (see Sect. 1.7). These equations form the starting point for our discussion of hysteresis phenomena in the TIM.

Eliminating $m_z(t)$ and $m_x(t)$ from the RHS of the equation for dm_y/dt, we find that $m_y(t)$ satisfies the equation of a damped harmonic oscillator:

$$\frac{d^2m_y}{dt^2} + 2\lambda\frac{dm_y}{dt} + \Omega^2 m_y(t) = 0 , \tag{7.57}$$

where

$$\Omega^2 = \lambda^2 + \frac{4H^2}{\hbar^2} . \tag{7.58}$$

The solution of (7.57) is

$$m_y(t) = \left[m_y(0)\cos\left(\frac{2H}{\hbar}t\right) + \frac{\hbar}{2H}\left[\lambda m_y(0) + \dot{m}_y(0)\right]\sin\left(\frac{2H}{\hbar}t\right)\right]e^{-\lambda t} . \tag{7.59}$$

Thus, $m_y(t)$ approaches its equilibrium value of zero exponentially at a rate λ. Correspondingly, the first two equations of (7.56) imply that $m_z(t)$ and $m_x(t)$ approach their respective equilibrium values:

$$\begin{aligned} m_z^{\mathrm{eq}} &= \frac{H}{h}\tanh(\beta h) , \\ m_x^{\mathrm{eq}} &= \frac{\Gamma}{h}\tanh(\beta h) . \end{aligned} \tag{7.60}$$

As expected, (7.60) is consistent with (7.33)-(7.34).

7.5 Quantum Glasses and Disordered Transverse Ising Model

An important generalization of the TIM, described by the Hamiltonian in (7.21), is obtained if the exchange interaction J_{ij} is taken to possess *quenched disorder*. The resultant model is usually referred to as the *disordered transverse Ising model* (DTIM). The DTIM is representative of a novel class of materials, which are a testing ground for several topical concepts of condensed matter physics. These materials are called *quantum glasses* as they exhibit various glass-like properties resulting from quenched disorder and frustration. Furthermore, they also display quantum behavior, particularly at low temperatures when tunneling becomes important. Before we discuss quantum glasses, it is relevant to briefly review *classical* (Ising) spin glasses, obtained from (7.21) in the limit $\Gamma = 0$.

Classical glasses are characterized by a complicated free-energy landscape, involving deep valleys and randomly-distributed barriers [172,173]. As a result, they exhibit novel phase transitions. The description of these requires the introduction of unusual order parameters. The theoretical treatment of the thermodynamics of such systems has necessitated the innovation of new statistical-mechanical techniques such as the replica method for evaluating the relevant partition function [174]. The time-dependent properties of spin glasses are of even greater interest. For example, spin auto-correlation functions decay in time as a *stretched exponential* of the Kohlrausch-Williams-Watts (KWW) form, with a characteristic relaxation time that obeys the Vogel-Fulcher (VF) relation [395].

Many of the important recent ideas in this field, e.g., replica-symmetry breaking, Edwards-Anderson order parameter, broken ergodicity, etc., first found their application to classical spin glasses. Examples of these are AuFe or CuMn, where a magnetic impurity (Fe or Mn) is substitutionally quenched in the host matrix of a metal like Au or Cu [396]. This results in an indirect interaction between the magnetic moments of the impurity, mediated by the conduction electrons of the host metal. The interaction is oscillatory (of the so-called Ruderman-Kittel-Kasuya-Yosida or RKKY form), with a length scale determined by the Fermi momentum of the electron. Furthermore, since the impurity atoms are randomly distributed, the net interaction between the magnetic moments fluctuates in both magnitude and direction. It is this competition between the ferromagnetic and antiferromagnetic bonds that is responsible for *frustration* which, in conjunction with disorder, gives rise to the novel effects mentioned earlier. However, we should stress that it is difficult to describe realistic systems by simple statistical-mechanical models because of a range of complicating factors, e.g., the metallic nature of the host, the vector nature of the coupling between spins, and the variable range of the interactions, etc. In this sense, theory and experiment seem to have evolved in divergent directions as far as classical spin glasses are concerned.

From this perspective, the situation appears to be far more satisfactory in the context of quantum glasses, as we shall discuss later.

If we turn our attention away from spin glasses and focus on real or structural glasses (e.g., window glass), we realize that analogous concepts apply there as well, even though the physical situations are quite different. Thus, real glasses have no quenched impurities, but certain degrees of freedom are *dynamically quenched* because of the manner of preparation (e.g., supercooling) of the system [397]. Apart from this distinction, real glasses also exhibit tunneling, especially at low temperatures, when the system makes transitions between neighboring configurations in the complicated free-energy landscape [398–401].

Therefore, it is of great interest to investigate new materials which are simple to treat theoretically, and share common properties of disorder, frustration and tunneling with spin glasses or real glasses. Quantum glasses, which are the focus of this section, are examples of such systems. In particular, we stress two important issues:
(i) Do quantum interactions preserve or destroy phase transitions in these glassy systems?
(ii) What, if any, are the influences of quantum transitions on relaxation phenomena?
In this section, we address the first question within MF theory. In Sect. 7.6, we will answer the second question.

As a simplifying step, we assume that the exchange couplings are infinite-ranged and randomly and independently distributed according to a Gaussian probability distribution:

$$P(J_{ij}) = \frac{1}{\sqrt{2\pi N^{-1} J^2}} \exp\left(-\frac{J_{ij}^2}{2N^{-1}J^2} \right) , \tag{7.61}$$

where $\overline{J_{ij}^2} = N^{-1}J^2$ (N being the number of spins) is the variance of the distribution. (We will use the "bar" to denote an average over the disorder distribution.) Such an assumption, in the context of the Ising model, was first made by Sherrington and Kirkpatrick (SK) [176]. As noted in (7.43), the single-spin Hamiltonian of the DTIM in the MF approximation can be written as

$$\mathcal{H}_i = -\left(H_i \sigma_{zi} + \Gamma \sigma_{xi} \right) , \tag{7.62}$$

where

$$H_i = \sum_{j=1}^{N} J_{ij} \langle \sigma_{zj} \rangle \equiv \sum_{j=1}^{N} J_{ij} m_j . \tag{7.63}$$

In accordance with (7.61), the effective *local fields* H_i can also be assumed to be Gaussian-distributed as follows:

$$P(H_i) = \frac{1}{\sqrt{2\pi \overline{H^2}}} \exp\left(-\frac{H_i^2}{2\overline{H^2}}\right), \tag{7.64}$$

where $\overline{H^2}$ is the appropriate variance in this case.

We have shown in Sect. 7.4 that the Hamiltonian in (7.62) can be diagonalized by transforming to a rotated frame in which the Hamiltonian assumes the form (cf. (7.46))

$$\tilde{\mathcal{H}}_i = -h_i \sigma_{zi}, \tag{7.65}$$

where

$$h_i = \sqrt{H_i^2 + \Gamma^2}. \tag{7.66}$$

The variance $\overline{H^2}$ is obtained from

$$\overline{H^2} = \overline{\left(\sum_{j=1}^{N} J_{ij} m_j\right)^2}. \tag{7.67}$$

Assuming that J_{ij} and m_j are independently distributed [174],

$$\overline{H^2} = \sum_{j=1}^{N} \overline{J_{ij}^2}\, \overline{m_j^2} = J^2 \overline{m_j^2}. \tag{7.68}$$

On the other hand, m_j is given by (7.33) as

$$m_j = \frac{H_j}{h_j} \tanh(\beta h_j). \tag{7.69}$$

Identifying $\overline{m_j^2}$ in (7.68) with the Edwards-Anderson order parameter q [174], and employing (7.69), we obtain the self-consistent equation:

$$q = \frac{1}{\sqrt{2\pi \overline{H^2}}} \int_{-\infty}^{\infty} dH \left[\frac{H}{h} \tanh(\beta h)\right]^2 e^{-H^2/(2\overline{H^2})}. \tag{7.70}$$

Introducing a dimensionless variable ξ as

$$\xi = \frac{H}{J\sqrt{q}}, \tag{7.71}$$

(7.70) yields

$$q = \frac{1}{\sqrt{2\pi}} \int_{-\infty}^{\infty} d\xi \left[\frac{H(\xi)}{h(\xi)} \tanh(\beta h(\xi))\right]^2 e^{-\xi^2/2}, \tag{7.72}$$

where we have used (7.66).

We can summarize the results as follows. The effective single-spin Hamiltonian in the MF approximation can be written as

$$\mathcal{H}_s = -\left[H(\xi)\sigma_z + \Gamma\sigma_x\right] , \tag{7.73}$$

where

$$H(\xi) = J\sqrt{q}\xi , \tag{7.74}$$

$$q = \frac{1}{\sqrt{2\pi}} \int_{-\infty}^{\infty} d\xi \; p(\xi)^2 e^{-\xi^2/2} . \tag{7.75}$$

In (7.75), $p(\xi)$ is the *local polarization* given by

$$p(\xi) = \frac{H(\xi)}{\sqrt{H(\xi)^2 + \Gamma^2}} \tanh\left(\beta\sqrt{H(\xi)^2 + \Gamma^2}\right) . \tag{7.76}$$

It is evident that, for $\Gamma = 0$, (7.73)-(7.76) reduce to the corresponding equations for the SK model [176].

The recipe for calculating physically observable quantities, e.g., the partition function, the free energy, correlation functions, etc., is as follows. We compute the corresponding quantity for a given realization of the disorder, and then perform an average over the underlying probability distribution of the disorder. In the MF approximation, the disorder is represented entirely in terms of the local polarization p (cf. (7.76)). Thus, all physical quantities are first calculated for a given value of p, and the results are then averaged over the distribution of p, denoted as $W(p)$. The latter is defined as

$$\begin{aligned} W(p) &= \frac{1}{N} \sum_{i=1}^{N} \delta\left(p - \langle \sigma_{zi} \rangle\right) \\ &= \int D\xi \delta\left[p(\xi) - p\right] , \end{aligned} \tag{7.77}$$

where the last term is a functional integral. We denote the argument of the delta function as $f(\xi)$, where

$$f(\xi) = \frac{H(\xi)}{\sqrt{H(\xi)^2 + \Gamma^2}} \tanh\left(\beta\sqrt{H(\xi)^2 + \Gamma^2}\right) - p , \tag{7.78}$$

and recall the identity [21]

$$\delta[f(\xi)] = \frac{\delta(\xi - \xi_0)}{f'(\xi_0)} , \tag{7.79}$$

where $\xi_0(p)$ is the solution of $f\left[\xi_0(p)\right] = 0$. Then, we obtain

$$W(p) = \frac{1}{\sqrt{2\pi}} \int_{-\infty}^{\infty} d\xi e^{-\xi^2/2} \delta[f(\xi)]$$
$$= \frac{1}{\sqrt{2\pi}} \frac{e^{-\xi_0^2/2}}{f'(\xi_0)} . \tag{7.80}$$

Using (7.74) in (7.78) and differentiating with respect to ξ, it is straightforward to show that

$$f'(\xi) = \beta J \sqrt{q} \left[\frac{H(\xi)^2}{H(\xi)^2 + \Gamma^2} + \frac{\Gamma^2}{\beta} \frac{p(\xi)}{H(\xi)(H(\xi)^2 + \Gamma^2)} - p(\xi)^2 \right] , \tag{7.81}$$

where we have employed the expression for $p(\xi)$ in (7.76). Hence, (7.80) yields

$$W(p) = \frac{e^{-\xi_0^2/2}}{\sqrt{2\pi q}} \frac{1}{\beta J} \left[\frac{H(\xi_0)^2}{H(\xi_0)^2 + \Gamma^2} + \frac{\Gamma^2}{\beta} \frac{p(\xi_0)}{H(\xi_0)(H(\xi_0)^2 + \Gamma^2)} - p(\xi_0)^2 \right]^{-1} , \tag{7.82}$$

where $\xi_0 = \xi_0(p)$ is the inverse function of $p(\xi)$ which satisfies (7.76). Again, in the limit $\Gamma = 0$, $W(p)$ reduces to the corresponding expression for the Ising glass [174]:

$$W^0(p) = \frac{e^{-\xi_0^2(p)/2}}{\sqrt{2\pi q}} \frac{1}{\beta J} \left(1 - p^2\right)^{-1} . \tag{7.83}$$

The steps specified above, leading to (7.82) follow from a generalization of the heuristic arguments of Southern (for the Ising-SK model) to the DTIM [402]. As has been discussed by Fischer and Hertz (in Chap. 3 of [174]), this argument is actually incorrect. For instance, the fields at different sites are correlated, hence the central-limit theorem implicit in (7.64) is not applicable. Furthermore, the site magnetizations are not independent of each other or of the J_{ij}'s, hence the factorization assumed in (7.68) is invalid. Finally, of course, the MF equations (7.69) are incorrect for a spin glass, even if they are reasonable for the corresponding ferromagnet. Yet, and quite remarkably, the above results are equivalent to those obtained from more rigorous analysis using, e.g., replica-symmetric theory [403], or the corresponding dynamical formulation of the SK model [404–406]. To summarize, the results obtained in crude MF theory are reasonable (even though their derivation is wrong) above the Almeida-Thouless (AT) line in the phase diagram [407]. Below the AT line, the physics is considerably more complicated as the assumption of replica symmetry breaks down, and analytical treatment becomes difficult.

We expect to encounter similar obstacles in the quantum glass problem, due to the presence of the additional transverse term in the SK model. In fact, the noncommutativity of operators appearing in the DTIM prohibits the use of the replica method [408–413]. A more formal treatment, using the *thermo-field dynamics* approach to many-body theory [414], has been given by Kopec [415]. Again, one finds that below an AT-like surface, there is no stable quantum glass solution. However, above the AT surface, one obtains the MF solutions given by (7.76) and (7.82). We shall assume here the approximate validity of these solutions near the quantum glass temperature T_g, and refrain from repeating the arguments of the thermofield dynamics method.

7.6 Relaxation Kinetics in Magnetic Glasses

The above discussion clarifies the static properties of the DTIM in MF theory. We now turn our attention to dissipative dynamics in the DTIM. Our reference system is the disordered version of $LiHoF_4$, already introduced in Sect. 7.2. We study the compound $LiHo_xY_{1-x}F_4$ in which non-magnetic Yttrium (Y) is randomly substituted in place of magnetic Holmium (Ho) [392,393]. The variance J^2 in (7.61) is then identified as $J^2 = x\tilde{J}^2$, where $\pm\tilde{J}$ measures the strength of bonds in the experimental system. As has been discussed earlier, the pure Holmium system ($x = 1$) has a Curie point ($T_c = 1.53$ K) of an underlying ferromagnetic transition. This transition has been found to be MF-like, with the relevant interaction corresponding to long-ranged dipolar coupling amongst the Holmium moments. Dilution results in a random, dipolar-coupled Ising system, and the theoretical understanding of this is as follows. A RG calculation shows that the critical exponents for the dipolar-coupled random Ising model are MF-like (with logarithmic corrections) in $d = 3$. However, these exponents are different from those in the undiluted case [391]. Moreover, the ordering temperature T_c varies with the spin concentration x; and studies suggest that spin-glass ordering may be favored over ferromagnetic ordering for $x < x_c$ [416,417]. Indeed, experiments show that for $x = 0.167$, there is a spin-glass transition from the paramagnetic phase at the transition temperature $T_g = 0.1$ K.

What about the effect of a transverse field, as was studied for the TIM by Bitko et al. [375]? The relevant experiment for the DTIM is due to Wu et al. [392,393], who investigated the transition from a classical glass to a quantum glass. For obvious reasons, we shall refer to $LiHo_xY_{1-x}F_4$ in a transverse field as a model *magnetic glass*, and focus on the measured absorptive susceptibility. We first note that the expression (in linear response theory) for the susceptibility due to an oscillatory magnetic field applied along the z-axis is [1]

$$\chi(\omega) = \beta \lim_{\delta \to 0} \lim_{s \to -i\omega+\delta} \left[1 - s\tilde{C}(s)\right] , \tag{7.84}$$

where $\tilde{C}(s)$ is the Laplace transform of the equilibrium correlation function

$$C(t) = \langle \sigma_z(0)\sigma_z(t)\rangle_0 \, . \tag{7.85}$$

In accordance with the prescription outlined in Sect. 7.5, we first compute $C(t)$ for a *given* local polarization p, and then average over the probability distribution $W(p)$ in (7.83). The analytical expression for $\chi(\omega)$ is then obtained from (7.84).

The correlation function $C(t)$ is evaluated in the equilibrium ensemble, so that (cf. (7.9))

$$C(t) = \mathrm{Tr}\left[\rho_{\mathrm{eq}}\sigma_z(0)e^{i\mathcal{H}t/\hbar}\sigma_z(0)e^{-i\mathcal{H}t/\hbar}\right], \tag{7.86}$$

where $\mathcal{H}$ is the total Hamiltonian for the spin subsystem plus the heat bath, as in (7.1). As before, we rewrite the argument of the "Trace" in the rotated frame (see Sect. 7.4). Therefore,

$$C(t) = \mathrm{Tr}\left[\tilde{\rho}_{\mathrm{eq}}\tilde{\sigma}_z(0)e^{i\tilde{\mathcal{H}}t/\hbar}\tilde{\sigma}_z(0)e^{-i\tilde{\mathcal{H}}t/\hbar}\right], \tag{7.87}$$

where $\tilde{\mathcal{H}}$ is given by (7.50) and (7.66) as follows:

$$\tilde{\mathcal{H}} = -h(\xi)\sigma_z + g\hat{b}\sigma_x + \mathcal{H}_B\,. \tag{7.88}$$

The operator $\tilde{\sigma}_z$ is identified as σ_z in the rotated frame (cf. (7.52)):

$$\tilde{\sigma}_z = \frac{1}{h(\xi)}\left[H(\xi)\sigma_z - \Gamma\sigma_x\right]. \tag{7.89}$$

Similarly,

$$\tilde{\rho}_{\mathrm{eq}} = S^{-1}\rho_{\mathrm{eq}}S\,, \tag{7.90}$$

the rotation operator S being given by (7.45). Now, from (7.11),

$$\tilde{\rho}_{\mathrm{eq}} = \rho_B \otimes \tilde{\rho}_s^{\mathrm{eq}}\,, \tag{7.91}$$

where

$$\tilde{\rho}_s^{\mathrm{eq}} = \frac{1}{Z_s}\exp\left(-\beta\tilde{\mathcal{H}}_s\right), \tag{7.92}$$

$\tilde{\mathcal{H}}_s$ being the subsystem Hamiltonian in the rotated frame, $\tilde{\mathcal{H}}_s = -h(\xi)\sigma_z$, and Z_s the corresponding partition function:

$$Z_s = 2\cosh\left[\beta h(\xi)\right]. \tag{7.93}$$

Therefore, following the prescription in (7.9)-(7.14), we may write from (7.87):

$$\tilde{C}(s) = \frac{1}{Z_s}\mathrm{Tr}_s\left[e^{-\beta\tilde{\mathcal{H}}_s}\left(\frac{H}{h}\sigma_z - \frac{\Gamma}{h}\sigma_x\right)\left\{\left(\tilde{U}(s)\right)_{\mathrm{av}}\left(\frac{H}{h}\sigma_z - \frac{\Gamma}{h}\sigma_x\right)\right\}\right], \tag{7.94}$$

where $\left(\tilde{U}(s)\right)_{\text{av}}$ is to be computed using the total Hamiltonian $\tilde{\mathcal{H}}$ in (7.88). Furthermore, we have dropped the argument ξ for the sake of brevity. Writing out the trace over the subsystem, we obtain from (7.94)

$$\tilde{C}(s) = \frac{1}{Z_s}\frac{1}{h^2}\sum_{\mu\nu,\mu'\nu'} e^{\beta h\mu}\langle\mu|\left(H\sigma_z - \Gamma\sigma_x\right)|\nu\rangle \left(\nu\mu\right|\left(\tilde{U}(s)\right)_{\text{av}}\left|\nu'\mu'\right)$$
$$\times\langle\nu'|\left(H\sigma_z - \Gamma\sigma_x\right)|\mu'\rangle\,, \tag{7.95}$$

where we have used the properties of Liouville operators as well as the notation introduced in (1.A.16). Now,

$$\langle\mu|\left(H\sigma_z - \Gamma\sigma_x\right)|\nu\rangle = H\mu\delta_{\mu\nu} - \Gamma\left(\delta_{\mu,\nu+2} + \delta_{\mu,\nu-2}\right)\,. \tag{7.96}$$

Hence, (7.95) yields

$$\tilde{C}(s) = \frac{1}{Z_s h^2}\sum_{\mu\nu,\mu'\nu'} e^{\beta h\mu}\left(\nu\mu\right|\left(\tilde{U}(s)\right)_{\text{av}}\left|\nu'\mu'\right)\left[H^2\mu\mu'\delta_{\mu\nu}\delta_{\mu'\nu'}\right.$$
$$+\Gamma^2\left(\delta_{\mu,\nu+2}\delta_{\mu',\nu'-2} + \delta_{\mu,\nu-2}\delta_{\mu',\nu'-2} + \delta_{\mu,\nu+2}\delta_{\mu',\nu'+2} + \delta_{\mu,\nu-2}\delta_{\mu',\nu'+2}\right)$$
$$\left.-H\Gamma\left(\mu'\delta_{\mu'\nu'}\delta_{\mu,\nu+2} + \mu'\delta_{\mu'\nu'}\delta_{\mu\nu-2} + \mu\delta_{\mu\nu}\delta_{\mu',\nu'-2} + \mu\delta_{\mu\nu}\delta_{\mu',\nu'+2}\right)\right]\,. \tag{7.97}$$

In order to make further progress, we need to compute the matrix elements of $\left(\tilde{U}(s)\right)_{\text{av}}$. As this quantity is a super-operator in the space of the subsystem alone, it is characterized by 16 elements as the subsystem is restricted to a 2-dimensional Hilbert space. Recall that, in second-order perturbation theory, and in the Markovian limit, $\left(\tilde{U}(s)\right)_{\text{av}}$ is given by (cf. (7.19)-(7.20))

$$\left(\tilde{U}(s)\right)_{\text{av}} = \left(s - iL_s + \tilde{\Sigma}(s=0)\right)^{-1}\,. \tag{7.98}$$

Using (7.46) and properties of the Liouville operators, the matrix elements of L_s can be written as (see Appendix 7.A)

$$\left(\mu\nu|L_s|\mu'\nu'\right) = \frac{h}{\hbar}(\nu-\mu)\delta_{\mu\mu'}\delta_{\nu\nu'}\,. \tag{7.99}$$

In a more explicit form, we have

$$L_s = \frac{2h}{\hbar}\begin{pmatrix} 0 & 0 & 0 & 0 \\ 0 & 0 & 0 & 0 \\ 0 & 0 & -1 & 0 \\ 0 & 0 & 0 & 1 \end{pmatrix}\,, \tag{7.100}$$

where the rows and columns labeled by $|\mu\nu)$ take the values $|++)$, $|--)$, $|+-)$ and $|-+)$, respectively. The next step is the evaluation of the matrix elements of $\tilde{\Sigma}(s=0)$. This is also done in Appendix 7.A, where we find

$$\tilde{\Sigma}(s=0) = \begin{pmatrix} \lambda p^{\rm eq}(-1) & -\lambda p^{\rm eq}(-1) & 0 & 0 \\ -\lambda p^{\rm eq}(1) & \lambda p^{\rm eq}(1) & 0 & 0 \\ 0 & 0 & \lambda/2 & -\lambda/2 \\ 0 & 0 & -\lambda/2 & \lambda/2 \end{pmatrix}, \tag{7.101}$$

where $p^{\rm eq}(\pm 1)$ is given by (1.92) (with H replaced by h), and λ is given by (1.98).

It is clear from (7.98) that, in order to obtain the matrix elements of $\left(\tilde{U}(s)\right)_{\rm av}$, we have to invert the 4 ×4 matrix M, given by

$$M = \begin{pmatrix} s+\lambda p^{\rm eq}(-1) & -\lambda p^{\rm eq}(-1) & 0 & 0 \\ -\lambda p^{\rm eq}(1) & s+\lambda p^{\rm eq}(1) & 0 & 0 \\ 0 & 0 & s+\lambda/2+2ih/\hbar & -\lambda/2 \\ 0 & 0 & -\lambda/2 & s+\lambda/2-2ih/\hbar \end{pmatrix}. \tag{7.102}$$

This matrix is easily inverted, yielding

$$\left(\tilde{U}(s)\right)_{\rm av} = \begin{pmatrix} \frac{s+\lambda p^{\rm eq}(1)}{s(s+\lambda)} & \frac{\lambda p^{\rm eq}(-1)}{s(s+\lambda)} & 0 & 0 \\ \frac{\lambda p^{\rm eq}(1)}{s(s+\lambda)} & \frac{s+\lambda p^{\rm eq}(-1)}{s(s+\lambda)} & 0 & 0 \\ 0 & 0 & \frac{s+\lambda/2-2ih/\hbar}{s(s+\lambda)+4h^2/\hbar^2} & \frac{\lambda/2}{s(s+\lambda)+4h^2/\hbar^2} \\ 0 & 0 & \frac{\lambda/2}{s(s+\lambda)+4h^2/\hbar^2} & \frac{s+\lambda/2+2ih/\hbar}{s(s+\lambda)+4h^2/\hbar^2} \end{pmatrix}. \tag{7.103}$$

It is clear from the structure of (7.97) that the cross-terms involving the product of H and Γ vanish, yielding

$$\begin{aligned} \tilde{C}(s) &= \frac{1}{Z_s h^2}\left\{\frac{H^2}{s(s+\lambda)}\left[e^{\beta h}\left(s+\lambda\tanh(\beta h)\right)+e^{-\beta h}\left(s-\lambda\tanh(\beta h)\right)\right]\right. \\ &\quad \left. + \frac{\Gamma^2}{s(s+\lambda)+4h^2/\hbar^2}\left[e^{\beta h}\left(s+\lambda+\frac{2ih}{\hbar}\right)+e^{-\beta h}\left(s+\lambda-\frac{2ih}{\hbar}\right)\right]\right\} \\ &= \frac{1}{h^2}\left\{\frac{H^2}{s(s+\lambda)}\left[s+\lambda\tanh^2(\beta h)\right]\right. \\ &\quad \left. + \frac{\Gamma^2}{s(s+\lambda)+4h^2/\hbar^2}\left[(s+\lambda)+\frac{2ih}{\hbar}\tanh(\beta h)\right]\right\}, \end{aligned} \tag{7.104}$$

where we have used (7.103). We have also used the relation

$$p^{\rm eq}(1) - p^{\rm eq}(-1) = \tanh(\beta h) = \frac{h}{H}p, \tag{7.105}$$

where p is the local polarization given by (7.76). Hence, (7.104) yields

$$\tilde{C}(s) = \frac{1}{s(s+\lambda)}\left(\frac{sH^2}{h^2}+\lambda p^2\right) + \frac{\Gamma^2}{s(s+\lambda)+4h^2/\hbar^2}\left(\frac{s+\lambda}{h^2}+\frac{2ip}{\hbar H}\right). \tag{7.106}$$

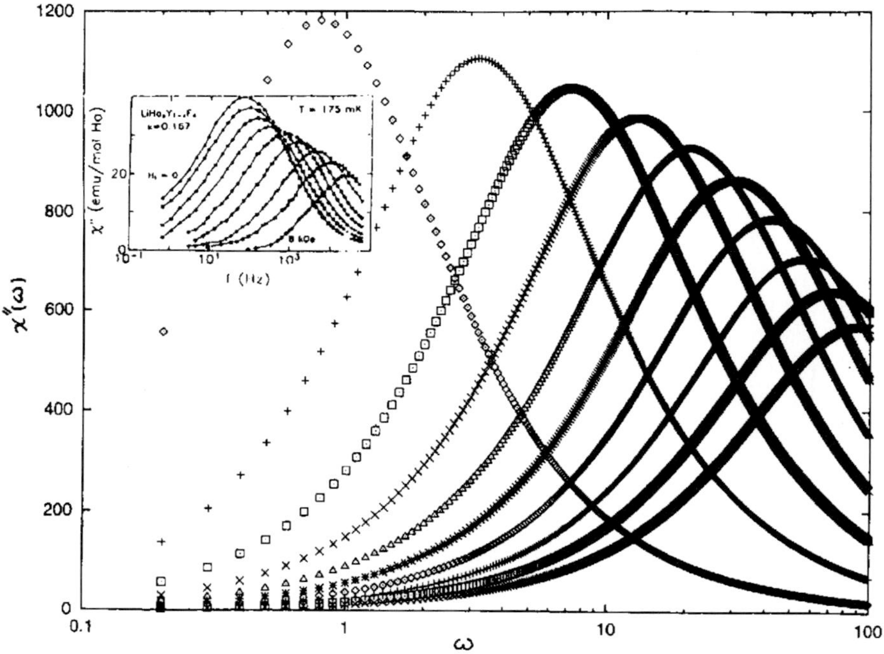

Fig. 7.2. Imaginary component of the dynamic susceptibility $\chi''(\omega)$ vs. ω for the DTIM. The parameter values are specified in the text. The inset figure shows the corresponding experimental results of Wu et al. [392,393]

With the correlation function at hand, the measured susceptibility is obtained from (7.84) as

$$\chi(\omega) = \frac{\beta}{4\pi} \int_{-1}^{1} dp W(p) \left\{ 1 - \frac{\left(-i\omega \frac{H^2}{\hbar^2} + \lambda p^2\right)}{\lambda - i\omega} - \frac{i\omega \frac{\Gamma^2}{\hbar^2} \left[\lambda - i\left(\omega - \frac{2p}{\hbar}\frac{h^2}{H}\right)\right]}{\left(\omega^2 - \frac{4h^2}{\hbar^2}\right) + i\omega\lambda} \right\} . \tag{7.107}$$

In particular, the absorptive susceptibility is obtained from the imaginary part of (7.107) as

$$\chi''(\omega) = \frac{\beta}{4\pi} \int_{-1}^{1} dp W(p) \left[\frac{\omega\lambda}{\lambda^2 + \omega^2} \left(\frac{H^2}{h^2} - p^2 \right) + \frac{4\omega\lambda \frac{\Gamma^2}{\hbar^2} \left(1 - \frac{\hbar\omega}{2H} p\right)}{\left(\omega^2 - \frac{4h^2}{\hbar^2}\right)^2 + \omega^2\lambda^2} \right] . \tag{7.108}$$

In (7.107) and (7.108), $W(p)$ is given by (7.82).

Figure 7.2 plots the imaginary component of the dynamic susceptibility $\chi''(\omega)$ vs. ω for the experimentally relevant value of x, viz., x =0.167 in $\mathrm{LiHo}_x\mathrm{Y}_{1-x}\mathrm{F}_4$ [418]. The graphs were obtained by a numerical integration of (7.108) at a temperature $k_B T = 1.1$ and a relaxation rate $\lambda = 50$, both in units of $\tilde{J}$. The curves, in descending order of amplitude, correspond to different values of $\Gamma/\tilde{J}$ – ranging from 0.1 (the largest-amplitude curve) to 1.0 (the smallest-amplitude curve) in steps of 0.1. The dynamic-susceptibility response around the peak value indicates a shift in the frequency as $\Gamma/\tilde{J}$ is increased. In addition to the reduction in amplitude, there is a clear indication of a broadening of peaks as quantum effects are enhanced. It can also be observed that the low-frequency tails of $\chi''(\omega)$ are greatly suppressed, implying that quantum routes to relaxation affect the long-time dynamics of the system. These features are qualitatively similar to experimental data (shown in the inset of Fig. 7.2) obtained by Wu et al. [392,393]. We have checked that the results are not reproducible when λ is small, hence the heat-bath coupling assumed by us is critical in clarifying the essential features of the quantum and relaxational dynamics in the compound $LiHo_xY_{1-x}F_4$.

7.7 Dielectric Relaxation in Proton Glasses

Proton glasses, like magnetic glasses, have the structure of a crystal but are characterized by a random mix of ferroelectric and antiferroelectric interactions. The most studied system in this context is $Rb_{1-x}(NH_4)_xH_2PO_4$, in which the Rb-system is ferroelectric while the NH_4-system is antiferroelectric [419]. This compound is called a proton glass as the basic random interaction is between protons, each of which can occupy two sites in the O-H...O bond. The *left* and *right* positions in an O-H...O bond are mapped onto an Ising spin variable σ_z, which assumes eigenvalues $+1$ and -1. The ferroelectric and antiferroelectric transitions in these systems are known to be driven by Ising-like interactions between the pseudospins. This interaction is actually between *electric dipoles*, which are induced in the medium by the distortion of the material caused by hydrogen, and hence the underlying Ising model is long-ranged. Glassiness arises from the random mixing of the ferroelectric and antiferroelectric bonds. Finally, tunneling can be incorporated by adding a term that couples to the transverse component of the pseudospin with respect to the easy axis of the Ising interaction.

Therefore, as in the magnetic glass problem, one has a disordered transverse Ising model (DTIM) for proton glasses. However, there is one additional term. This involves the coupling between the pseudospin and a local random field seen by it due to the presence of strain fields generated by the random substitute, e.g., NH_4 in place of Rb. The influence of the random field (hitherto ignored in our analysis of magnetic glasses) on the relaxational behavior of proton glasses is the topic of our present discussion.

The Hamiltonian for a proton glass is a generalized version of (7.21), and is given by

$$\mathcal{H} = -\sum_{i>j=1}^{N} J_{ij}\sigma_{zi}\sigma_{zj} - \Gamma\sum_{i=1}^{N} f_i\sigma_{xi}\,, \qquad (7.109)$$

where f_i is the local random field mentioned above. It is customary to assume that the distributions of both J_{ij} and f_i are Gaussian and independent, i.e.,

$$P(J_{ij}) = \frac{1}{\sqrt{2\pi N^{-1}J^2}} \exp\left[-\frac{(J_{ij}-J_0)^2}{2N^{-1}J^2}\right], \tag{7.110}$$

$$P(f_i) = \frac{1}{\sqrt{2\pi\Delta_f}} \exp\left(-\frac{f_i^2}{2\Delta_f}\right). \tag{7.111}$$

In contrast to (7.61), the mean and the variance of the J_{ij}-distribution are now given by

$$J_0 = (1-2x)N^{-1/2}\tilde{J}, \quad J^2 = 4x(1-x)\tilde{J}^2, \tag{7.112}$$

where x and $(1-x)$ are the concentrations of antiferroelectric and ferroelectric bonds, respectively [420,421]. As before, $\pm\tilde{J}$ measures the strength of these bonds. A similar (but independent) concentration dependence can be assumed for the variance Δ_f of the distribution for the local random field.

In MF theory, the single-spin Hamiltonian is again given by (7.73), but because of the presence of the local random field as well as the ferromagnetic order J_0, $H(\xi)$ is now given by (in contrast to (7.74))

$$H(\xi) = \xi J\sqrt{q+\tilde{\Delta}_f} + NJ_0 p, \tag{7.113}$$

$$\tilde{\Delta}_f = \frac{\Delta_f}{J^2}. \tag{7.114}$$

However, the MF equations for the Edwards-Anderson order parameter q and the local polarization $p(\xi)$ are identical to (7.75)-(7.76), obtained for magnetic glasses.

As an example of an experimentally relevant quantity, let us focus upon the dynamic dielectric permittivity $\epsilon(\omega)$. In linear response theory, this is given by $\chi(\omega)$, defined in (7.84). Thus, the result in (7.107), derived earlier for the magnetic glass, applies once again – with appropriate modifications in the probability distribution of the local polarization $W(p)$. We first study the effect of the local random field $\tilde{\Delta}_f$. As mentioned before, this additional term, which is an integral part of the proton-glass Hamiltonian, was absent in our earlier discussion of magnetic glasses. Figure 7.3(a) shows the effect of the absence of the local random field (i.e., $\tilde{\Delta}_f = 0$) on the imaginary part of the dielectric susceptibility for different values of the frequency ($\omega = 1, ...5$), as the transverse field Γ is varied. The corresponding results with the random field ($\tilde{\Delta}_f = 1.0$) are shown in Fig. 7.3(b) [422]. As discussed earlier, Γ is a disordering field, analogous to the temperature. The values of $\lambda/\tilde{J}$ and x were selected to be 1.0 and 0.5, respectively. Note that for the chosen value of x, J_0 vanishes (cf. (7.112)) and the distribution $P(J_{ij})$ in (7.110) becomes identical to that for magnetic glasses.

It is clearly seen that the peak value of the amplitude of the response decreases in the presence of the field. In the absence of the random field, the

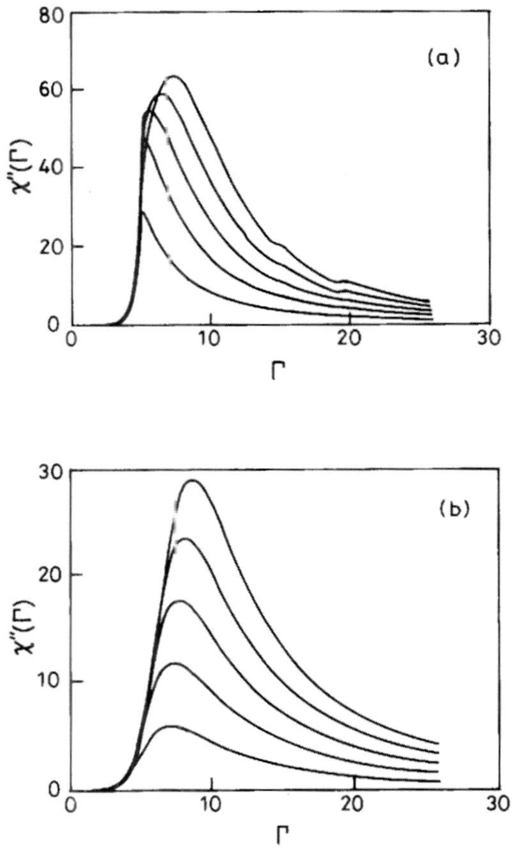

Fig. 7.3. Imaginary part of the dielectric susceptibility $\chi''(\Gamma)$ as a function of transverse field strength Γ (measured in units of $\tilde{J}$). We show results for different frequencies $\omega = 1, 2, 3, 4, 5$. The lowest-amplitude curve corresponds to $\omega = 1$, and the highest-amplitude curve corresponds to $\omega = 5$. The results correspond to (**a**) $\tilde{\Delta}_f = 0$, (**b**)$\tilde{\Delta}_f = 1.0$

dielectric susceptibility shows a sharp increase from its value at $\Gamma = 0$ to its peak value (see Fig. 7.3(a)). This is not so in the presence of the random field, where a broadening of the response function is observed (see Fig. 7.3(b)). We point out that this behavior is qualitatively similar to that observed in experiments on mixed crystals of $Rb_{1-x}(NH_4)_xH_2PO_4$ for $x = 0.35$ [423–425]. This emphasizes the importance of the random-field term in our model.

Finally, we study the effect of quantum tunneling on the imaginary part of the dielectric susceptibility. In Fig. 7.4, we plot $\chi''(\omega)$ vs. ω for different strengths of the tunneling term $\Gamma(\Gamma/\tilde{J} = 0.1, ...0.5)$. As before, the values of $\lambda/\tilde{J}$ and x were fixed as 1.0 and 0.5, respectively. Proton glasses are known to exhibit a shift in the frequency of the peak value of the response towards

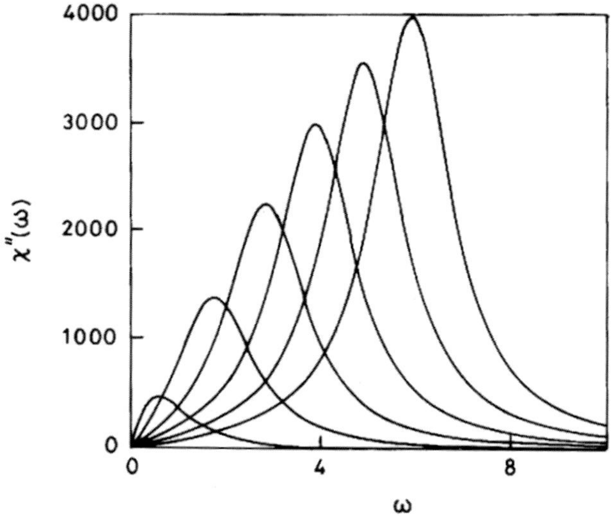

Fig. 7.4. Imaginary part of the dielectric susceptibility $\chi''(\omega)$ as a function of frequency ω. We show results for different values of $\Gamma/\tilde{J} = 0.1, ...0.5$. The lowest-amplitude curve corresponds to $\Gamma/\tilde{J} = 0.1$, and the highest-amplitude curve corresponds to $\Gamma/\tilde{J} = 0.5$

higher values of ω with increasing temperature. These shifts are seen to occur with increasing quantum effects, quite akin to those encountered in magnetic glasses [392,393]. Due to the glassy nature of the system, the dielectric relaxation is evidently of the non-Debye type. The low-frequency tails appear to be quenched, suggesting that quantum effects are important at longer times.

7.A Appendix: Matrix Elements of L_s and $\tilde{\Sigma}(s=0)$

In the Markovian approximation, we have

$$\left(\tilde{U}(s)\right)_{\mathrm{av}} = \left(s - iL_s + \tilde{\Sigma}(s=0)\right)^{-1} . \tag{7.A.1}$$

We compute matrix elements of L_s and $\tilde{\Sigma}(s=0)$, as described below.

Matrix Elements of L_s

The matrix elements of L_s are as follows:

$$\begin{aligned}(\mu\nu|L_s|\mu'\nu') &= \frac{1}{\hbar}\left[\langle\mu|\tilde{\mathcal{H}}_s|\mu'\rangle\delta_{\nu\nu'} - \langle\nu'|\tilde{\mathcal{H}}_s|\nu\rangle\delta_{\mu\mu'}\right] \\ &= \frac{h(\xi)}{\hbar}(\nu-\mu)\delta_{\mu\mu'}\delta_{\nu\nu'} .\end{aligned} \tag{7.A.2}$$

Matrix elements of $\tilde{\Sigma}(s=0)$

From (1.A.51), we have (using the Heisenberg picture)

$$\begin{aligned}&\left(\mu\nu|\tilde{\Sigma}(s=0)|\mu'\nu'\right) \\ &= \int_0^\infty dt \left(\mu\nu|\left(L_I e^{i(L_s+L_B)t}L_I\right)_{\mathrm{av}}|\mu'\nu'\right) \\ &= \int_0^\infty dt \mathrm{Tr}_B\left[\rho_B\left(\mu\nu|L_I e^{i(L_s+L_B)t}L_I|\mu'\nu'\right)\right] \\ &= \int_0^\infty dt \sum_{i_B i'_B}\langle i_B|\rho_B|i_B\rangle\left(\mu i_B\nu i_B|L_I e^{i(L_s+L_B)t}L_I|\mu' i'_B\nu' i'_B\right) .\end{aligned} \tag{7.A.3}$$

Furthermore,

$$\begin{aligned}&\left(\mu i_B\nu i_B|L_I e^{i(L_s+L_B)t}L_I|\mu' i'_B\nu' i'_B\right) \\ &= \sum_{\mu_1\nu_1 j_B j'_B}(\mu i_B\nu i_B|L_I|\mu_1 j_B\nu_1 j'_B)\left(\mu_1 j_B\nu_1 j'_B|e^{i(L_s+L_B)t}|\mu_1 j_B\nu_1 j'_B\right) \\ &\times(\mu_1 j_B\nu_1 j'_B|L_I|\mu' i'_B\nu' i'_B) \\ &= \frac{1}{\hbar^2}\sum_{\mu_1\nu_1 j_B j'_B}\left(\langle\mu i_B|\mathcal{H}_I|\mu_1 j_B\rangle\delta_{\nu\nu_1}\delta_{i_B j'_B} - \langle\nu_1 j'_B|\mathcal{H}_I|\nu i_B\rangle\delta_{\mu\mu_1}\delta_{i_B j_B}\right) \\ &\times\left(\mu_1 j_B\nu_1 j'_B|e^{i(L_s+L_B)t}|\mu_1 j_B\nu_1 j'_B\right) \\ &\times\left(\langle\mu_1 j_B|\mathcal{H}_I|\mu' i'_B\rangle\delta_{\nu_1\nu'}\delta_{i'_B j'_B} - \langle\nu' i'_B|\mathcal{H}_I|\nu_1 j'_B\rangle\delta_{\mu_1\mu'}\delta_{j_B i'_B}\right)\end{aligned}$$

$$
\begin{aligned}
&= \frac{1}{\hbar^2}\Bigg[\sum_{\mu_1 j_B} \langle \mu i_B|\mathcal{H}_I|\mu_1 j_B\rangle\langle \mu_1 j_B|\mathcal{H}_I|\mu' i_B\rangle \delta_{\nu\nu'}\delta_{i_B i'_B} \\
&\times \left(\mu_1 j_B \nu i_B\middle|e^{i(L_s+L_B)t}\middle|\mu_1 j_B \nu i_B\right) \\
&+ \sum_{\nu_1 j'_B} \langle \nu' i'_B|\mathcal{H}_I|\nu_1 j'_B\rangle\langle \nu_1 j'_B|\mathcal{H}_I|\nu i_B\rangle \delta_{\mu\mu'}\delta_{i_B i'_B} \\
&\times \left(\mu i_B \nu_1 j'_B\middle|e^{i(L_s+L_B)t}\middle|\mu i_B \nu_1 j'_B\right) \\
&- \langle \mu i_B|\mathcal{H}_I|\mu' i'_B\rangle\langle \nu' i'_B|\mathcal{H}_I|\nu i_B\rangle \left(\mu' i'_B \nu i_B\middle|e^{i(L_s+L_B)t}\middle|\mu' i'_B \nu i_B\right) \\
&- \langle \mu i_B|\mathcal{H}_I|\mu' i'_B\rangle\langle \nu' i'_B|\mathcal{H}_I|\nu i_B\rangle \left(\mu i_B \nu' i'_B\middle|e^{i(L_s+L_B)t}\middle|\mu i_B \nu' i'_B\right)\Bigg] .
\end{aligned}
\tag{7.A.4}
$$

Therefore, we calculate the non-zero matrix elements as

$$
\begin{aligned}
&\left(++|\tilde{\Sigma}|++\right) \\
&= \frac{g^2}{\hbar^2}\int_0^\infty dt \sum_{i_B i'_B} \langle i_B|\rho_B|i_B\rangle \Bigg[\sum_{j_B} \langle +|\sigma_+|-\rangle\langle -|\sigma_-|+\rangle\langle i_B|\hat{b}|j_B\rangle \\
&\times \langle j_B|\hat{b}|i_B\rangle \delta_{i_B i'_B} e^{2iht/\hbar}\langle j_B|e^{i\mathcal{H}_B t/\hbar}|j_B\rangle\langle i_B|e^{-i\mathcal{H}_B t/\hbar}|i_B\rangle \\
&+ \sum_{j'_B} \langle +|\sigma_+|-\rangle\langle -|\sigma_-|+\rangle\langle i_B|\hat{b}|j'_B\rangle \\
&\times \langle j'_B|\hat{b}|i_B\rangle \delta_{i_B i'_B} e^{-2iht/\hbar}\langle i_B|e^{i\mathcal{H}_B t/\hbar}|i_B\rangle\langle j'_B|e^{-i\mathcal{H}_B t/\hbar}|j'_B\rangle\Bigg] \\
&= \frac{g^2}{\hbar^2}\int_0^\infty dt \left[\langle \hat{b}(0)\hat{b}(t)\rangle e^{2iht/\hbar} + \langle \hat{b}(t)\hat{b}(0)\rangle e^{-2iht/\hbar}\right] \\
&= W(1|-1)\,, \quad \text{from (1.81)} \\
&= \lambda p^{\mathrm{eq}}(-1)\,, \quad \text{from (1.100)} .
\end{aligned}
\tag{7.A.5}
$$

Similarly,

$$
\begin{aligned}
&\left(++|\tilde{\Sigma}|--\right) \\
&= -\frac{g^2}{\hbar^2}\int_0^\infty dt \sum_{i_B i'_B} \langle i_B|\rho_B|i_B\rangle \\
&\times \Big[\langle i_B|\hat{b}|i'_B\rangle\langle i'_B|\hat{b}|i_B\rangle e^{2iht/\hbar}\langle i'_B|e^{i\mathcal{H}_B t/\hbar}|i'_B\rangle\langle i_B|e^{-i\mathcal{H}_B t/\hbar}|i_B\rangle \\
&+ \langle i_B|\hat{b}|i'_B\rangle\langle i'_B|\hat{b}|i_B\rangle e^{-2iht/\hbar}\langle i_B|e^{i\mathcal{H}_B t/\hbar}|i_B\rangle\langle i'_B|e^{-i\mathcal{H}_B t/\hbar}|i'_B\rangle\Big]
\end{aligned}
$$

$$
\begin{aligned}
&= -\frac{g^2}{\hbar^2}\int_0^\infty dt\left[\langle\hat{b}(0)\hat{b}(t)\rangle e^{2iht/\hbar} + \langle\hat{b}(t)\hat{b}(0)\rangle e^{-2iht/\hbar}\right] \\
&= -W(1|-1) \\
&= -\lambda p^{\mathrm{eq}}(-1)\,.
\end{aligned} \tag{7.A.6}
$$

Next, we consider

$$
\begin{aligned}
&\left(--|\tilde{\Sigma}|--\right) \\
&= \frac{g^2}{\hbar^2}\int_0^\infty dt\sum_{i_B i'_B}\langle i_B|\rho_B|i_B\rangle\left[\sum_{j_B}\langle i_B|\hat{b}|j_B\rangle\langle j_B|\hat{b}|i_B\rangle\delta_{i_B i'_B}\right. \\
&\times e^{-2iht/\hbar}\langle j_B|e^{i\mathcal{H}_B t/\hbar}|j_B\rangle\langle i_B|e^{-i\mathcal{H}_B t/\hbar}|i_B\rangle \\
&+\sum_{j'_B}\langle i_B|\hat{b}|j'_B\rangle\langle j'_B|\hat{b}\, i_B\rangle\delta_{i_B i'_B} \\
&\left.\times e^{2iht/\hbar}\langle i_B|e^{i\mathcal{H}_B t/\hbar}|i_B\rangle\langle j'_B|e^{-i\mathcal{H}_B t/\hbar}|j'_B\rangle\right] \\
&= \frac{g^2}{\hbar^2}\int_0^\infty dt\left[\langle\hat{b}(0)\hat{b}(t)\rangle e^{-2iht/\hbar} + \langle\hat{b}(t)\hat{b}(0)\rangle e^{2iht/\hbar}\right] \\
&= W(-1|1)\,, \quad \text{from (1.81)} \\
&= \lambda p^{\mathrm{eq}}(1)\,, \quad \text{from (1.99)}\,.
\end{aligned} \tag{7.A.7}
$$

Next,

$$
\begin{aligned}
&\left(--|\tilde{\Sigma}|++\right) \\
&= -\frac{g^2}{\hbar^2}\int_0^\infty dt\sum_{i_B i'_B}\langle i_B|\rho_B|i_B\rangle \\
&\times\left[\langle i_B|\hat{b}|i'_B\rangle\langle i'_B|\hat{b}|i_B\rangle e^{-2iht/\hbar}\langle i'_B|e^{i\mathcal{H}_B t/\hbar}|i'_B\rangle\langle i_B|e^{-i\mathcal{H}_B t/\hbar}|i_B\rangle\right. \\
&\left.+\langle i_B|\hat{b}|i'_B\rangle\langle i'_B|\hat{b}|i_B\rangle e^{2iht/\hbar}\langle i_B|e^{i\mathcal{H}_B t/\hbar}|i_B\rangle\langle i'_B|e^{-i\mathcal{H}_B t/\hbar}|i'_B\rangle\right] \\
&= -\frac{g^2}{\hbar^2}\int_0^\infty dt\left[\langle\hat{b}(0)\hat{b}(t)\rangle e^{-2iht/\hbar} + \langle\hat{b}(t)\hat{b}(0)\rangle e^{2iht/\hbar}\right] \\
&= -W(-1|1) \\
&= -\lambda p^{\mathrm{eq}}(1)\,.
\end{aligned} \tag{7.A.8}
$$

Next,

$$
\begin{aligned}
&\left(+-|\tilde{\Sigma}|+-\right) \\
&= \frac{g^2}{\hbar^2}\int_0^\infty dt\sum_{i_B i'_B}\langle i_B|\rho_B|i_B\rangle
\end{aligned}
$$

$$\times \left[\sum_{j_B} \langle i_B|\hat{b}|j_B\rangle\langle j_B|\hat{b}|i_B\rangle \delta_{i_B i'_B} \langle j_B|e^{i\mathcal{H}_B t/\hbar}|j_B\rangle\langle i_B|e^{-i\mathcal{H}_B t/\hbar}|i_B\rangle \right.$$
$$\left. + \sum_{j'_B} \langle i_B|\hat{b}|j'_B\rangle\langle j'_B|\hat{b}|i_B\rangle \delta_{i_B i'_B} \langle i_B|e^{i\mathcal{H}_B t/\hbar}|i_B\rangle\langle j'_B|e^{-i\mathcal{H}_B t/\hbar}|j'_B\rangle \right]$$
$$= \frac{g^2}{\hbar^2}\int_0^\infty dt \left[\langle\hat{b}(0)\hat{b}(t)\rangle + \langle\hat{b}(t)\hat{b}(0)\rangle \right]$$
$$\simeq \frac{1}{2}[W(-1|1) + W(1|-1)]\,, \quad \text{from (1.94)}$$
$$= \frac{\lambda}{2}\,, \quad \text{from (1.97)}\,. \tag{7.A.9}$$

Next,

$$\left(-+|\tilde{\Sigma}|-+\right)$$
$$= \frac{g^2}{\hbar^2}\int_0^\infty dt \sum_{i_B i'_B} \langle i_B|\rho_B|i_B\rangle$$
$$\times \left[\sum_{j_B} \langle i_B|\hat{b}|j_B\rangle\langle j_B|\hat{b}|i_B\rangle \delta_{i_B i'_B} \langle i_B|e^{i\mathcal{H}_B t/\hbar}|i_B\rangle\langle j_B|e^{-i\mathcal{H}_B t/\hbar}|j_B\rangle \right.$$
$$\left. + \sum_{j'_B} \langle i_B|\hat{b}|j'_B\rangle\langle j'_B|\hat{b}|i_B\rangle \delta_{i_B i'_B} \langle i_B|e^{i\mathcal{H}_B t/\hbar}|i_B\rangle\langle j'_B|e^{-i\mathcal{H}_B t/\hbar}|j'_B\rangle \right]$$
$$= \frac{g^2}{\hbar^2}\int_0^\infty dt \left[\langle\hat{b}(0)\hat{b}(t)\rangle + \langle\hat{b}(t)\hat{b}(0)\rangle \right]$$
$$= \frac{\lambda}{2}\,. \tag{7.A.10}$$

Next,

$$\left(+-|\tilde{\Sigma}|-+\right)$$
$$= -\frac{g^2}{\hbar^2}\int_0^\infty dt \sum_{i_B i'_B} \langle i_B|\rho_B|i_B\rangle$$
$$\times \left[\langle i_B|\hat{b}|i'_B\rangle\langle i'_B|\hat{b}|i_B\rangle\langle i'_B|e^{i\mathcal{H}_B t/\hbar}|i'_B\rangle\langle i_B|e^{-i\mathcal{H}_B t/\hbar}|i_B\rangle \right.$$
$$\left. + \langle i_B|\hat{b}|i'_B\rangle\langle i'_B|\hat{b}|i_B\rangle\langle i_B|e^{i\mathcal{H}_B t/\hbar}|i_B\rangle\langle i'_B|e^{-i\mathcal{H}_B t/\hbar}|i'_B\rangle \right]$$
$$= -\frac{g^2}{\hbar^2}\int_0^\infty dt \left[\langle\hat{b}(0)\hat{b}(t)\rangle + \langle\hat{b}(t)\hat{b}(0)\rangle \right]$$
$$= -\frac{\lambda}{2}\,. \tag{7.A.11}$$

Finally,

$$\begin{aligned}
&\left(-+|\tilde{\Sigma}|+-\right) \\
&= -\frac{g^2}{\hbar^2}\int_0^\infty dt \sum_{i_B i'_B} \langle i_B|\rho_B|i_B\rangle \\
&\times \Big[\langle i_B|\hat{b}|i'_B\rangle\langle i'_B|\hat{b}|i_B\rangle\langle i'_B|e^{i\mathcal{H}_B t/\hbar}|i'_B\rangle\langle i_B|e^{-i\mathcal{H}_B t/\hbar}|i_B\rangle \\
&+\langle i_B|\hat{b}|i'_B\rangle\langle i'_B|\hat{b}|i_B\rangle\langle i_B|e^{i\mathcal{H}_B t/\hbar}|i_B\rangle\langle i'_B|e^{-i\mathcal{H}_B t/\hbar}|i'_B\rangle\Big] \\
&= -\frac{g^2}{\hbar^2}\int_0^\infty dt \left[\langle\hat{b}(0)\hat{b}(t)\rangle + \langle\hat{b}(t)\hat{b}(0)\rangle\right] \\
&= -\frac{\lambda}{2}\,.
\end{aligned} \tag{7.A.12}$$

8 Dissipative Two-State Systems

8.1 Spin-Boson Model

In Chap. 1, we discussed the dynamics of a single quantum spin in a dissipative environment using the system-bath-approach (SBA). Although the bath was described by a Hamiltonian, no explicit modeling of the latter was considered. In fact, the bath properties were incorporated in a phenomenological manner through parameters such as relaxation rates. It is therefore of interest, in our search for simple models of quantum dissipative systems, to extend the discussion of Chap. 1 to an explicit quantum treatment of the bath Hamiltonian. This is achieved by introducing the *spin-boson Hamiltonian* [426]:

$$\mathcal{H} = \mathcal{H}_s + V + \mathcal{H}_B\,, \tag{8.1}$$

$$\mathcal{H}_s = -H\sigma_z\,, \tag{8.2}$$

$$V = \sigma_x \sum_{\boldsymbol{k}} g_{\boldsymbol{k}}(b_{\boldsymbol{k}} + b_{\boldsymbol{k}}^+)\,, \tag{8.3}$$

$$\mathcal{H}_B = \sum_{\boldsymbol{k}} \omega_{\boldsymbol{k}} b_{\boldsymbol{k}}^+ b_{\boldsymbol{k}}\,. \tag{8.4}$$

The model described by (8.1)-(8.4) has several physical realizations as detailed below. (For notational convenience, we will work in units with $\hbar = 1$ in this and subsequent chapters.)

8.1.1 Spin-Lattice Relaxation in Solids

In this context, the subsystem Hamiltonian $\mathcal{H}_s$ corresponds to a Zeeman coupling between an external magnetic field of strength H applied along the z-axis. The term V (written in second-quantized notation [22]) describes spin-phonon interactions via a coupling constant $g_{\boldsymbol{k}}$ for the $\boldsymbol{k}^{\text{th}}$ phonon mode. Finally, the term $\mathcal{H}_B$ corresponds to a phonon bath with a prescribed phonon dispersion governed by the $\boldsymbol{k}$-dependence of $\omega_{\boldsymbol{k}}$. The external field fixes the quantization axis (z) with respect to which $\mathcal{H}_s$ is represented by a 2×2 diagonal matrix. The interaction term, being proportional to σ_x, is off-diagonal in this representation, and causes quantum transitions between the energy levels of $\mathcal{H}_s$ (see Chap. 7). The strength of these transitions is proportional to a

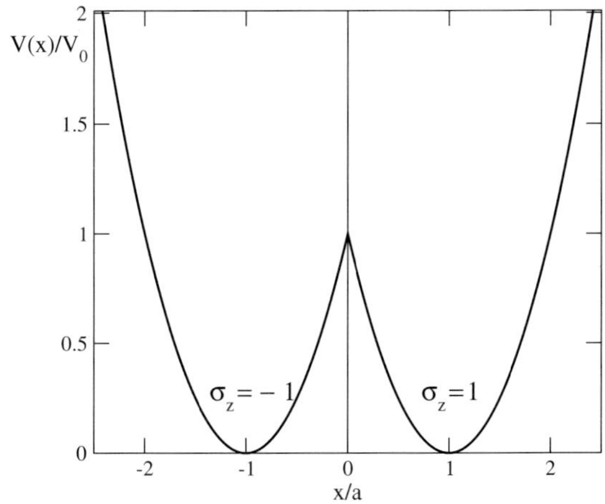

Fig. 8.1. Schematic of symmetric double-well potential with minima at $\pm\, a$, e.g., $V(x) = k(|x| - a)^2/2$, where k is a parameter. The barrier height is $V_0 = ka^2/2$

phonon displacement field, which is then modulated by the bath Hamiltonian $\mathcal{H}_B$.

Since $\mathcal{H}_s$ and $\mathcal{H}_B$ operate in two disjoint Hilbert spaces, one could study the problem in the *interaction picture* of $(\mathcal{H}_s + \mathcal{H}_B)$, where the phonon displacement field acquires a time-varying character. From this perspective, the coupling term V generalizes (1.74) in which the phonons were treated through a single stochastic field $\hat{b}$.

The model described by (8.1)-(8.4) is widely used for discussing spin-lattice relaxation processes in solids, which constitute the basic mechanism for magnetization of a system under the influence of an external magnetic field. The study of spin-relaxation dynamics is of great importance in the context of *nuclear magnetic resonance* and *electron paramagnetic resonance* [376].

8.1.2 Dissipative Tunneling in a Symmetric Double Well

Consider a quantum-mechanical particle (e.g., electron, positron, proton, muon) constrained to move in a symmetric double-well potential (see Fig. 8.1). This physical situation is often encountered in ultra-low-temperature defect physics in the solid state [427]. A similar situation also arises in molecular rotations in solids [428], or in the ammonia-maser problem where the nitrogen atom tunnels between two equivalent positions [429]. The two lowest energy states of this double-well potential are symmetric and anti-symmetric about the origin, respectively. These are referred to as the *tunneling states* and their energy difference is known as the *tunnel splitting* [22]. (We will discuss these states further in Chap. 10.)

We denote the states associated with the two minima of the double-well potential, which are the two *physical* locations of the tunneling particle, by the eigenstates of the Pauli operator σ_z. In that case, the tunneling Hamiltonian can be written as

$$\mathcal{H}_s = \Delta\sigma_x\,, \tag{8.5}$$

where Δ is the tunnel splitting. Thus, when the eigenvalue of σ_z is $+1$ (associated with the state $|+\rangle$), the particle is taken to be localized in the right well. Similarly, when the eigenvalue is -1 (associated with the state $|-\rangle$), the particle is localized in the left well. The Hamiltonian $\mathcal{H}_s$ is clearly off-diagonal in this chosen representation. This implies that, if the initial preparation is such that the particle is in the right well, the *stay-put* probability $P(t)$ that the particle will continue to stay in the right well is an oscillatory function of time:

$$P(t) = \cos(\Delta t)\,. \tag{8.6}$$

The tunneling particle is then a *quantum clock* with frequency Δ, as the particle makes transitions between the two wells.

On the other hand, when the tunneling is influenced by quantum dissipation due to coupling with a heat bath, the system can be described by the following Hamiltonian:

$$\mathcal{H} = \Delta\sigma_x + \sigma_z \sum_{\boldsymbol{k}} g_{\boldsymbol{k}} \left(b_{\boldsymbol{k}} + b_{\boldsymbol{k}}^{+}\right) + \sum_{\boldsymbol{k}} \omega_{\boldsymbol{k}} b_{\boldsymbol{k}}^{+} b_{\boldsymbol{k}}\,. \tag{8.7}$$

The Hamiltonian in (8.7) is equivalent to that in (8.1), and is obtained from the latter by a counter-clockwise $\pi/2$-rotation of the coordinate system about the y-axis.

This Hamiltonian describes a hydrogen atom, proton or a positive muon that is trapped amongst two equivalent interstitial sites in a solid. The interstitial particle distorts the site it occupies and carries the distortion field, described by phonons, as it tunnels between the two equivalent sites. This is the picture adopted in the *polaron hopping* theory [430]. Clearly, the nature of the phonons involved in this process, acoustic or optic, will be dictated by the chosen form of the spectral density of phonons. In a different context, we may also consider tunneling centers in metals at low temperatures. Here, the distortion fields are not phonons but are caused by the conduction electrons, due to Coulomb interactions. If we describe electron-hole excitations from the Fermi surface of a metal as bosons, the Hamiltonian in (8.7) is applicable again, though with a different form of the spectral density of bosonic excitations [431–434].

In yet another realization of the Hamiltonian in (8.1), consider the Kondo problem in a metal. In this case, the eigenstates of σ_z now represent the *actual* spin states of the impurity (rather than the pseudospin states of an interstitial particle), whereas the bosonic field again represents the electron-hole pairs of the host metal [435].

The Hamiltonian in (8.1) or (8.7) has been referred to by Leggett et al. [426] as the spin-boson Hamiltonian in the context of quantum tunneling in SQUID junctions. These authors have made extensive use of functional-integral techniques to study the static and dynamic properties of the spin-boson Hamiltonian, which have been reviewed by Weiss [20]. They find that dissipation can transform the probability function $P(t)$ from an oscillatory form to an exponential one – thereby changing the tunneling behavior from *coherent* to *incoherent*. An extreme form of incoherent tunneling occurs when the particle does not tunnel at all, i.e., it becomes *localized* in one of the wells, corresponding to a *broken-symmetry* transition [436,437]. This transition is analogous to the so-called *watched-pot effect* [438,439], where the two minima of the double well depict the *boiling* and *non-boiling* states of the pot – localization means the pot does not boil at all if it is continually watched.

In our subsequent discussion, we shall be interested in the calculation of two physically relevant quantities. These are as follows:
(a) *the survival probability*

$$P(t) = \frac{1}{2}[1 + \langle\sigma_z(t)\rangle] , \tag{8.8}$$

with the initial condition

$$\langle\sigma_z(0)\rangle = 1 . \tag{8.9}$$

(b) *the symmetrized equilibrium auto-correlation function*

$$C(t) = \frac{1}{2}\left[\langle\sigma_z(0)\sigma_z(t)\rangle_0 + \langle\sigma_z(t)\sigma_z(0)\rangle_0\right] , \tag{8.10}$$

with

$$C(0) = 1 . \tag{8.11}$$

The quantity $P(t)$ is directly measurable by the technique of muon-spin rotation, whereas $C(t)$ is related to the cross-section of neutron scattering from hydrogen in metals [1].

8.2 Dilute Bounce Gas Approximation (DBGA)

As mentioned earlier, the spin-boson Hamiltonian in (8.7), and its equivalent in (8.1)-(8.4), has been extensively studied by Leggett et al. [426] using functional-integral methods of quantum mechanics [440,441]. A major advantage of this approach is that it is nonperturbative. Therefore, strong-coupling cases, which are relevant for the Kondo-type problems mentioned above, can be effectively dealt with. Thus, the approach of Leggett et al. has provided a new direction to quantum dissipative phenomena – distinct from most quantum-optics applications, where the weak-coupling approximation is valid [33]. In the weak-coupling limit, there has been extensive usage of the master equation approach discussed in Appendix 1.A. However, we show below how the master equation method can be adapted to handle strong-coupling cases as well.

In the functional-integral approach, a useful approximation is the so-called *dilute bounce gas approximation* (DBGA) [426,442]. In order to appreciate the physical meaning of the DBGA, it is useful to consider the example of dissipative tunneling in a double well as in Sect. 8.1.2. If the tunneling particle is initially localized in the right well, it transits to the left well and then bounces back again into the right well, after a time Δ^{-1}. This is a coherent clock-like motion that occurs when there is no heat-bath induced dissipation, i.e., the coupling term proportional to $g_{\boldsymbol{k}}$ in (8.7) is absent. Therefore, one return journey to the left well (i.e., a *sojourn*) is like a bounce. The bounces are rare or dilute if Δ is small compared to bath-induced relaxation rates. This condition will be properly quantified below.

The first step in our study of the strong-coupling case is a unitary transformation of the Hamiltonian in (8.7) [443]. This transformation is defined by the operator

$$\mathcal{S} = \exp\left[-\sum_{\boldsymbol{q}} \frac{g_{\boldsymbol{q}}}{2\omega_{\boldsymbol{q}}}\left(b_{\boldsymbol{q}} - b_{\boldsymbol{q}}^{+}\right)\sigma_z\right] . \tag{8.12}$$

This transforms the Hamiltonian $\mathcal{H}$ to $\tilde{\mathcal{H}}$, where

$$\begin{aligned}\tilde{\mathcal{H}} &= \mathcal{S}\mathcal{H}\mathcal{S}^{-1} \\ &= \frac{1}{2}\Delta(\sigma^{+}B_{-} + \sigma^{-}B_{+}) + \sum_{\boldsymbol{k}} \omega_{\boldsymbol{k}} b_{\boldsymbol{k}}^{+} b_{\boldsymbol{k}} ,\end{aligned} \tag{8.13}$$

with

$$B_{\pm} = \exp\left[\pm\sum_{\boldsymbol{q}} \frac{g_{\boldsymbol{q}}}{2\omega_{\boldsymbol{q}}}\left(b_{\boldsymbol{q}} - b_{\boldsymbol{q}}^{+}\right)\right] . \tag{8.14}$$

It may be noted that the transformation in (8.12) also generates a counter-term, which has been ignored as it has no consequence for the dynamics. In the spirit of the SBA, we have on comparison of $\tilde{\mathcal{H}}$ with (8.1)-(8.4),

$$\begin{aligned}\mathcal{H}_s &= 0 , \\ V &= \frac{1}{2}\Delta(\sigma^{+}B_{-} + \sigma^{-}B_{+}) , \\ \mathcal{H}_B &= \sum_{\boldsymbol{k}} \omega_{\boldsymbol{k}} b_{\boldsymbol{k}}^{+} b_{\boldsymbol{k}} .\end{aligned} \tag{8.15}$$

In accordance with the decomposition in (1.A.29), we may identify

$$\begin{aligned} S_1 &= \sigma^+ , \\ S_2 &= \sigma^- , \\ B_1 &= \frac{\Delta}{2} B_- , \\ B_2 &= \frac{\Delta}{2} B_+ . \end{aligned} \tag{8.16}$$

From Appendix 1.A, recall that the interaction term V is treated in the Born–Markov approximation in the master equation approach. A similar perturbative treatment in the context of (8.13) implies that certain types of terms upto *all* orders in the coupling constant $g_{\boldsymbol{q}}$ are incorporated in the analysis. This is borne out by using (8.16) to obtain an equation for $\langle \sigma_z(t) \rangle$, defined as

$$\langle \sigma_z(t) \rangle = \mathrm{Tr}\left[\rho_s(t) \sigma_z \right] . \tag{8.17}$$

The resultant equation is [444]

$$\frac{d}{dt} \langle \sigma_z(t) \rangle = -\Delta^2 \int_0^t d\tau \left[\Phi(\tau) + \Phi(-\tau) \right] \langle \sigma_z(t-\tau) \rangle , \tag{8.18}$$

where $\Phi(\tau)$ is the correlation function involving bath operators:

$$\Phi(\tau) = \langle B_-(0) B_+(\tau) \rangle_{\mathrm{av}} = \langle B_+(0) B_-(t) \rangle_{\mathrm{av}} . \tag{8.19}$$

In obtaining (8.18), we have used the following properties of $B_\pm$ as defined in (8.14):

$$\begin{aligned} \langle B_\pm \rangle_{\mathrm{av}} &= 0 , \\ \langle B_\pm(t) B_\pm(0) \rangle_{\mathrm{av}} &= 0 . \end{aligned} \tag{8.20}$$

It has been shown by Aslangul et al. [444] that the solution of (8.18) is consistent with the solution obtained from the path-integral approach within the DBGA.

Before concluding this section, it is relevant to obtain (for the sake of completeness) the explicit structure of the correlation function $\Phi(\tau)$ in the context of the *Ohmic dissipation model*, first discussed in Sect. 1.3.1. The Ohmic model ascribes a specific frequency-dependence to the spectral density $J(\omega)$, defined as

$$J(\omega) = 2 \sum_{\boldsymbol{q}} g_{\boldsymbol{q}}^2 \delta(\omega - \omega_{\boldsymbol{q}}) . \tag{8.21}$$

In general, $J(\omega)$ can be written as

$$J(\omega) = K \omega^s e^{-\omega/\omega_c} , \tag{8.22}$$

where K is a damping parameter (or a relaxation rate) which measures the strength of the coupling with the heat bath; ω_c is a suitably chosen cut-off frequency; and the index s characterizes the spectral nature of the bath modes. The Ohmic case corresponds to

$$s = 1\,. \tag{8.23}$$

Using (8.14), the correlation function $\Phi(\tau)$ can be derived from (8.19) as [445]

$$\Phi(\tau) = \exp\left\{-\sum_{\boldsymbol{q}} \frac{4g_{\boldsymbol{q}}^2}{\omega_{\boldsymbol{q}}^2}\left[\coth\left(\frac{\beta\omega_{\boldsymbol{q}}}{2}\right)[1-\cos(\omega_{\boldsymbol{q}}\tau)] + i\sin(\omega_{\boldsymbol{q}}\tau)\right]\right\}. \tag{8.24}$$

The summation over $\boldsymbol{q}$ can be converted into an integral over a continuum of modes using (8.21), thus yielding

$$\Phi(\tau) = \exp\left\{-2\int_0^\infty d\omega \frac{J(\omega)}{\omega^2}\left[\coth\left(\frac{\beta\omega}{2}\right)[1-\cos(\omega\tau)] + i\sin(\omega\tau)\right]\right\}. \tag{8.25}$$

Using the Ohmic form of the spectral density (cf. (8.22)-(8.23)), we obtain

$$\Phi(\tau) = \exp\left\{-2K\int_0^\infty \frac{d\omega}{\omega} e^{-\omega/\omega_c}\left[\coth\left(\frac{\beta\omega}{2}\right)[1-\cos(\omega\tau)] + i\sin(\omega\tau)\right]\right\}. \tag{8.26}$$

For $\beta\omega_c \gg 1$, this can be reduced to the closed form

$$\Phi(\tau) = \exp[i\pi K \mathrm{sgn}(\tau)]\left[\frac{\pi}{\beta\omega_c \sinh(\pi|\tau|/\beta)}\right]^{2K}. \tag{8.27}$$

8.3 Beyond the DBGA

It is clear from (8.18) that the DBGA result, though valid in the strong-coupling regime, is actually of second order in the tunneling frequency Δ. Furthermore, because the DBGA is based on the unitary transformation in (8.12), the systematic part of the Hamiltonian $\mathcal{H}_s$ does not contain the tunneling term at all. However, in the SBA, the asymptotic (i.e., $t \to \infty$) properties are governed by $\mathcal{H}_s$ alone, as the system is expected to be equilibrated at the bath temperature. Thus, it is an inherent defect of the DBGA that correct thermal equilibrium results, which must contain the tunneling term Δ, are not recovered from the time-dependent solutions in the appropriate asymptotic limit [445]. One way of circumventing this problem, within the path-integral formulation, is to invoke interacting bounces as has been demonstrated using quantum Monte Carlo techniques by Weiss and Wollensak [446].

We follow here a simpler method of improving the DBGA by adding and subtracting the free-tunneling term in (8.13) [447], i.e.,

$$\tilde{\mathcal{H}} = \frac{\Delta}{2}(\sigma^+ + \sigma^-) + \frac{1}{2}\Delta[\sigma^+(B_- - 1) + \sigma^-(B_+ - 1)] + \sum_{\boldsymbol{k}} \omega_{\boldsymbol{k}} b_{\boldsymbol{k}}^+ b_{\boldsymbol{k}} . \tag{8.28}$$

The idea behind the above decomposition is that, in any perturbative treatment of the second term in (8.28), the first term (i.e., the free-tunneling term) is dealt with exactly. Therefore, in accordance with the separation indicated in (8.1), $\mathcal{H}_s$ is now given by

$$\mathcal{H}_s = \frac{\Delta}{2}(\sigma^+ + \sigma^-) . \tag{8.29}$$

In this case, the master equation (1.A.39) leads to the following (closed) equations of motion [448]:

$$\frac{d}{dt}\langle\sigma_z(t)\rangle = -\int_0^t d\tau \left[K_{11}(\tau)\langle\sigma_z(t-\tau)\rangle + K_{13}(\tau)\langle\sigma^+(t-\tau) - \sigma^-(t-\tau)\rangle\right] , \tag{8.30}$$

$$\begin{aligned} \frac{d}{dt}\left(\langle\sigma^+(t)\rangle - \langle\sigma^-(t)\rangle\right) = -\int_0^t d\tau \, [4K_{13}(\tau)\langle\sigma_z(t-\tau)\rangle \\ +K_{33}(\tau)\langle\sigma^+(t-\tau) - \sigma^-(t-\tau)\rangle] , \end{aligned} \tag{8.31}$$

where

$$K_{11}(t) = \frac{\Delta^2}{2}[1 + \cos(2\Delta t)][\Phi(-t) + \Phi(t)] , \tag{8.32}$$

$$K_{13}(t) = i\frac{\Delta^2}{2} \sin(2\Delta t)[\Phi(-t) + \Phi(t)] , \tag{8.33}$$

$$K_{33}(t) = \Delta^2 \cos(2\Delta t)[\Phi(-t) + \Phi(t)] . \tag{8.34}$$

It is interesting to note that the DBGA equation (8.18) is recovered from (8.30)-(8.34) on setting $\cos(2\Delta t) = 1$ and $\sin(2\Delta t) = 0$. This observation provides a physical interpretation to the DBGA, viz., it is valid over time-scales much shorter than the inverse tunneling frequency. It is therefore not surprising that the DBGA is tantamount to assuming that the tunneling frequency is effectively small.

It is instructive to compare the results of the DBGA and extensions thereof by computing the *survival probability* $P(t)$, defined in (8.8). In the context of the double-well problem, the condition (8.9) implies that the tunneling particle is initially localized in the right well. The *leakage* in $P(t)$ is

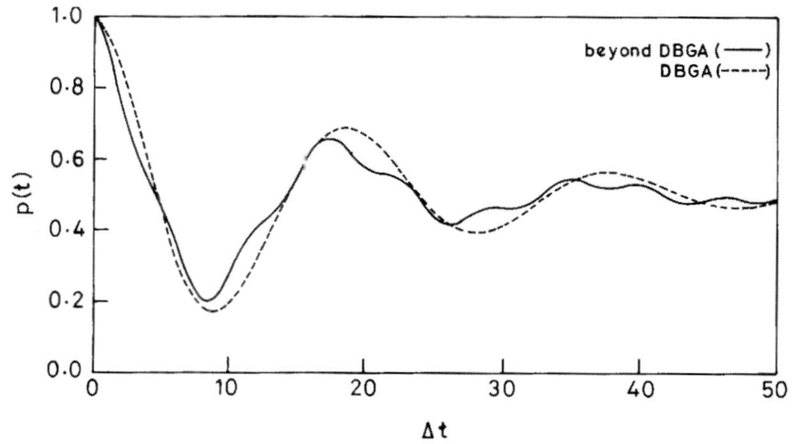

Fig. 8.2. Survival probability $P(t)$ vs. Δt, obtained using the DBGA method (denoted by a dashed line) and the extended-DBGA method (denoted by a solid line)

then a measure of transport across the barrier. The results obtained from a numerical solution of (8.18) and (8.30)-(8.31) are shown in Fig. 8.2. It can be seen from the figure that there are many more coherent oscillations in the survival probability as a function of time in the extended-DBGA result from (8.30)-(8.31).

8.4 Dissipative Tunneling in an Asymmetric Double Well

For practical applications of dissipative tunneling in a double-well potential, it is important to add an asymmetry term to the Hamiltonian in (8.7), which is then modified to

$$\mathcal{H} = \epsilon\sigma_z + \Delta\sigma_x + \sigma_z \sum_{\boldsymbol{k}} g_{\boldsymbol{k}}(b_{\boldsymbol{k}} + b_{\boldsymbol{k}}^+) + \sum_{\boldsymbol{k}} \omega_{\boldsymbol{k}} b_{\boldsymbol{k}}^+ b_{\boldsymbol{k}} \,. \tag{8.35}$$

The presence of the first term implies that the left and right wells have different energies. In the context of spin-lattice relaxation in solids (discussed in Sect. 8.1.1), (8.35) can be interpreted to define the Hamiltonian dynamics of a spin-1/2 particle, subjected to magnetic fields along the z- and x-axis of strengths ϵ and Δ, respectively. Furthermore, the spin is assumed to be in contact with a phononic heat bath, described in terms of the boson operators $b_{\boldsymbol{k}}$ and $b_{\boldsymbol{k}}^+$.

We first study this Hamiltonian in the DBGA, as discussed in Sect. 8.2. For this, we perform the unitary transformation defined in (8.12), which reduces the Hamiltonian in (8.35) to

$$\tilde{\mathcal{H}} = \epsilon\sigma_z + \frac{1}{2}\Delta(\sigma^+ B_- + \sigma^- B_+) + \sum_{\boldsymbol{k}} \omega_{\boldsymbol{k}} b_{\boldsymbol{k}}^+ b_{\boldsymbol{k}} \,. \tag{8.36}$$

Note that the asymmetry term remains unaffected by the unitary transformation as $\mathcal{S}$ commutes with σ_z. The SBA decomposition then implies that

$$\mathcal{H}_s = \epsilon\sigma_z \,, \tag{8.37}$$

whereas the interaction term V and the bath term $\mathcal{H}_B$ remain the same as before, and are given by (8.15).

In what follows, we shall calculate the correlation function $C(t)$ defined in (8.10), rather than the survival probability $P(t)$, computed in Sect. 8.3. The motivation for this comes from neutron-scattering studies of quantum diffusion of hydrogen in metals [449,450]. In that context, the structure factor for scattering can be expressed directly in terms of $C(t)$. In addition, since scattering data are usually analyzed in the frequency domain (as indeed is the case in most spectroscopy experiments) it will be relevant to consider the Laplace transform of $C(t)$:

$$\tilde{C}(s) = \int_0^\infty dt e^{-st} C(t) \,, \tag{8.38}$$

where

$$\mathrm{Re}(s) > 0 \,. \tag{8.39}$$

Note that the evaluation of $C(t)$ requires the calculation of averages such as (cf. (7.86) with $\hbar = 1$)

$$\langle \sigma_z(0)\sigma_z(t)\rangle_0 = \mathrm{Tr}\left[\rho_{\mathrm{eq}}\sigma_z(0)e^{i\mathcal{H}t}\sigma_z(0)e^{-i\mathcal{H}t}\right] \,, \tag{8.40}$$

where ρ_{eq}, the overall density matrix in equilibrium, is given by

$$\rho_{\mathrm{eq}} = \frac{1}{Z}\exp(-\beta\mathcal{H}) \,, \tag{8.41}$$

Z being the overall partition function. As σ_z commutes with the unitary operator $\mathcal{S}$, the correlation function in (8.40) can be rewritten as

$$\langle \sigma_z(0)\sigma_z(t)\rangle_0 = \mathrm{Tr}\left[\tilde{\rho}_{\mathrm{eq}}\sigma_z(0)e^{i\mathcal{H}t}\sigma_z(0)e^{-i\mathcal{H}t}\right] \,, \tag{8.42}$$

upon using the cyclic property of the trace. In (8.42), $\tilde{\rho}_{\mathrm{eq}}$ is obtained from (8.41) by replacing $\mathcal{H}$ with $\tilde{\mathcal{H}}$.

As mentioned earlier, the interaction term V is treated perturbatively in the DBGA. Thus, the term V can be ignored in $\tilde{\rho}_{\mathrm{eq}}$, yielding a *factorized* equilibrium density matrix:

$$\tilde{\rho}_{\mathrm{eq}} \simeq \frac{1}{2}\left[1 - \tanh(\beta\epsilon)\sigma_z\right]\rho_B \,, \tag{8.43}$$

where

$$\rho_B = \frac{1}{Z_B} \exp\left(-\beta \sum_{\boldsymbol{k}} \omega_{\boldsymbol{k}} b_{\boldsymbol{k}}^{+} b_{\boldsymbol{k}}\right), \tag{8.44}$$

Z_B being the partition function associated with the bath Hamiltonian $\mathcal{H}_B$. This allows us to decompose $C(t)$ as [445]

$$C(t) = \frac{1}{2} \sum_{\mu\mu'} (\mu + \langle\sigma_z\rangle)\mu' \, (\mu\mu|[U(t)]_{\text{av}}|\mu'\mu') , \tag{8.45}$$

where $\mu = \pm 1$ refer to the two eigenvalues of σ_z. Here, the quantity $\langle\sigma_z\rangle$ is the equilibrium expectation value given by

$$\langle\sigma_z\rangle = -\tanh(\beta\epsilon) , \tag{8.46}$$

and $U(t)$ is the *time-development* operator, defined by

$$U(t) = \exp(i\tilde{L}t) , \tag{8.47}$$

$\tilde{L}$ being the Liouvillian associated with $\tilde{\mathcal{H}}$ [1]. (Clearly, (8.45) is incomplete as it neglects the tunneling term Δ, for reasons discussed in Sect. 8.3.) Recall that $[U(t)]_{\text{av}}$ is a *partial* average of the time-development operator over the bath states alone (cf. (7.12)), given by

$$[U(t)]_{\text{av}} = \sum_{n_b n_b'} (n_b n_b|U(t)|n_b' n_b')\langle n_b'|\rho_B|n_b'\rangle , \tag{8.48}$$

$|n_b\rangle, |n_b'\rangle$ being the eigenstates of the bath Hamiltonian $\mathcal{H}_B$ (cf. (1.A.16)).

As mentioned earlier, we are interested in calculating the Laplace transform of $[U(t)]_{\text{av}}$. Towards this objective, it is convenient to employ the resolvent expansion of Appendix 1.A.2 and Sect. 7.1. From (7.19), we recall

$$\left[\tilde{U}(s)\right]_{\text{av}} = \frac{1}{s - iL_s + \tilde{\Sigma}(s)} , \tag{8.49}$$

where the self-energy $\tilde{\Sigma}(s)$ is approximated by (7.20). The result in (8.49) applies for interactions V which vanish when averaged over the bath density matrix, as is indeed the case for the interaction in (8.36) (see (8.20)).

As in Sect. 7.6, the super-operator $[\tilde{U}(s)]_{\text{av}}$ has a 4×4 matrix representation for the two-level system. In order to obtain the elements of this matrix, our first task is to tabulate the matrix of the resolvent in (8.49) and then invert it. The enumeration of the 4 elements of L_s is simple, as it involves the operator $\mathcal{H}_s$ that is diagonal in the basis states $|\mu\rangle$, etc. (see Appendix 7.A). The next step is to obtain the matrix elements of the self-energy in (8.49). Some algebra yields [445]

$$\tilde{\Sigma}(s) \simeq \Delta^2 \begin{pmatrix} a_{11} & a_{12} & 0 & 0 \\ a_{21} & a_{22} & 0 & 0 \\ 0 & 0 & a_{33} & a_{34} \\ 0 & 0 & a_{43} & a_{44} \end{pmatrix} . \tag{8.50}$$

Here, the rows and columns are labeled by $++$, $--$, $+-$ and $-+$, respectively, and

$$\begin{aligned} a_{11} &= \tilde{\Phi}_{-+}(s_+) + \tilde{\Phi}_{-+}(s_-) , \\ a_{12} &= -\tilde{\Phi}_{-+}(s_+) - \tilde{\Phi}'_{-+}(s_-) , \\ a_{21} &= -\tilde{\Phi}_{+-}(s_-) - \tilde{\Phi}'_{+-}(s_+) , \\ a_{22} &= \tilde{\Phi}_{+-}(s_-) + \tilde{\Phi}'_{+-}(s_+) , \\ a_{33} &= \tilde{\Phi}_{-+}(s) + \tilde{\Phi}'_{+-}(s) , \\ a_{34} &= -\tilde{\Phi}_{--}(s) - \tilde{\Phi}'_{--}(s) , \\ a_{43} &= -\tilde{\Phi}_{++}(s) - \tilde{\Phi}'_{++}(s) , \\ a_{44} &= \tilde{\Phi}_{+-}(s) + \tilde{\Phi}'_{-+}(s) , \\ s_\pm &= s \pm 2i\epsilon , \end{aligned} \tag{8.51}$$

with

$$\begin{aligned} \Phi_{+-}(t) &= \langle B_+(0) B_-(t) \rangle_{\text{av}} , \\ \Phi_{-+}(t) &= \langle B_-(0) B_+(t) \rangle_{\text{av}} , \\ \Phi_{++}(t) &= \langle B_+(0) B_+(t) \rangle_{\text{av}} , \\ \Phi_{--}(t) &= \langle B_-(0) B_-(t) \rangle_{\text{av}} . \end{aligned} \tag{8.52}$$

Furthermore, all primed quantities in (8.51) are obtained by replacing the argument t by $-t$ in (8.52). It may be stressed again that the angular brackets in (8.52) denote thermal averages governed by ρ_B, and the time-development of $B_\pm(t)$ is dictated by $\mathcal{H}_B$ alone.

The correlation functions in (8.52) have been listed in Sect. 8.2 for the Ohmic dissipation model. Recall that $\Phi_{++}(t) = \Phi_{--}(t) = 0$, whereas $\Phi_{+-}(t) = \Phi_{-+}(t) = \Phi(t)$, given in (8.27). Since we require the Laplace transform of $\Phi(t)$, we obtain it explicitly from (8.27) as

$$\tilde{\Phi}(s) = \frac{e^{i\pi K}}{\omega_c} \left(\frac{2\pi}{\beta\omega_c} \right)^{2K-1} \frac{\Gamma(1-2K)\Gamma(K + s\beta/2\pi)}{\Gamma(1-K+s\beta/2\pi)} , \tag{8.53}$$

where $\Gamma(x)$ is the gamma function of argument x. Similarly, $\tilde{\Phi}'(s)$ is obtained by replacing i with $-i$ in (8.53) above. Substituting (8.53) in (8.51), we obtain the matrix in (8.50). This yields the matrix elements of $[\tilde{U}(s)]_{\text{av}}$, and thereby the following form for $\tilde{C}(s)$ from (8.45):

$$\tilde{C}(s) = \frac{s - i\langle\sigma_z\rangle \sin(\pi K)[\tilde{J}(s+2i\epsilon) - \tilde{J}(s-2i\epsilon)]}{s\{s + \cos(\pi K)[\tilde{J}(s+2i\epsilon) + \tilde{J}(s-2i\epsilon)]\}} , \tag{8.54}$$

where

$$\tilde{J}(s) = \frac{2\Delta^2}{\omega_c} \left(\frac{2\pi}{\beta\omega_c} \right)^{2K-1} \frac{\Gamma(1-2K)\Gamma(K+s\beta/2\pi)}{\Gamma(1-K+s\beta/2\pi)} . \tag{8.55}$$

For further discussion and clarification of the limiting cases of (8.54), it is useful to introduce an effective tunneling frequency as

$$\tilde{\Delta} = [\cos(\pi K)\Gamma(1-2K)]^{1/(2-2K)} \Delta \left(\frac{2\Delta}{\omega_c} \right)^{K/(1-K)} , \tag{8.56}$$

so that

$$\tilde{J}(s) = \tilde{\Delta} \left(\frac{\beta\tilde{\Delta}}{\pi} \right)^{1-2K} \frac{\Gamma(K+s\beta/2\pi)}{\cos(\pi K)\Gamma(1-K+s\beta/2\pi)} . \tag{8.57}$$

First, we discuss the symmetric case ($\epsilon = 0$) of Sect. 8.2 for which (8.54) simplifies to

$$\tilde{C}(s) = [s + 2\cos(\pi K)\tilde{J}(s)]^{-1} . \tag{8.58}$$

This is identical to the DBGA result for dissipative tunneling in a symmetric double well [426].

Furthermore, in the zero-damping limit ($K = 0$) we have

$$\tilde{J}(s) = \frac{2\tilde{\Delta}^2}{s} , \quad \tilde{\Delta} = \Delta , \tag{8.59}$$

and (8.54) yields

$$\tilde{C}(s) = \frac{s^2 + 4\epsilon^2}{s(s^2 + 4\epsilon^2 + 4\Delta^2)} . \tag{8.60}$$

Finally, in order to verify the asymptotic form for the correlation function, we consider

$$\lim_{s\to 0} s\tilde{C}(s) = -\langle\sigma_z\rangle \tanh(\beta\epsilon) . \tag{8.61}$$

Now, an independent argument based on the idea of *mixing* yields

$$\lim_{s\to 0} s\tilde{C}(s) = \lim_{t\to\infty} C(t) = \langle\sigma_z\rangle^2 . \tag{8.62}$$

Therefore, internal consistency demands that the only admissible form of $\langle\sigma_z\rangle$ is the one given by (8.46). This again emphasizes the limitations of the DBGA, in that the tunneling frequency Δ is absent from the asymptotic result.

8.5 Asymmetric Double Well: The Weak-Coupling Limit

In Sects. 8.2-8.4, we have clarified the physical meaning and the limitations of the DBGA in the path-integral formulation of dissipative tunneling. We have emphasized that the DBGA is valid for arbitrary damping, because the underlying unitary transformation (8.12) incorporates the original coupling constants $g_{\boldsymbol{q}}$ in the exponent of the perturbation term (cf. (8.14)). However, the tunneling term is effectively treated only upto second order and *inter-bounce interactions* are neglected.

It is physically plausible that the DBGA is a good approximation in the limit of either strong coupling (damping) or high temperatures, where coherent tunneling motion is impeded and the tunneling frequency is effectively small. However, it turns out that the DBGA is a systematic approximation at *all* temperatures even for weak coupling, provided that the asymmetry parameter $\epsilon = 0$. In that case, the neglected inter-bounce interactions vanish as K^2 in the limit $K \to 0$ [446]. On the other hand, for an asymmetric system ($\epsilon \neq 0$) and in the limit of weak coupling, the DBGA breaks down at low temperatures, as the inter-bounce interactions are no longer negligible. Additionally, the DBGA yields an incorrect asymptotic expression for both $P(t)$ and $C(t)$, as discussed earlier.

Against this background, it is important to consider an alternate treatment of the weak-coupling limit, distinct from the DBGA – especially as the damping parameter K for interstitial tunneling in metals is usually *small* [451]. This treatment should take cognizance of the fact that, if the coupling is weak, the two terms in $\mathcal{H}_s$ proportional to ϵ and Δ (cf. (8.35)) should be treated on the same footing. Hence, it is appropriate to develop a perturbation theory using the original Hamiltonian in (8.35), and not the transformed one in (8.36), thereby treating $\mathcal{H}_s$ exactly. To this end, it is useful to diagonalize $\mathcal{H}_s$ by performing a rotation in *spin-space* about the y-axis by an angle $\theta = \tan^{-1}(\Delta/\epsilon)$. In the rotated frame, the Hamiltonian in (8.35) becomes

$$\mathcal{H}' = \Omega\sigma_z + \frac{1}{\Omega}(\epsilon\sigma_z + \Delta\sigma_x)\sum_{\boldsymbol{k}} g_{\boldsymbol{k}}(b_{\boldsymbol{k}}^+ + b_{\boldsymbol{k}}) + \sum_{\boldsymbol{k}} \omega_{\boldsymbol{k}} b_{\boldsymbol{k}}^+ b_{\boldsymbol{k}}, \tag{8.63}$$

where

$$\Omega = (\epsilon^2 + \Delta^2)^{1/2}\,. \tag{8.64}$$

Correspondingly,

$$\langle\sigma_z(0)\sigma_z(t)\rangle_0 = \frac{1}{\Omega^2}\mathrm{Tr}\left\{\rho'_{\mathrm{eq}}\left[\epsilon\sigma'_z(0) + \Delta\sigma'_x(0)\right]\left[\epsilon\sigma'_z(t) + \Delta\sigma'_x(t)\right]\right\}\,, \tag{8.65}$$

where ρ'_{eq} is obtained from (8.41) by replacing $\mathcal{H}$ with $\mathcal{H}'$, and

$$\sigma'_z(t) = e^{i\mathcal{H}'t}\sigma_z(0)e^{-i\mathcal{H}'t}\,. \tag{8.66}$$

A similar expression holds for $\sigma'_x(t)$.

Equation (8.63) has the generic SBA form. As is customary, we then make a factorization approximation on the density matrix as

$$\rho'_{\rm eq} \simeq \rho_s \otimes \rho_B \,, \tag{8.67}$$

where

$$\rho_s = \frac{1}{2}\left[1 - \tanh(\beta\Omega)\sigma_z\right] , \tag{8.68}$$

and ρ_B is given by (8.44). Using (8.65)-(8.68), the Laplace transform of the correlation function can be written from (8.10) as

$$\tilde{C}(s) = \frac{1}{\Omega^2}\mathrm{Tr}\left\{\rho_B[\epsilon\sigma_z(0) + \Delta\sigma_x(0) - \epsilon\tanh(\beta\Omega)][\tilde{U}(s)(\epsilon\sigma_z(0) + \Delta\sigma_x(0))]\right\} . \tag{8.69}$$

Here, $\tilde{U}(s)$ is the Laplace transform of the time-development operator defined in (8.47), appropriate to the Hamiltonian $\mathcal{H}'$. Writing out the trace explicitly using the spin states (denoted by $|\mu\rangle$), (8.69) reduces to

$$\begin{aligned}\tilde{C}(s) = &\sum_{\mu\nu\mu'\nu'} \langle\mu|\left[\epsilon\sigma_z(0) + \Delta\sigma_x(0) - \epsilon\tanh(\beta\Omega)\right]|\nu\rangle \\ &\times \langle\nu'|\left[\epsilon\sigma_z(0) + \Delta\sigma_x(0)\right]|\mu'\rangle \left(\nu\mu\right|\left[\tilde{U}(s)\right]_{\rm av}\left|\nu'\mu'\right) ,\end{aligned} \tag{8.70}$$

where $\left[\tilde{U}(s)\right]_{\rm av}$ is given by (8.49).

The evaluation of the matrix elements of $[\tilde{U}(s)]_{\rm av}$ proceeds similarly as in Sect. 8.4. However, as the original coupling is now directly treated as a perturbation, the relevant quantity is the Laplace transform of the following bath correlation function:

$$\xi(t) = \langle X(0)X(t)\rangle_0 \,, \tag{8.71}$$

where

$$X(t) = \sum_{\boldsymbol{k}} g_{\boldsymbol{k}}\left[b_{\boldsymbol{k}}(t) + b^+_{\boldsymbol{k}}(t)\right] . \tag{8.72}$$

Using the harmonic nature of the bath, it can easily be shown that

$$\xi(t) = \sum_{\boldsymbol{k}} g^2_{\boldsymbol{k}}\left[\coth\left(\frac{\beta\omega_{\boldsymbol{k}}}{2}\right)\cos(\omega_{\boldsymbol{k}}t) + i\sin(\omega_{\boldsymbol{k}}t)\right] . \tag{8.73}$$

Using the Ohmic dissipation model (cf. (8.21)-(8.23)), this reduces to

$$\xi(t) = \frac{K}{2}\int_0^\infty d\omega\omega e^{-\omega/\omega_c}\left[\coth\left(\frac{\beta\omega}{2}\right)\cos(\omega t) + i\sin(\omega t)\right] . \tag{8.74}$$

The Laplace transform of $\xi(t)$, in the relevant temperature regime for which $\beta\omega_c \gg 1$, is

$$\tilde{\xi}(s) = \frac{K}{2}\left[sF(s) + i\left(\omega_c - \frac{\pi s}{2}\right)\right], \tag{8.75}$$

where

$$F(s) = -\psi\left(1 + \frac{s\beta}{2\pi}\right) - \ln\left(\frac{2\pi}{\beta\omega_c}\right) + \frac{\pi}{s\beta} - \psi(1), \tag{8.76}$$

ψ being Euler's polygamma function [452]. In order to obtain the elements of $[\tilde{U}(s)]_{\mathrm{av}}$, we need to invert a 4×4 matrix. This is done using the algebraic manipulation program REDUCE [453]. We present here the final result for the correlation function, valid upto order K:

$$\tilde{C}(s) = \frac{s^2 + 4\epsilon^2 + K[G_1(s) + G_2(s)\tanh(\beta\Omega)]}{s(s^2 + 4\Omega^2) + KR(s)}, \tag{8.77}$$

where

$$\begin{aligned} G_1(s) &= \frac{1}{\Omega^2}\left\{2\epsilon^2 s_+ s_-[F(s_-) + F(s_+)] + s\Delta^2[s_+F(s_+) + s_-F(s_-)]\right. \\ &\quad \left. +2s^2\Delta^2 F(s)\right\}, & (8.78) \\ G_2(s) &= \frac{4\epsilon^2\Delta^2}{s\Omega^3}\left[\frac{\pi}{2}(s^2 + 2\Omega^2) - s\omega_c\right], & (8.79) \\ R(s) &= \frac{\epsilon^2}{\Omega^2} s_+ s_-[s_+F(s_-) + s_-F(s_+)] + s_+ s_-[s_+F(s_+) + s_-F(s_-)] \\ &\quad +2\frac{\Delta^2}{\Omega^2}s^3 F(s), & (8.80) \\ s_\pm &= s \pm 2i\Omega. & (8.81) \end{aligned}$$

The expression in (8.77) is evaluated numerically, and the results are plotted in Figs. 8.3 and 8.4. These figures also show the corresponding DBGA result of Sect. 8.4 as dashed lines, for the sake of comparison. The breakdown of the DBGA at low temperatures is demonstrated by these results.

It is also useful to calculate $P(t)$, the *stay-put* probability defined in (8.8). We obtain the following expression for the Laplace transform of $P(t)$ [451]:

$$\tilde{P}(s) = \frac{1}{2s} + \frac{s^2 + 4\epsilon^2 + K\left[G(s) + \frac{i\Omega}{\epsilon}\left(\omega_c - \frac{\pi s}{2}\right)\right]}{2[s(s^2 + 4\Omega^2) + KR(s)]}. \tag{8.82}$$

Equation (8.82), derived here from the resolvent-expansion method, allows for a direct comparison with the result obtained using the functional-integral approach, but going beyond DBGA by incorporating interacting bounces [446]. The relevant comparison is presented in Figs. 8.5 and 8.6, where the two sets of results are found to be numerically indistinguishable. As before,

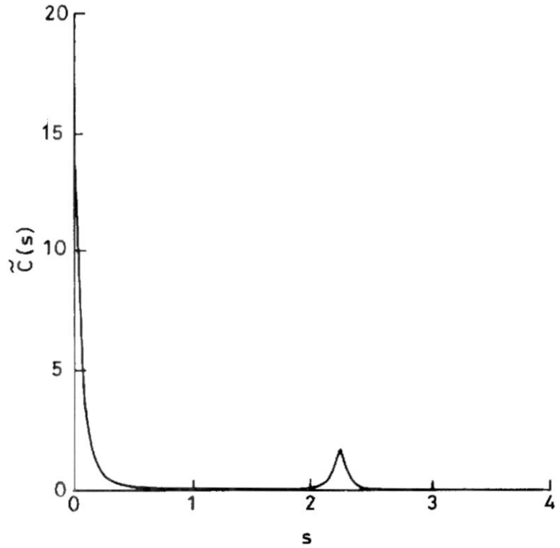

Fig. 8.3. Plot of Laplace-transformed correlation function $\tilde{C}(is + \Gamma/2)$ vs. s, obtained by numerical evaluation of (8.77). The parameter values (in units of Δ) were $\Gamma = 0.1, \epsilon = 2, \omega_c = 100$ and $T = 10^{-3}$. The dimensionless coupling constant was taken to be $K = 10^{-3}$. For these parameters, the DBGA result is numerically indistinguishable from the result in (8.77)

the DBGA result (indicated by a dashed line) shows considerable departure at low temperatures.

Before concluding, it is instructive to discuss a few limiting cases in order to have a clearer understanding of the validity of (8.77). First, note that in the zero-damping case ($K = 0$) we recover (8.60), as expected. Second, we find for the asymptotic case, *before* taking the $\beta\omega_c \gg 1$ limit,

$$\lim_{t\to\infty} C(t) = \lim_{s\to 0} s\tilde{C}(s) = \left[\frac{\epsilon}{\Omega}\tanh(\beta\Omega)\right]^2 . \tag{8.83}$$

The RHS of (8.83) has the expected form, i.e., it is the square of the expectation value of σ_z in the equilibrium ensemble defined by ρ_s in (8.68). The present treatment therefore removes a major deficiency of the DBGA, as discussed in Sect. 8.4. Finally, we consider the symmetric case with $\epsilon = 0$. In that case, (8.77) yields

$$\tilde{C}(s) = \left[s + \frac{4\Delta^2}{s(1 + 4KF(s))}\right]^{-1} . \tag{8.84}$$

Interestingly, this result turns out to be identical to the small-K limit of the DBGA result given in (8.58) [445]. This observation underscores the fact, as mentioned earlier, that inter-bounce interactions are negligible for a symmetric system, *even* for weak coupling.

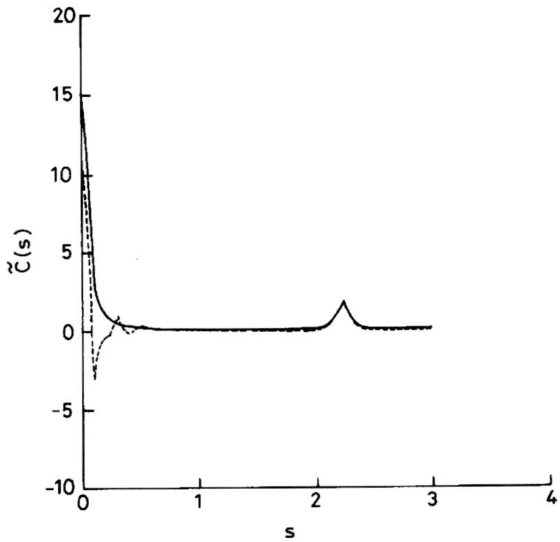

Fig. 8.4. Analogous to Fig. 8.3, except the temperature (in units of Δ) was $T = 5\times10^{-7}$. The DBGA result (dashed line) is now unphysical, and $\tilde{C}$ becomes negative for moderately small values of s

8.6 Dynamics of an Impurity Spin Coupled to a Spin-Boson System

One of the physical realizations of the spin-boson model introduced in (8.1)-(8.4) is spin-lattice relaxation in solids, as discussed in Sect. 8.1.1. In this case, the system part of the Hamiltonian $\mathcal{H}_s$ describes a Zeeman coupling between a spin (representing, e.g., an electronic moment) and an external magnetic field H. We would like to consider now a situation in which H is an *internal* field rather than an *external* field, say that arising from a nuclear spin at the electronic site because of the magnetic hyperfine interaction:

$$\mathcal{H}_s = -aI_z\sigma_z \,. \tag{8.85}$$

Equation (8.85) corresponds to a uniaxial hyperfine interaction with coupling constant a, and the z-axis being the anisotropy axis. The rest of the Hamiltonian remains the same as in (8.3)-(8.4). The field H $(= aI_z)$ is now dynamical and is governed by a quantum operator I_z. Furthermore, it can induce time-evolution in another quantum operator, say I_x, the x-component of the nuclear angular momentum. Thus, in the present context, we consider the following *transverse* correlation function relevant for nuclear spin dynamics:

$$C_{xx}(t) = \langle I_x(0)I_x(t)\rangle \,, \tag{8.86}$$

where $\langle\ldots\rangle$, as usual, denotes the statistical average.

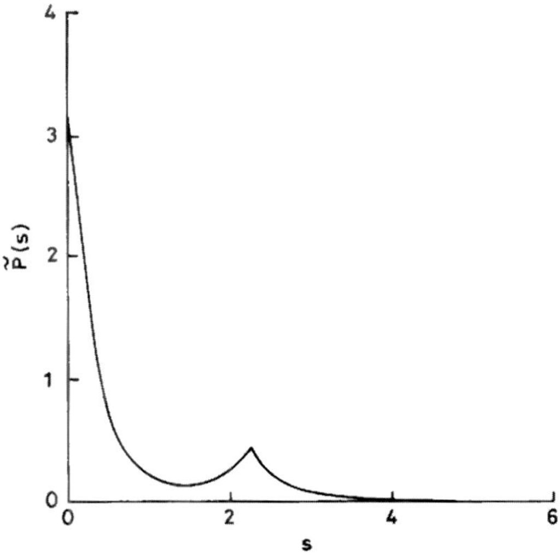

Fig. 8.5. Comparison of results for $\tilde{P}(is+\Gamma/2)$ vs. s, obtained using the resolvent-expansion method and the Weiss-Wollensak approach [446]. The two results coincide on the scale of the plot. The parameter values (in units of Δ) were $\Gamma = 0.5, \epsilon = 2, \omega_c = 100$ and $T = 10^{-2}$. The coupling constant was $K = 10^{-3}$. At these parameter values, the DBGA result is also numerically indistinguishable from the other results

In our earlier calculation of the correlation function in (8.10), the probe spin (which is electronic) was *directly* in contact with the dissipative bath. In this case, the probe spin (which is nuclear) is driven by the hyperfine coupling (cf. (8.85)) which, in turn, is influenced by the bath. Therefore, the effect of the bath on the nuclear spin is *indirect*, and is mediated by the electronic spin.

A motivating factor behind formulating the Hamiltonian in (8.85), in conjunction with the terms in (8.3)-(8.4), is the Kondo problem. In that context, $\boldsymbol{\sigma}$ represents the electronic spin of an impurity atom and $\boldsymbol{I}$ represents the nuclear spin of the impurity atom. The spin-boson part is then taken to model the interaction between the localized spin and the conduction electrons. This interaction has a bosonic representation in the Ohmic-dissipation limit for a highly anisotropic Kondo system [454]. A quantity such as $C_{xx}(t)$ is relevant in interpreting resonance experiments, e.g., NMR. A related quantity is also useful for analyzing hyperfine line-shapes, which can be measured by the Mössbauer and angular-correlation techniques [1].

Another motivation for studying the present Hamiltonian arises from the conceptual interest in examining the interplay of quantum and dissipative effects for the original spin-boson problem itself. A detailed analysis is expected to elucidate the underlying basis of certain equivalent Hamiltonians, which are often introduced from purely stochastic considerations [1]. These

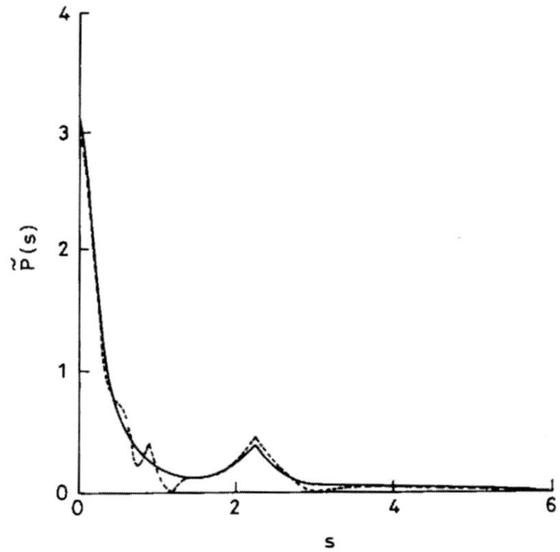

Fig. 8.6. Analogous to Fig. 8.5, except the temperature (in units of Δ) was $T = 10^{-7}$. The corresponding DBGA result is denoted by a dashed line

arise from the observation that an *effective* time-dependent Hamiltonian can be constructed by going to the interaction picture which, in the present case, reads

$$\mathcal{H}_s(t) = -aI_z\sigma_z(t)\,. \tag{8.87}$$

In order to simulate dissipative effects, it is often useful to replace $\mathcal{H}_s(t)$ by the following stochastic Hamiltonian:

$$\mathcal{H}_s(t) = -aI_z h(t)\,. \tag{8.88}$$

As σ_z can take values $+1$ or -1, it is appropriate to regard $h(t)$ as a two-level jump process (i.e., a telegraph process) in which $h(t)$ jumps randomly between two values $+1$ and -1. The underlying stochastic process, assumed to be Markovian, is specified completely in terms of the rate λ at which $h(t)$ jumps from $+1$ to -1 and vice versa. The stochastic model in (8.88) enables a straightforward calculation of $C_{xx}(t)$ [1]. The resultant expression can then be compared with that obtained from the more detailed quantum-mechanical analysis of the dynamics. Such a comparison provides a microscopic interpretation for the phenomenological parameter λ in the stochastic model.

With this background, we present next a calculation of $C_{xx}(t)$, based on the DBGA with Ohmic dissipation. It is useful to first perform a rotation in the space of $\boldsymbol{\sigma}$ *alone*, as indicated after (8.7), such that $x \to z$ and $z \to -x$. This rotation has no effect on I_x and hence $C_{xx}(t)$. Therefore, the rotated Hamiltonian becomes

$$\mathcal{H} = aI_z\sigma_x + \sigma_z \sum_{\boldsymbol{k}} g_{\boldsymbol{k}}(b_{\boldsymbol{k}} + b_{\boldsymbol{k}}^+) + \sum_{\boldsymbol{k}} \omega_{\boldsymbol{k}} b_{\boldsymbol{k}}^+ b_{\boldsymbol{k}} \,. \tag{8.89}$$

In this Hamiltonian, the tunneling term Δ is replaced by the dynamical term aI_z, as is clear upon comparison with (8.7). We now perform the unitary transformation in (8.12), which yields (cf. (8.13))

$$\tilde{\mathcal{H}} = \frac{1}{2} aI_z(\sigma^- B_- + \sigma^- B_+) + \sum_{\boldsymbol{k}} \omega_{\boldsymbol{k}} b_{\boldsymbol{k}}^+ b_{\boldsymbol{k}} \,. \tag{8.90}$$

In terms of $\tilde{\mathcal{H}}$, the transverse correlation function $C_{xx}(t)$ in (8.86) may be written as

$$C_{xx}(t) = \mathrm{Tr}\left[\tilde{\rho}\tilde{I}_x(0)\tilde{I}_x(t)\right] , \tag{8.91}$$

where $\tilde{\rho}$ is the density matrix appropriate to the Hamiltonian $\tilde{\mathcal{H}}$, and

$$\begin{aligned} \tilde{I}_x(t) &= e^{i\tilde{\mathcal{H}}t} I_x(0) e^{-i\tilde{\mathcal{H}}t} \\ &= e^{iLt} I_x(0) \\ &= U(t) I_x(0) \,. \end{aligned} \tag{8.92}$$

Since we are interested in analyzing spectroscopy experiments, it is convenient to consider the Laplace transform of $C_{xx}(t)$, as discussed in Sect. 8.4. We find

$$\tilde{C}_{xx}(s) = \frac{1}{2(2I+1)} \mathrm{Tr}_s \left\{ I_x \left([\tilde{U}(s)]_{\mathrm{av}} I_x \right) \right\} , \tag{8.93}$$

where we have made the usual factorization of the density matrix:

$$\tilde{\rho} \simeq \frac{1}{2(2I+1)} \rho_B \,. \tag{8.94}$$

In (8.93), $\mathrm{Tr}_s\{\ldots\}$ denotes the trace over the degrees of freedom for the system governed by the Hamiltonian $\mathcal{H}_s$. The prefactor on the RHS of (8.94) is valid for $a \ll k_B T$, as is often the case for the hyperfine interaction in metals. In a more explicit form, (8.93) becomes

$$\begin{aligned} \tilde{C}_{xx}(s) = {} & \frac{1}{2(2I+1)} \sum_{m_0 m_1} |\langle m_1 | I_x | m_0 \rangle|^2 \\ & \times \sum_{\mu\mu'} \left(m_0\mu, m_1\mu | [\tilde{U}(s)]_{\mathrm{av}} | m_0\mu', m_1\mu' \right) , \end{aligned} \tag{8.95}$$

where $|m_0\rangle, |m_1\rangle$, etc., are the eigenstates of I_z while $|\mu\rangle, |\mu'\rangle$, etc., are the eigenstates of σ_z, as before. Note that the super-operator $[\tilde{U}(s)]_{\mathrm{av}}$, defined in (8.48), is diagonal in the nuclear angular-momentum representation because the underlying Hamiltonian involves I_z alone.

We again employ the resolvent expansion (cf. (8.49)) and enumerate the 4×4 matrix $[\tilde{U}(s)]_{\rm av}$ (in the space of σ_z), in which each matrix element depends either on m_0 or m_1. The final result for the transverse correlation function is obtained as [454]

$$\tilde{C}_{xx}(s) = \frac{1}{2(2I+1)} \sum_{m_0 m_1} |\langle m_1|I_x|m_0\rangle|^2$$
$$\times \left\{ s + \frac{\tilde{J}(s)}{4\Delta^2} \left[(a_0 m_0 - a_1 m_1)^2 \cos(\pi K) + i(a_0^2 m_0^2 - a_1^2 m_1^2) \sin(\pi K) \right] \right\}^{-1} , \tag{8.96}$$

where $\tilde{J}(s)$ is given by (8.55), corresponding to the case with Ohmic dissipation. At this stage, two remarks on (8.96) are in order. First, we have deliberately employed the same notation as in Sect. 8.4, so as to facilitate comparison with the results presented there. Second, we have allowed for the possibility that the hyperfine constant a takes two distinct values, a_0 and a_1, depending on whether the nuclear state is m_0 or m_1. This is particularly relevant in an experiment involving the Mössbauer effect, where the states m_0 and m_1 refer to the ground and excited states of the nucleus. Having said that, we now specialize to the NMR case for which $a_0 = a_1 = a$. Further, for the sake of simplicity, we consider the case $I = 1/2$. Then, (8.96) reduces to

$$\tilde{C}_{xx}(s) = \left[s + 2\tilde{J}(s) \left(\frac{a}{2\Delta} \right)^2 \cos(\pi K) \right]^{-1} . \tag{8.97}$$

Notice that (8.97) is identical to (8.58), if we identify $a = 2\Delta$ – see the remarks following (8.89). This is a rather fortuitous result, given the difference in the two physical situations.

We now return to the task of comparing our results in (8.96)-(8.97) with the predictions of the stochastic model introduced in (8.88). The stochastic Hamiltonian, as mentioned earlier, constitutes a drastic simplification in that a quantum many-body problem is reduced to a classical one. Using expressions derived in [1], we obtain

$$\tilde{C}_{xx}(s) = \frac{1}{(2I+1)} \sum_{m_0 m_1} |\langle m_1|I_x|m_0\rangle|^2 \left[s + \frac{(a_0 m_0 - a_1 m_1)^2}{s + \lambda} \right]^{-1} . \tag{8.98}$$

For the spin $I = 1/2$ NMR case ($a_0 = a_1 = a$), this simplifies to

$$\tilde{C}_{xx}(s) = \left(s + \frac{a^2}{s+\lambda} \right)^{-1} . \tag{8.99}$$

When we compare (8.99) with the result given in (8.97), it is evident that the latter has a richer structure. First, the relaxation parameters in

(8.97), described by the second term inside the braces, depend strongly on the frequency s. Such non-Markovian features are expected, of course, when the heat bath is a quantum system. Second, the parameters also depend on the hyperfine constant a, indicating the importance of the *feedback* of the system into the heat bath. We see that the term which plays the role of $(s+\lambda)$ is $\gamma(K,s)$, where (using (8.55))

$$\gamma(K,s) = \omega_c \left(\frac{\beta\omega_c}{2\pi}\right)^{2K-1} \frac{\Gamma(1-K+s\beta/2\pi)\sec(\pi K)}{\Gamma(1-2K)\Gamma(K+s\beta/2\pi)} . \tag{8.100}$$

Making a weak-coupling expansion, we may write

$$\begin{aligned} \gamma(K,s) &= \gamma(0,s) + K\frac{\partial\gamma(K,s)}{\partial K}\bigg|_{K=0} + \ldots \\ &\simeq s + K\frac{\partial\gamma(K,s)}{\partial K}\bigg|_{K=0} . \end{aligned} \tag{8.101}$$

Thus, the relaxation rate λ is given by the second term on the RHS of (8.101), which is still frequency-dependent. Finally, we may consider the *motional-narrowing* limit in which λ is so large that $s+\lambda \simeq \lambda$ on the RHS of (8.99). In this limit, $\lambda = \gamma(K, s=0)$, which can be obtained from (8.100). The approximation $s=0$ in (8.100) is equivalent to assuming $s\beta \ll 1$. Therefore, the quantum analysis shows that the motional-narrowing limit is applicable only at high temperatures. We shall return to a further discussion of the validity of the motional-narrowing limit in Sect. 8.7 below.

8.7 Quantum Diffusion of Muons in Metals

The dynamics of light interstitial particles such as hydrogen (H^+) or muons (μ^+) in metals is a problem of considerable interest. At very low temperatures, the dynamics is governed by quantum laws. Nevertheless, the influence on this motion of damping caused by metallic electrons leads to fascinating effects, characteristic of a quantum system in contact with a fermionic bath [455]. Both hydrogen and the muons are positively-charged light interstitials, and therefore characterized by similar behavior in metals. However, the experimental methods for probing their dynamics are quite distinct. Hydrogen is mostly studied by incoherent neutron scattering, and the theoretical interest lies in the evaluation of the structure factor. In contrast, the muon is probed through its spin precession in local magnetic fields that vary from site to site. The relevant theoretical issue is how this coherent precessional motion gets affected by the hopping of the muon from one site to another.

In Sect. 8.8, we will calculate the structure factor for tunneling of hydrogen in Nb with trapping impurities. In this section, we will analyze muon-spin rotation (μSR), where the basic issues can be clarified by a simple example.

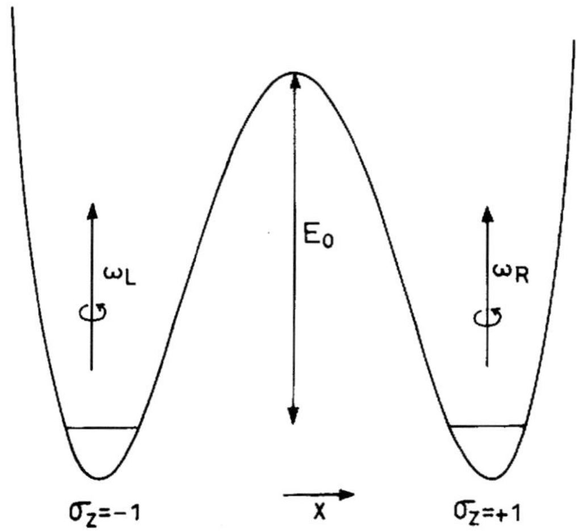

Fig. 8.7. Muon in a symmetric double-well potential, with different magnetic fields in the two wells. The quantities ω_L and ω_R denote the corresponding Larmor precession frequencies in the left and right wells, respectively. The barrier height is denoted as E_0

Consider a muon moving in a symmetric double-well potential, in which it encounters different magnetic fields (e.g., H and $-H$, acting along the z-axis) in the two wells (see Fig. 8.7). If the motion of the muon can be described by thermally-activated stochastic jumps from one well to the other across the intervening barrier E_0, the μSR line-shape can be described by $\tilde{C}_{xx}(s)$, given by (8.99) with the hyperfine constant $a = H$. The parameter λ is the mean rate of jumps of the muon across the barrier which, in the classical domain, is given by the Arrhenius formula:

$$\lambda = \lambda_0 e^{-\beta E_0}\,, \tag{8.102}$$

λ_0 being the *attempt frequency*. The main difficulty in the quantum case arises from two considerations. First, we require a modification of (8.102), which incorporates the surrounding fermionic system. Second, and perhaps more important, it is not even clear whether the general formulation in (8.102) is valid. A critical assessment of these issues was already undertaken in the context of NMR in Sect. 8.6. We would like to perform a similar analysis for μSR in this section.

It may seem that restricting the motion of the muon between only two sites is a gross simplification of the actual physical situation. However, this model does capture the essential features of the issues mentioned above. In addition, this model can be extended to cover certain experimental situations in which the muon is trapped by impurities and performs localized motion over a finite number of sites. In the present treatment, we have also simplified the discussion by assuming that the local magnetic fields at the two sites are oriented in opposite directions.

With the preceding survey of the scope of the present analysis in the context of quantum dissipative phenomena, we write the Hamiltonian describing the dynamics of the muon as follows:

$$\mathcal{H} = HI_z\sigma_z + \Delta\sigma_x - \sigma_z \sum_{\boldsymbol{k}} g_{\boldsymbol{k}}(b_{\boldsymbol{k}} + b_{\boldsymbol{k}}^+) + \sum_{\boldsymbol{k}} \omega_{\boldsymbol{k}} b_{\boldsymbol{k}}^+ b_{\boldsymbol{k}} \,. \tag{8.103}$$

Here, I_z (for $I = 1/2$) is the spin angular momentum of the muon in the direction in which the internal fields act. As usual, σ_z is the pseudospin operator whose eigenvalues (± 1) label the two wells, as described in Sect. 8.1.2. This Hamiltonian generalizes the spin-boson Hamiltonian for tunneling in an asymmetric double well. The asymmetry parameter ϵ is now replaced by the dynamic field HI_z (cf. (8.35)). As mentioned earlier, the bosonic bath, in conjunction with an Ohmic spectral density of bath excitations, adequately describes the electron-hole pairs of the surrounding metallic system.

We follow the DBGA approach of Sect. 8.4, but now use it to calculate the transverse correlation function $\tilde{C}_{xx}(s)$, defined by (8.93). The calculation is similar to that described earlier. For brevity, we only present the final result here [455]:

$$\tilde{C}_{xx}(s) = \left[s + \frac{H^2}{s + 2\cos(\pi K)\tilde{J}(s)} \right]^{-1} , \tag{8.104}$$

where $\tilde{J}(s)$ is again given by (8.55). The derivation of (8.104) assumes that the Zeeman energy associated with the internal field H is negligible compared to the thermal energy. Note that the above result is different from the one derived for $\tilde{C}_{xx}(s)$ in Sect. 8.6 (cf. (8.97)). This is because the Hamiltonian in (8.103) is quite distinct from the Hamiltonian in (8.89). In particular, the Zeeman coupling in (8.103) is not directly influenced by the bath as it commutes with the interaction term.

Comparing (8.104) with (8.99), we identify $\lambda = 2\cos(\pi K)\tilde{J}(s)$. This quantity is frequency-dependent and hence would yield non-Lorentzian lineshapes. This clarifies the question raised in the beginning regarding the validity of (8.99). In order to obtain an expression for λ and compare it with the Arrhenius form in (8.102), it is necessary to consider the high-temperature limit of $\tilde{J}(s)$, yielding

$$\lambda(K, \beta) \simeq \frac{4\Delta^2}{\omega_c} \left(\frac{2\pi}{\beta\omega_c} \right)^{2K-1} \frac{\Gamma(1-2K)\Gamma(K)\cos(\pi K)}{\Gamma(1-K)} \,. \tag{8.105}$$

The dependence of the μSR relaxation rate on the coupling constant K and the temperature T ($\propto \beta^{-1}$) are explicitly seen in (8.105). In particular,

$$\lambda(K, \beta) \sim T^{2K-1} \,, \tag{8.106}$$

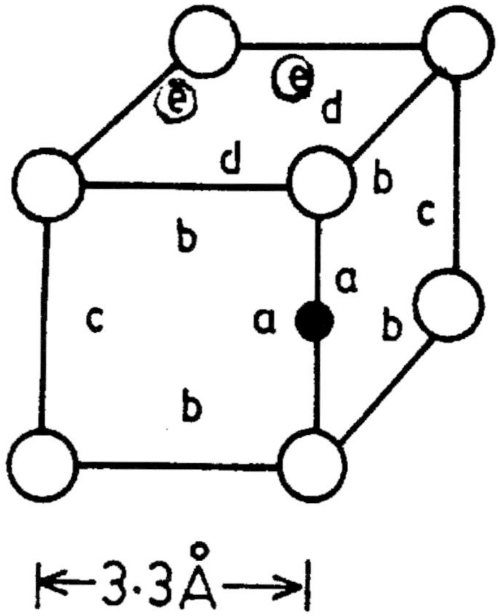

Fig. 8.8. The bcc structure of Nb. The big open circles are Nb sites, and the filled circle denotes an immobile oxygen atom on an octahedral interstitial site. The sites a→e denote tetrahedral interstitial sites on the visible surfaces of the unit cell. The trapped hydrogen occupies one of the two sites labeled by e, which have been encircled

which is a power law, quite distinct from the classical Arrhenius expression of (8.102). This power law was first derived by Kondo [456], and was shown to be a direct consequence of the fermionic nature and Ohmic spectrum of bath excitations. To conclude, the present analysis demonstrates that the μSR line-shape cannot be simply described by an expression based on the picture of a muon stochastically jumping from one site to another with a mean jump rate, especially at temperatures where quantum effects are important.

8.8 Neutron Scattering Study of H-Tunneling in Niobium

Having discussed the quantum diffusion of muons in metals within the framework of the spin-boson model, we consider next another successful application of this model. We will use it to analyze neutron-scattering experiments on the dynamics of hydrogen, trapped by oxygen in niobium (Nb). The niobium metal has a BCC structure and becomes superconducting at low temperatures. However, we shall consider only those experiments in which superconductivity is eliminated by an applied magnetic field, so that all electrons

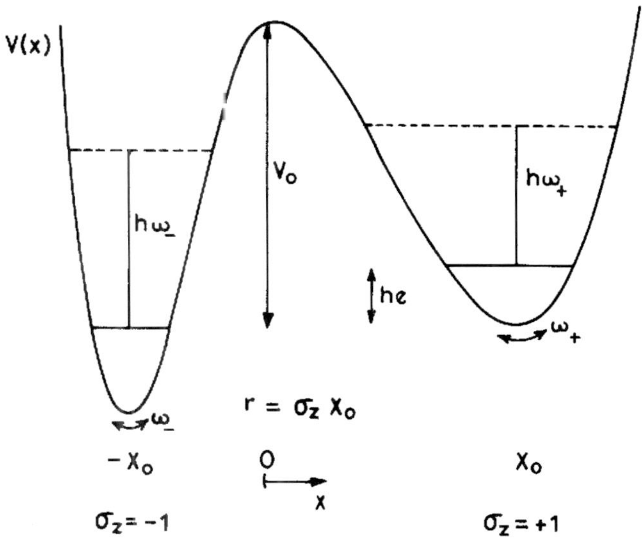

Fig. 8.9. The asymmetric double-well potential ($V(x)$ vs. x) for H^+ in $Nb(OH)_x$. The energies associated with small-oscillation frequencies $\omega_\pm$ are much larger than the asymmetry energy ϵ, the tunneling energy Δ, and the thermal energy k_BT. As usual, the left and right minima of the well are mapped into pseudospin states $\sigma_z = -1$ and $+1$, respectively

are in the normal state. Furthermore, our focus of interest will be the dilute system of hydrogen interstitials which occupy tetrahedral sites, indicated by a→e in Fig. 8.8. These can be trapped by oxygen interstitials, which occupy octahedral sites in Fig. 8.8. From a large body of experimental data involving X-ray and channeling measurements, it has been ascertained that the H^+ (which are essentially isolated) can occupy one of the two sites encircled in Fig. 8.8. This is reminiscent of a double-well potential in which H^+, being a light quantum particle, tunnels back and forth. However, this double well is asymmetric (see Sect. 8.4) because of long-range elastic interactions between OH-pairs (see Fig. 8.9). Of course, the asymmetry is random at different hydrogen sites and has to be averaged out in the final analysis.

At this stage, it is relevant to specify the energy scales in the problem. It turns out that the tunneling energy Δ, the asymmetry energy ϵ and the thermal energy k_BT (at temperatures of experimental interest) are all of the same order, viz., a few meV. On the other hand, the harmonic-oscillator-like energy separation (in each well), the Debye energy (associated with phonons), and the Fermi energy are all at least two orders of magnitude higher. Therefore, the reduction to a double-well potential, in which the subsystem Hamiltonian can be described in terms of spin operators alone, is indeed a valid model for the tunneling center.

As has been already mentioned, our treatment will focus on neutron-scattering measurements. It is well known that neutrons have a large scattering cross-section with hydrogen atoms, resulting in a structure factor which is the Fourier transform of the Van Hove correlation function [457,458]:

$$S(\boldsymbol{k},\omega) = \int_{-\infty}^{\infty} dt\; e^{i\omega t} \left\langle e^{-i\boldsymbol{k}.\boldsymbol{r}(0)} e^{i\boldsymbol{k}.\boldsymbol{r}(t)} \right\rangle . \tag{8.107}$$

In (8.107), $\boldsymbol{k}$ is the wave-vector transfer in the scattering process, and ω is the corresponding frequency transfer. The positions of the scatterer (hydrogen, in this case) are denoted by $\boldsymbol{r}(0)$ and $\boldsymbol{r}(t)$ at times 0 and t, respectively. Note that $\boldsymbol{r}(t)$, which is obtained from the Heisenberg evolution of $\boldsymbol{r}(0)$, is a quantum operator that *does not* commute with $\boldsymbol{r}(0)$. Finally, the angular brackets $\langle \ldots \rangle$ denote a quantum statistical average over the entire system of interest, maintained at a fixed temperature T.

The system consists of the tunneling center (i.e., the subsystem) plus the environment with which it is coupled. Now, as the hydrogen (which is normally stripped of its electron and therefore behaves as a proton) moves, it is dragged by the cloud of conduction electrons of Nb. Therefore, the environment is a heat bath of conduction electrons which cause *quantum friction.* Since the thermal energy is much smaller than the Fermi energy, the relevant excitations are those of electron-hole pairs off the Fermi surface. These can be represented by bosons, as mentioned earlier. Motivated by these considerations, the system of interest may be described by the Hamiltonian in (8.35).

The details of the calculation for the correlation function $\tilde{C}(s)$ have already been presented in Sects. 8.4 and 8.5. We now discuss the relation of $\tilde{C}(s)$ to $S(\boldsymbol{k},\omega)$ and the neutron-scattering data. In view of the two-level abstraction of the double well, the position operator $\boldsymbol{r}$ can be mapped into the pseudospin operator σ_z (see Fig. 8.9):

$$\boldsymbol{r} = \hat{i} x_0 \sigma_z , \tag{8.108}$$

where $\hat{i}$ is the unit vector along the x-axis, and $2x_0$ is the separation between the two minima of the double well. Because of the special property of σ_z (viz., $\sigma_z^2 = 1$), the operator $\exp(-i\boldsymbol{k}\cdot\boldsymbol{r})$ can be simplified drastically, reducing the expression for the structure factor to [1]

$$S(\boldsymbol{k},\omega) = f_1(kx_0)\delta(\omega) + \frac{f_2(kx_0)}{1+e^{\beta\omega}} \mathrm{Re}\; \tilde{C}(i\omega) . \tag{8.109}$$

Here, f_1 and f_2 are the so-called *form factors* which depend on the dimensionless quantity kx_0.

Figure 8.10 plots the experimental structure factors for H-tunneling [459–463] as a function of the neutron energy gain ω (in meV) for fixed

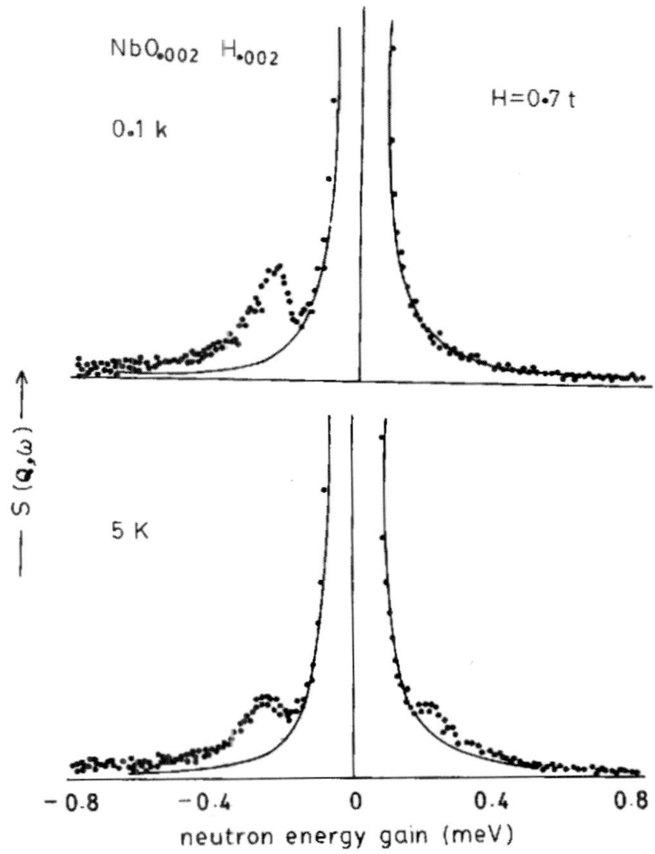

Fig. 8.10. Experimental structure factors $S(\boldsymbol{k},\omega)$ vs. ω for fixed $\boldsymbol{k}$. Data sets are shown for two different temperatures, viz., 0.1 K and 5 K [462]

$\boldsymbol{k}$, when the sample is placed in a magnetic field of 0.7 Tesla. The data for $T = 0.1$ K clearly shows an *inelastic* peak located at energy $-\sqrt{\Delta^2 + \epsilon^2}$. This is a consequence of the *coherent* tunneling motion of the hydrogen, picked up by the scattered neutron. The other inelastic peak on the right shoulder, located at energy $\sqrt{\Delta^2 + \epsilon^2}$, is suppressed at $T = 0.1$ K because of the detailed-balance factor, but shows its presence at the higher temperature of 5 K. Figure 8.10 exhibits signatures of coherence, as far as neutron scattering is concerned. Now, as the temperature increases, there is more phase space available for the tunneling center to interact with the heat-bath excitations. The coherent clock-like motion of the hydrogen is then impeded and decoherence sets in. As a result, neutron scattering changes over from the *inelastic* regime to the *quasi-elastic* regime, yielding a central line (see Fig. 8.11). The latter, for a fixed $\boldsymbol{k}$, can be fitted to a sum of Lorentzians:

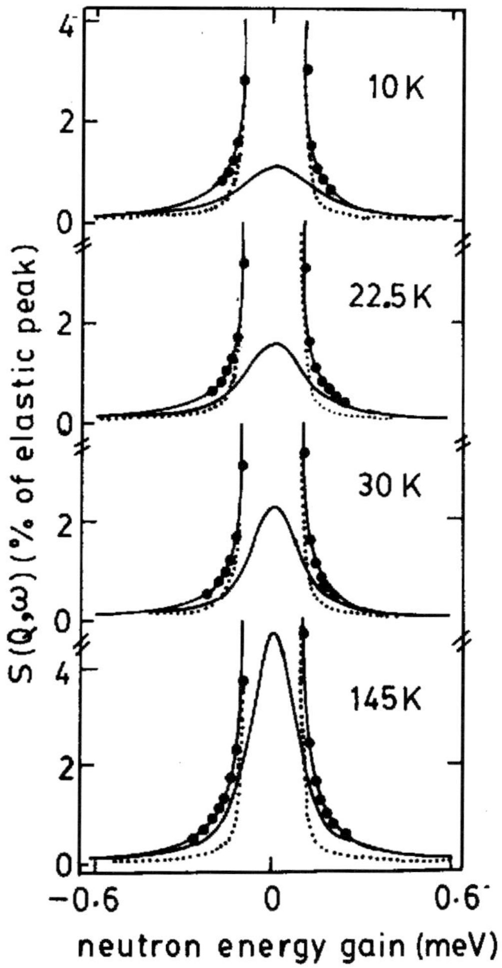

Fig. 8.11. Analogous to Fig. 8.10, but at higher temperatures, where the H-motion is incoherent [463]

$$S(\boldsymbol{k},\omega) = \sum_j \frac{a_j \Gamma_j}{\omega^2 + \Gamma_j^2} \,, \tag{8.110}$$

where a_j's are constant coefficients and Γ_j's are widths related to the jump rate of hydrogen atoms. We should emphasize that these jumps are associated with the *quantum* diffusion of hydrogen, analogous to the case of muons (cf. Sect. 8.7).

The structure factor has been analytically calculated using the DBGA results of Sect. 8.4 for $\tilde{C}(s)$ [449]. The result is reasonable for most parameter regimes, but fails for weak damping at low temperatures for reasons eluci-

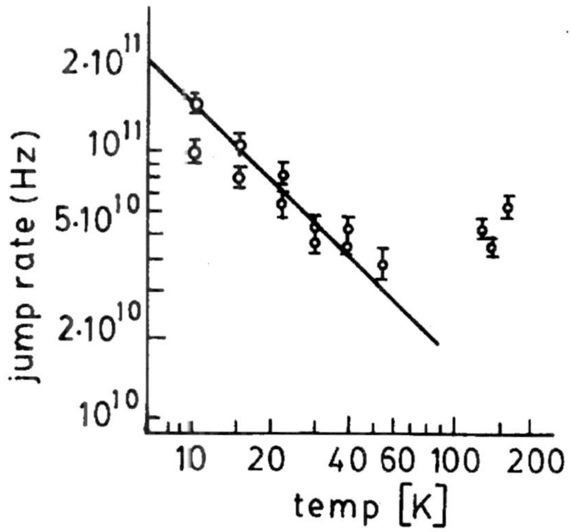

Fig. 8.12. Dependence of the jump rate λ on the temperature T for the $Nb(OH)_x$ system with $x = 0.002$ [463]. The experimental data is plotted on a log-log scale, and the solid line denotes the best linear fit to the data for $T < 60\,K$

dated earlier. Indeed, μSR studies in Nb suggest that the damping parameter $K \simeq 0.05$ [459–463]. Hence, it is more appropriate to use the weak-coupling results for $\tilde{C}(s)$ obtained in Sect. 8.5 for proper interpretation of the low-temperature data and coherent tunneling. An intriguing result is the dependence of the line-width (or jump rate) on temperature, shown in Fig. 8.12 [463,464]. It is seen that, in the temperature range of 10-60 K, the jump rate λ *decreases* with temperature, following a power law $\lambda \propto T^{2K-1}$. This power law is identical to the one discussed earlier in the context of μSR (see (8.106)). It is a consequence of the following facts: (a) the heat bath is composed of fermions, and (b) the electron-hole excitations can be characterized by an Ohmic spectral density.

For further analysis, it is convenient to introduce dimensionless quantities as follows:

$$\nu = \frac{\omega}{\Delta}, \qquad \tau = \frac{k_B T}{\Delta}. \tag{8.111}$$

In terms of these quantities, the inelastic part of the structure factor is proportional to

$$j(\nu) = \frac{2\Delta}{1 + e^{\nu/\tau}} \mathrm{Re}\, \tilde{C}(i\nu\Delta), \tag{8.112}$$

where $\tilde{C}(i\nu\Delta)$ can be obtained from (8.77). The plots for $j(\nu)$ vs. ν from (8.112) and (8.77) are shown in Fig. 8.13, and should be qualitatively com-

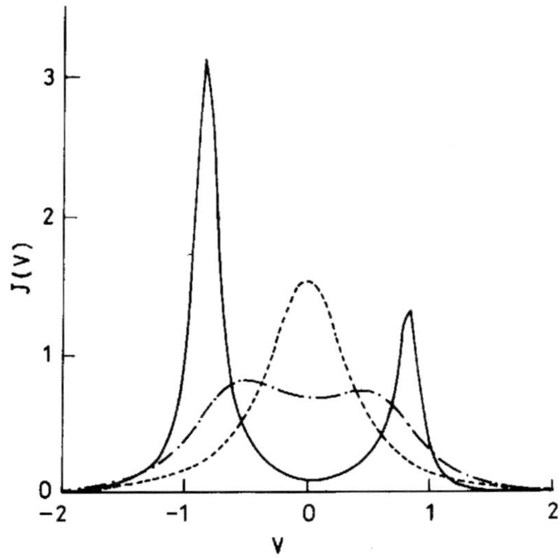

Fig. 8.13. Inelastic part of the structure factor $j(\nu)$ vs. ν from (8.112). The parameter values are $K = 0.055$ and $\epsilon/\Delta = 0.1$. We show results for $k_BT/\Delta = 1.0$ (solid line), 4.5 (dot-dashed line), 10 (dashed line)

pared with the experimental results in Figs. 8.10-8.11. As expected, the low-temperature structure factor is characterized by two inelastic peaks near $\nu = \pm\Omega/\Delta$, the relative intensity of which is determined by the detailed-balance factor. As the temperature increases, the resonances shift towards the origin – at a characteristic temperature, the two peaks merge into a quasi-elastic peak centered at $\nu = 0$. With further increase of temperature, the quasi-elastic peak narrows. We have calculated the numerical values of $j(\nu)$ at different values of ν around $\nu = 0$, and fitted the data to a Lorentzian line-shape function to extract the line-width or jump rate [449]. The analytical results for λ vs. T are consistent with the power-law behavior seen in Fig. 8.12.

The upturn in the curve at temperatures beyond 60 K in Fig. 8.12 is associated with phonon-assisted tunneling of hydrogen. It turns out that the Hamiltonian in (8.35) can account for dissipation due to both conduction electrons and phonons, provided we modify the spectral density defined in (8.22) as [465,466]

$$J(\omega) = \left(K\omega + G\omega^{2s+1}\right) e^{-\omega/\omega_c} , \tag{8.113}$$

where G is the coupling constant appropriate for phonons. The value of the index s depends on the type of phonons (acoustic or optic) that are relevant in a given situation. If we use (8.113), and focus only on the quantum diffusion of hydrogen, we can concentrate on the *incoherent* tunneling regime in which the correlation function can be expressed in a Markovian form:

$$\tilde{C}(s) = \left[s + \tilde{\phi}(s=0) \right]^{-1} , \tag{8.114}$$

where $\tilde{\phi}(s)$ is the appropriate memory kernel. The jump rate of the hydrogen atoms can then be obtained from

$$\lambda = \tilde{\phi}(s=0) . \tag{8.115}$$

We summarize now the results for the jump rate λ, which are presented here for the symmetric case ($\epsilon = 0$) only, as the expressions for the asymmetric case ($\epsilon \neq 0$) are cumbersome. The quantity λ can be expressed in terms of a time-dependent *decay function* $\Lambda(t)$ [467,468]:

$$\lambda = \frac{\Delta^2}{4} \int_{-\infty}^{\infty} dt e^{-\Lambda(t)} , \tag{8.116}$$

where

$$\Lambda(t) = 2 \int_0^{\infty} d\omega \frac{J(\omega)}{\omega^2} \left[\coth\left(\frac{\beta\omega}{2} \right) - \frac{\cos(\omega t)}{\sinh(\beta\omega/2)} \right] . \tag{8.117}$$

Using (8.113), the decay function can be split into *electronic* and *phononic* parts:

$$\Lambda(t) = \Lambda_{\mathrm{el}}(t) + \Lambda_{\mathrm{ph}}(t) , \tag{8.118}$$

resulting in (after a number of simplifying approximations [465]):

$$\lambda_{\mathrm{el}} = \frac{\Gamma(K)}{\Gamma(1-K)} \frac{\tilde{\Delta}}{2} \left(\frac{2\pi}{\beta\tilde{\Delta}} \right)^{2K-1} , \tag{8.119}$$

and

$$\lambda_{\mathrm{ph}} = \left(\frac{\pi\beta}{16E_c} \right)^{1/2} \tilde{\Delta}^2 e^{-\beta E_c} , \tag{8.120}$$

where $\tilde{\Delta}$ is the renormalized tunneling frequency in (8.56), and the parameter E_c will be clarified shortly. In obtaining (8.120), we have used $s = 3$, corresponding to acoustic phonons. The electronic contribution has been already analyzed. The phononic contribution is of the Arrhenius form, but has a quantum-mechanical origin. In fact, the expression for λ_{ph} can also be derived from the *golden rule* of perturbation theory in which the tunneling term is treated upto second order [466]. The parameter E_c is referred to as the *coincidence energy*, and is defined in terms of the phononic part of the spectral density as

$$E_c = \frac{C}{2} \int_0^{\infty} d\omega \omega^{2s} e^{-\omega/\omega_c} . \tag{8.121}$$

Equations (8.116)-(8.121) can be used to interpret the diffusion data for λ over a wide range of temperatures, from 10-60 K in the electron-dominated regime, and from 60-150 K in the phonon-dominated regime.

Before concluding this section, we remark that for $T > 150$ K, the two-level picture of tunneling breaks down. Experimentally, a break in the Arrhenius curve for hydrogen diffusion in Nb and Ta is observed [469]. Computer simulations have shown that this is due to tunneling from excited states of the H-vibration. The asymmetry of the double well further accentuates the tunneling of the H from its ground state in one well to an excited state in the second well. We have demonstrated how a generalized spin-boson model, with pseudospin-1 instead of 1/2, can account for these additional effects [470].

8.9 Spectroscopic Data in Kondo Systems

Next, we would like to consider an important application of the spin-boson model to a fundamental problem in the study of correlated electron systems, viz., the Kondo effect [471]. In particular, we would like to understand spectroscopic data in Kondo systems, obtained from various absorption spectroscopies [1]. This analysis would place in proper perspective the interrelated topics of spectroscopy, correlation functions and relaxation phenomena, especially in the context of quantum dissipative systems. Our present discussion will be based on the material in Sects. 8.6 and 8.7, with the impurity now being interpreted as a Kondo magnetic impurity. We will first demonstrate how the Kondo Hamiltonian can be reduced to the generic spin-boson form, using a formulation due to Anderson and Yuval [472–474]. We will then incorporate the hyperfine interaction, in the presence of a Zeeman coupling, which will act as the probe for studying nuclear relaxation effects.

The conventional Kondo Hamiltonian may be written as

$$\mathcal{H}_K = J_K \boldsymbol{\sigma} \cdot \boldsymbol{s}(0) + \sum_{\boldsymbol{k},\sigma} \epsilon(\boldsymbol{k}) c^+_{\boldsymbol{k}\sigma} c_{\boldsymbol{k}\sigma} \, , \tag{8.122}$$

where the spin $\boldsymbol{\sigma}$ represents the impurity spin localized at $\boldsymbol{r} = 0$. This spin interacts with $\boldsymbol{s}(0)$, the effective spin of the assembly of conduction electrons at $\boldsymbol{r} = 0$, with strength J_K (> 0). The operators $c^+_{\boldsymbol{k}\sigma}$ (and $c_{\boldsymbol{k}\sigma}$) create (and annihilate) electrons of wave-vector $\boldsymbol{k}$, with energy $\epsilon(\boldsymbol{k})$ and spin index $\sigma = \pm 1$. The Anderson-Yuval formulation considers the Kondo coupling to be *anisotropic* with two exchange constants $J_\parallel$ and $J_\perp$. They then rewrite the Hamiltonian with the aid of fermion operators:

$$\begin{aligned}\mathcal{H}_K = {} & \frac{J_\parallel}{2} \sigma_z \sum_{\boldsymbol{k}\boldsymbol{k}'} \left(c^+_{\boldsymbol{k}'\uparrow} c_{\boldsymbol{k}\uparrow} - c^+_{\boldsymbol{k}'\downarrow} c_{\boldsymbol{k}\downarrow} \right) + \frac{J_\perp}{2} \sum_{\boldsymbol{k}\boldsymbol{k}'} \left(\sigma^+ c^+_{\boldsymbol{k}'\downarrow} c_{\boldsymbol{k}\uparrow} - \sigma^- c^+_{\boldsymbol{k}'\uparrow} c_{\boldsymbol{k}\downarrow} \right) \\ & + \sum_{\boldsymbol{k},\sigma} \epsilon(\boldsymbol{k}) c^+_{\boldsymbol{k}\sigma} c_{\boldsymbol{k}\sigma} \, .\end{aligned} \tag{8.123}$$

This Hamiltonian can be represented in terms of spin-density and charge-density excitations which are bosonic in nature [426]. Ignoring the charge-density excitations, we have

$$\begin{aligned}\mathcal{H}_K &= \frac{J_\parallel}{4}\sigma_z \sum_{k>0} e^{-k/(2k_c)}\sqrt{\frac{k}{\pi L}}\left(b^+_{\boldsymbol{k}} + b_{\boldsymbol{k}}\right) + \frac{J_\perp k_c}{4\pi}\left(\sigma^+ e^{\xi} + \sigma^- e^{-\xi}\right) \\ &+ v_F \sum_{k>0} k b^+_{\boldsymbol{k}} b_{\boldsymbol{k}} \,. \end{aligned} \tag{8.124}$$

Here, $b^+_{\boldsymbol{k}}, b_{\boldsymbol{k}}$ are Bose operators for spin-density excitations, k_c is a cut-off wave-vector, L is the length of a normalization box, and v_F is the Fermi speed. Furthermore,

$$\xi = \sum_{k>0} e^{-k/(2k_c)}\sqrt{\frac{4\pi}{kL}}\left(b_{\boldsymbol{k}} - b^+_{\boldsymbol{k}}\right) \,. \tag{8.125}$$

We now make a unitary transformation $\mathcal{R}^{-1}\mathcal{H}_K\mathcal{R}$ [426] with

$$\mathcal{R} = \exp(\sigma_z \xi/2) \,, \tag{8.126}$$

which results in the transformed Hamiltonian:

$$\begin{aligned}\mathcal{H}'_K &= \frac{J_\perp k_c}{4\pi}\sigma_x + \left(\frac{J_\parallel}{4\pi v_F} - 1\right) v_F \sigma_z \sum_{k>0} e^{-k/(2k_c)}\sqrt{\frac{\pi k}{L}}\left(b_{\boldsymbol{k}} + b^+_{\boldsymbol{k}}\right) \\ &+ v_F \sum_{k>0} k b^+_{\boldsymbol{k}} b_{\boldsymbol{k}} \,. \end{aligned} \tag{8.127}$$

Introducing the quantities

$$\begin{aligned} \Delta &= \frac{J_\perp k_c}{4\pi} \,, \\ g_{\boldsymbol{k}} &= -\left(\frac{K\pi v_F \omega_{\boldsymbol{k}}}{L}\right)^{1/2} e^{-\omega_{\boldsymbol{k}}/(2\omega_c)} \,, \\ K &= \left(1 - \frac{J_\parallel}{4\pi v_F}\right)^2 \,, \\ \omega_{\boldsymbol{k}} &= v_F k \,, \\ \omega_c &= v_F k_c \,. \end{aligned} \tag{8.128}$$

(8.127) is identified as the spin-boson Hamiltonian of (8.7) with Ohmic dissipation.

We now incorporate in (8.127) the hyperfine coupling between the angular momentum of the nucleus (I_z) and the electron (σ_z) of the impurity atom. We also include the corresponding Zeeman couplings with an externally-applied magnetic field H, so that the complete Hamiltonian is as follows [435]:

$$\mathcal{H} = -\omega_N I_z + \omega_e \sigma_z + a I_z \sigma_z + \Delta \sigma_x + \sigma_z \sum_{\boldsymbol{k}} g_{\boldsymbol{k}} (b_{\boldsymbol{k}} + b_{\boldsymbol{k}}^{+}) + \sum_{\boldsymbol{k}} \omega_{\boldsymbol{k}} b_{\boldsymbol{k}}^{+} b_{\boldsymbol{k}} \,, \tag{8.129}$$

where ω_N and ω_e are, respectively, the Larmor frequencies of the nuclear and electronic spins in the external magnetic field. Note that the additional terms (i.e., the first three terms on the RHS of (8.129)) could have been added to the original Hamiltonian $\mathcal{H}_K$ itself because they are unaffected by the unitary transformation $\mathcal{R}$. This assumes that the hyperfine coupling has a uniaxial form, which is reasonable in the presence of a large Zeeman field which suppresses the off-diagonal terms of the hyperfine interaction. One final remark is in order, i.e., at the end of the calculation leading to (8.127), we may reset $J_{\parallel} = J_{\perp} = J_K$, and further write the coupling constant K in terms of the density of states at the Fermi surface (ρ):

$$K = \left(1 - \frac{1}{2}\rho J_K\right)^2 , \quad \rho = (2\pi v_F)^{-1} \,. \tag{8.130}$$

Recall that the Hamiltonian in (8.129) (except for the first two terms) is identical to the one discussed in Sect. 8.7, if we assume the nuclear spin $I = 1/2$ (cf. (8.103)). The additional terms, being diagonal with respect to z as the quantization axis, are not expected to create complications in adapting the results of Sect. 8.7 to the situation at hand. The only alteration required is to replace $\tilde{J}(s)$, occurring in the expression for the transverse correlation function in (8.104), as follows:

$$\tilde{J}(s) \to \frac{1}{2}\left[\tilde{J}(s_+) + \tilde{J}(s_-)\right] , \tag{8.131}$$

where

$$s_{\pm} = s \pm i\omega_e \,. \tag{8.132}$$

The Kondo relaxation rate can then be obtained directly from (8.104) as

$$\lambda = \mathrm{Re}\left[\tilde{J}(i\omega_{0+}) + \tilde{J}(i\omega_{0-})\right] \cos(\pi K) \,, \tag{8.133}$$

where

$$\omega_{0\pm} = \omega_r - \omega_N \pm \omega_e \,, \tag{8.134}$$

ω_r being the appropriate resonance frequency.

There are two relevant parameters which characterize the Kondo effect: (a) the dimensionless quantity ρJ_K, and (b) the high-energy cut-off ω_c. We will subsequently use ρJ_K as an expansion parameter. Therefore, it is useful to rewrite the coupling constant K as (cf. (8.130))

$$K = 1 - \delta \,, \tag{8.135}$$

where

$$\delta = \rho J_K - \left(\frac{\rho J_K}{2}\right)^2 . \tag{8.136}$$

Furthermore, from (8.128), we have

$$\Delta = \frac{\omega_c \rho J_K}{2} . \tag{8.137}$$

Therefore, using (8.55),

$$\begin{aligned}\lambda &= \alpha T^{-2\delta} (\rho J_K)^2 \cos(\pi\delta) \Gamma(2\delta - 1) \\ &\times \left[\frac{\Gamma(1 - \delta + i\Theta_+/T)}{\Gamma(\delta + i\Theta_+/T)} + \frac{\Gamma(1 - \delta + i\Theta_-/T)}{\Gamma(\delta + i\Theta_-/T)} \right] ,\end{aligned} \tag{8.138}$$

where we have introduced

$$\begin{aligned}\alpha &= 2\pi k_B T_0^{2\delta} \,, \\ T_0 &= \frac{\omega_c}{2\pi k_B} \,, \\ \Theta_\pm &= \frac{\omega_{0\pm}}{2\pi k_B} .\end{aligned} \tag{8.139}$$

Recalling the expression for δ from (8.136), a Taylor-series expansion in ρJ_K yields

$$\lambda \simeq \pi k_B T (\rho J_K)^2 \left[1 - 2\rho J_K \ln\left(\frac{T}{T_0}\right) + \ldots \right] . \tag{8.140}$$

In arriving at (8.140), we have kept in mind the NMR case in which $\omega_0 \simeq \omega_e$, the electronic Larmor frequency, as $\omega_N \ll \omega_e$. The first term in (8.140), of order $k_B T (\rho J_K)^2$, has the so-called Korringa form [475–477], and the second term is the first *Kondo correction*.

We next analyze the correlation function $\tilde{C}_{xx}(s)$ in the limit $T \to 0$. Because $\delta \ll \Theta/T$, where $\Theta = \omega_e/(2\pi k_B)$, we can make the following approximation:

$$\frac{\Gamma(1 - \delta + i\Theta/T)}{\Gamma(\delta + i\Theta/T)} \simeq \exp\left[(1 - 2\delta)\psi\left(\frac{i\Theta}{T}\right) \right] , \tag{8.141}$$

where ψ represents Euler's digamma function [452]. As $T \to 0, \psi(i\Theta/T) \simeq \ln(i\Theta/T)$, which finally yields for λ:

$$\lambda \simeq \pi k_B T (\rho J_K)^2 \left[1 - 2\rho J_K \ln\left(\frac{\Theta}{T_0}\right) \right] . \tag{8.142}$$

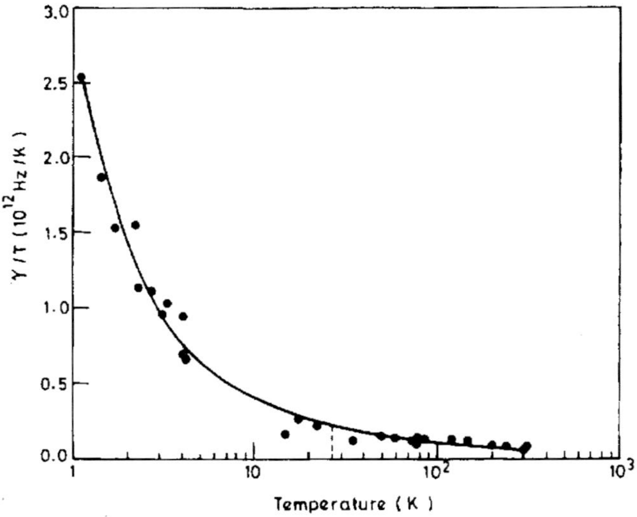

Fig. 8.14. Best-fit of (8.138) (denoted by a solid line) to the NMR longitudinal relaxation rates derived by Alloul [478] from satellite NMR data on Cu:Fe. The external field $\Theta = 0.9834$, and the best-fit line corresponds to $\rho J_K \simeq 0.28$ and $\alpha \simeq 3.4$. The Kondo temperature for Cu:Fe is approximately 27.6 K, and is indicated by a dashed line

This is referred to as *saturation* of the Kondo correction, and is observed in Mössbauer spectroscopy at very low temperatures [475–477].

The expression in (8.138) has been fitted to the NMR longitudinal relaxation rate derived by Alloul [478] from satellite NMR data on Cu:Fe (see Fig. 8.14). For this purpose, we have used the experimental value of the magnetic moment $\mu_{\text{eff}} = 2.74\mu_B$ for Cu:Fe. From the best-fit, the value of ρJ_K turns out to be slightly larger than 0.2, which is normally assumed for the Cu:Fe system. In the Mössbauer case, there is no external magnetic field, and ω_0 should be replaced by $2\pi a/k_B$, where a is the hyperfine constant. The resulting equation (8.138) is fitted to the data obtained by Bonville et al. [479] in Fig. 8.15.

In conclusion, we have derived an expression for the longitudinal impurity-spin-relaxation rate relevant for NMR and Mössbauer experiments. The analysis is based on the DBGA of a generalized spin-boson model. In our calculations, the exchange constant $J_{\|}$ has been treated exactly in order to remove the Kondo divergence. The DBGA has its limitations at low temperatures and long times, i.e., at low frequencies, as stated earlier. However, because the Zeeman or the hyperfine field is ubiquitous in resonance spectroscopy, the invalidity of the DBGA at low frequencies does not pose serious problems. In addition to the longitudinal relaxation rate, the transverse relaxation rate, the NMR Knight shift and Mössbauer hyperfine field can also be computed [435].

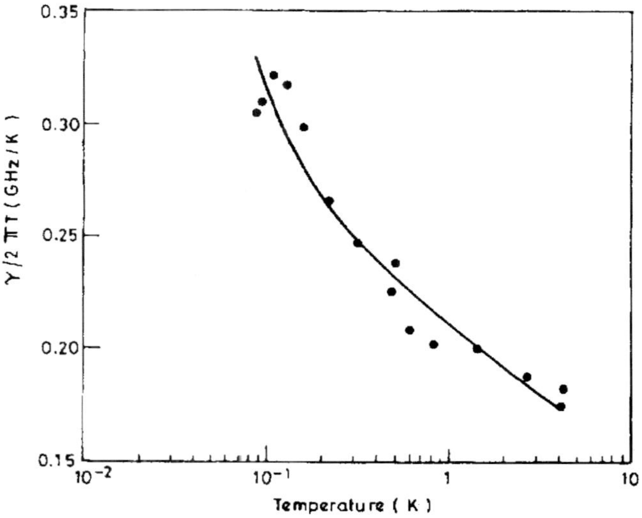

Fig. 8.15. Best-fit of (8.138) (denoted by a solid line) to the impurity spin-relaxation rates derived by Bonville et al. [479] from Mössbauer experiments on Au:Yb. The external field $\Theta = 0$, and the best-fit line corresponds to $\rho J_K \simeq 0.069$ and $\alpha \simeq 7.821$

8.10 Two-level Systems in Glasses

The quantum diffusion of hydrogen in $Nb(OH)_x$, as discussed in Sect. 8.8, provides an example of the low-temperature dynamics of two-level systems in metals. Fortunately, as elaborated earlier, a large body of unambiguous experimental results is available for this system, which can be subjected to detailed theoretical analysis. We have further shown in Sect. 8.9 that the analysis is based upon certain recent advances made in the theory of tunneling systems, which also bears a close resemblance to the theory of the Kondo effect. Hence, the tunneling of hydrogen in $Nb(OH)_x$ is a critical testing ground for these theoretical developments. It is also appropriate to query whether the theory, developed so successfully for hydrogen in $Nb(OH)_x$, can be applied to other systems in which two-level excitations are relevant.

An important example of such systems in materials science is that of *structural glasses*. In pioneering work, Pohl and his collaborators demonstrated that numerous insulating glasses exhibit universal thermodynamic properties at low temperatures [398,399]. This is depicted in Fig. 8.16, where the specific heat $C(T)$ (divided by the Debye contribution proportional to T^3) is plotted against the temperature T for some typical materials. The specific-heat data for crystalline α-quartz and its amorphous version (vitreous silica) demonstrate that quartz exhibits Debye behavior ($C(T)T^{-3} \sim$ constant) for $T < 5$ K. However, the specific heat for vitreous silica differs

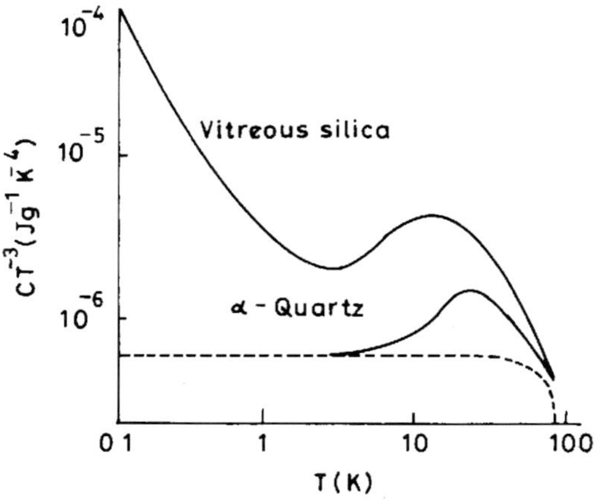

Fig. 8.16. Plot of specific heat $C(T)T^{-3}$ vs. T for vitreous silica and crystalline α-quartz [480]

significantly on two counts. Firstly, the magnitude of the low-temperature specific heat for the glassy system is two orders of magnitude larger than its crystalline counterpart. Secondly, and perhaps more significantly, the specific heat depends linearly on T between 0 and 1 K. Furthermore, this anomalous behavior of the specific heat is common to a wide variety of glassy materials. Similar universality is also exhibited in the thermal-conductivity data, which has a T^2-dependence at low temperatures.

Guided by these experimental results, Anderson et al. [400] and (independently) Phillips [401] postulated the existence of two-level systems in glasses. Recall that the free-energy landscape of a glass is characterized by deep valleys with multiple minima. Typically, the system becomes trapped in these local minima, and spends extended times in these metastable states. However, it is possible that there could be two minima close to each other in the configuration space of the glass, and a group of atoms occupying one of these two adjacent minima can tunnel to the other at low temperatures ($0 < T < 1$ K). Let $n(\Delta)d\Delta$ represent the number of two-level systems with tunnel splittings between Δ and $\Delta + d\Delta$. We make the simple assumption that $n(\Delta) \simeq n_0$, which is a constant. This results in the following expression for the specific heat [480]:

$$C(T) \simeq n_0 \int_0^\infty d\Delta \left(\frac{\Delta^2}{4k_B T^2} \right) \text{sech}^2 \left(\frac{\Delta}{2k_B T} \right) . \tag{8.143}$$

The integrand is simply the Schottky contribution which, on integrating over Δ, yields the requisite linear behavior at small T:

$$C(T) \simeq \frac{\pi^2}{6} n_0 k_B^2 T . \tag{8.144}$$

Recall that the specific heat is a thermodynamic quantity. On the other hand, the thermal conductivity κ is a transport property, and is given by the relation

$$\kappa = \frac{1}{3} C(T) v_s l \,, \tag{8.145}$$

where v_s is the velocity of sound, and l is the mean-free path of phonons. The calculation of κ is somewhat more complicated and requires a *golden-rule* evaluation of l, employing the coupling between tunneling centres and the phonons [480]. This coupling is analogous to that described in (8.35), and the subsequent analysis can be performed in parallel with the discussion in Sect. 8.4. In fact, the theory outlined earlier is applicable to *all* the dynamical properties of insulating glasses at low temperatures. We may further note that, as in the case of H in $Nb(OH)_x$, the double-well potentials in glasses also have a pronounced asymmetry.

Finally, we have discussed in Sect. 8.8 that the interaction between tunneling centres and phonons is characterized by the G-dependent term in the spectral density $J(\omega)$ in (8.113). On the other hand, the K-dependent (electronic) term is also relevant if we deal with metallic glasses instead of insulating glasses [481]. Thus, two-level systems in metallic glasses can also be treated using the methods outlined here for understanding nonadiabatic effects involving conduction electrons.

9 Quantum Diffusion

9.1 Introduction

Our discussion of dissipative phenomena has been based on two distinct examples of stochastic processes, viz., the discrete jump process and the continuous diffusion process. From a microscopic perspective, the jump process is modeled in terms of a single magnetic spin in contact with a heat bath (see Chap. 1). This model is generalizable to the case of an interacting many-body system, described by an Ising model (see Chap. 2). In Chaps. 3-6, we discussed various examples of phase ordering systems, which are described by generalizations of kinetic Ising models and their coarse-grained counterparts. In Chap. 7, we encountered a natural extension of the jump process to the quantum domain, i.e., the spin-boson model, where a single spin is subjected to mutually perpendicular magnetic fields in order to generate non-commuting quantum dynamics. Additionally, the bath is taken to be quantum-mechanical in Chap. 8, and is described by non-interacting bosons.

In the context of the above discussion, it is appropriate to next consider the quantum version of the diffusion process discussed in Chap. 1. A natural framework for this study is the phase-space dynamics of a single mechanical particle subjected to an arbitrary potential, and interacting with a bosonic bath. This bosonic bath is simply the second-quantized version of the non-interacting harmonic oscillators discussed in Chap. 1 [11,12]. Therefore, the system Hamiltonian is identical to that in (1.19). Furthermore, the bath Hamiltonian is the same as that in Chap. 8. However, an important difference from the spin-boson model of Chap. 8 is that the Hilbert space of the subsystem dynamics is infinite-dimensional in the present context.

It is relevant to highlight some important works in this context. Several authors have worked on models of a *quantum harmonic oscillator* in contact with a heat bath consisting of an infinite set of harmonic oscillators [482–488]. Furthermore, the Feynman-Vernon scheme [441] was extended by Caldeira and Leggett [19] to the case of a quantum particle moving in an *arbitrary potential*, and in contact with a bath of quantum harmonic oscillators. Caldeira and Leggett worked in the Schrödinger-picture representation of the density operator in a path-integral formalism. On the other hand, Ford et al. [489] used the Heisenberg picture to derive a *quantum Langevin equation* in a similar context. Their approach is analogous to that of Zwanzig [11,12], and is followed in our discussion in this chapter.

Our goal, as in Chap. 1, is to derive the quantum analog of the Langevin equation. However, as we will show subsequently, this quantum Langevin equation is governed by *non-white* noise even when the friction coefficient is a time-independent parameter, as appropriate to Ohmic dissipation. This result emphasizes the fact that a quantum bath is essentially *non-Markovian* over time-scales shorter than a *quantum time-scale* defined by $\beta\hbar$.

As mentioned above, our starting point is the Hamiltonian in (1.19), but we now use the Heisenberg equations of motion for position and momentum operators, as opposed to the classical Hamiltonian equations in Sect. 1.3. The resulting equations (presented below) are formally identical to the equations in Sect. 1.3, but it is necessary to interpret dynamical variables and their time-derivatives as quantum operators with a prescribed commutator algebra. For convenience, we use the canonically-transformed Hamiltonian in (1.22) in our subsequent discussion. Then, the relevant Langevin equation for the position operator of the particle is as follows (cf. (1.32)):

$$M\frac{d^2Q}{dt^2} + \int_0^t dt'\zeta(t-t')\frac{dQ(t')}{dt'} + \frac{\partial U(Q)}{\partial Q} = \theta(t)\,, \tag{9.1}$$

where, by choosing the *retarded* solution of the equation of motion as before, we have (tacitly) induced breaking of the time-reversal invariance. For the sake of ready reference, we quote again the expressions for the *memory function* $\zeta(t)$ and the *noise* $\theta(t)$ (cf. (1.33)-(1.34) in the canonically-transformed variables):

$$\zeta(t) = s(t)\sum_j m_j\omega_j^2\cos(\omega_j t)\,, \tag{9.2}$$

$$\theta(t) = \sum_j m_j\omega_j^2\left[q_j(0)\cos(\omega_j t) + \frac{p_j(0)}{m_j\omega_j}\sin(\omega_j t)\right]\,. \tag{9.3}$$

In (9.2), $s(t)$ denotes the step function. We should emphasize that the noise $\theta(t)$ is now a quantum operator as it depends explicitly on the dynamical variables $\{q_j\}$ and $\{p_j\}$ of the bath, which do not commute. We assume that the harmonic-oscillator bath is initially in thermal equilibrium at temperature T. Then, the spectral properties of the noise operator are characterized by the following symmetric and antisymmetric correlation functions:

$$\langle\{\theta(t),\theta(t')\}\rangle = \frac{2}{\pi}\int_0^\infty d\omega \mathrm{Re}\left[\tilde{\zeta}(\omega+i0^+)\right]\omega\coth\left(\frac{\beta\omega}{2}\right)\cos\left[\omega(t-t')\right]\,, \tag{9.4}$$

$$\langle[\theta(t),\theta(t')]\rangle = \frac{2}{i\pi}\int_0^\infty d\omega \mathrm{Re}\left[\tilde{\zeta}(\omega+i0^+)\right]\omega\sin\left[\omega(t-t')\right]\,. \tag{9.5}$$

Here, we have defined the "Laplace" transform of the memory function:

$$\tilde{\zeta}(s) = \int_{0}^{\infty} dt e^{ist}\zeta(t)\,, \quad \mathrm{Im} s > 0\,. \tag{9.6}$$

In this chapter, as in Chap. 8 we will work in units where $\hbar = 1$.

As in the case of the classical Langevin equation, it is useful to consider the case of constant *friction*, which is appropriate to Ohmic dissipation. Then, $\zeta(t-t') \to 2M\gamma\delta(t-t')$, and we retrieve the ordinary Langevin equation:

$$M\frac{d^2Q}{dt^2} + M\gamma\frac{dQ}{dt} + \frac{\partial U(Q)}{\partial Q} = \theta(t)\,, \tag{9.7}$$

where now (cf. (9.4)-(9.5))

$$\langle\{\theta(t), \theta(t')\}\rangle = \frac{2M\gamma}{\pi}\int_0^{\infty} d\omega\omega \coth\left(\frac{\beta\omega}{2}\right)\cos[\omega(t-t')]\,, \tag{9.8}$$

$$\langle[\theta(t), \theta(t')]\rangle = \frac{2M\gamma}{i\pi}\int_0^{\infty} d\omega\omega \sin[\omega(t-t')]\,. \tag{9.9}$$

Notice that the underlying stochastic process is non-Markovian (non-white) even though there is no memory, because of quantum fluctuations. Equations (9.7)-(9.9) will form the basis of our further analysis in this chapter.

9.2 Quantum Diffusion of a Free Particle

In the case of a free particle, the potential-energy term is absent, and (9.7) reduces to

$$\frac{d^2Q}{dt^2} + \gamma\frac{dQ}{dt} = \frac{\theta(t)}{M}\,. \tag{9.10}$$

The quantum dissipative motion of a free particle was discussed in detail by Hakim and Ambegaokar (HA), amongst others [490–492]. HA employed the functional-integral representation of the time-dependent density matrix of the system, following Caldeira and Leggett [19]. They evaluated the so-called *influence functional* by explicitly diagonalizing the Hamiltonian in the free-particle case. It is instructive to compare their results with those obtained from the quantum Langevin equation (9.10), as detailed below.

The solution of (9.10) is as follows:

$$Q(t) = Q(t_0) + \gamma^{-1}\left.\frac{dQ}{dt}\right|_{t_0}\left[1 - e^{-\gamma(t-t_0)}\right] + \int_{t_0}^{t} d\tau e^{-\gamma\tau}\int_{t_0}^{\tau} dt' e^{\gamma t'}\frac{\theta(t')}{M}\,, \tag{9.11}$$

where we have taken the initial time to be t_0. Without loss of generality, we choose $Q(t_0) = 0$. Taking the time-derivative of (9.11), we obtain

$$v(t) = v(t_0)e^{-\gamma(t-t_0)} + e^{-\gamma t} \int_{t_0}^{t} dt' e^{\gamma t'} \frac{\theta(t')}{M}, \tag{9.12}$$

where $v(t)$ is the velocity operator of the particle.

Equations (9.11)-(9.12) can be used to evaluate various correlation functions. For instance, the velocity auto-correlation in equilibrium is given by [493]

$$\langle v(t)v(0)\rangle_0 = \frac{1}{2M\pi} \int_{-\infty}^{\infty} d\omega \frac{\gamma}{\gamma^2+\omega^2} \, \omega \coth\left(\frac{\beta\omega}{2}\right) \cos(\omega t). \tag{9.13}$$

For a free particle, all three cartesian components of the velocity have the same correlation. Therefore, the vector version of (9.13) is

$$\langle \boldsymbol{v}(t) \cdot \boldsymbol{v}(0)\rangle_0 = \frac{3}{2M\pi} \int_{-\infty}^{\infty} d\omega \frac{\gamma}{\gamma^2+\omega^2} \, \omega \coth\left(\frac{\beta\omega}{2}\right) \cos(\omega t). \tag{9.14}$$

In particular, the equal-time correlation is as follows:

$$\langle \boldsymbol{v}^2\rangle_0 = \frac{3}{M\pi} \int_{0}^{\infty} d\omega \frac{\gamma}{\gamma^2+\omega^2} \, \omega \coth\left(\frac{\beta\omega}{2}\right). \tag{9.15}$$

In deriving (9.13)-(9.15), we have ignored the dependence on the initial time t_0. This assumes that, in the distant past, the particle is decoupled from the heat bath, which is maintained in thermal equilibrium at a fixed temperature T. Equation (9.15) agrees with the corresponding result due to HA [490].

On the other hand, if we retain the dependence on the initial time t_0 $(= 0$, say), (9.12) yields

$$\begin{aligned}\langle \boldsymbol{v}^2(t)\rangle = {} & \frac{3\gamma}{M\pi} \int_{0}^{\infty} d\omega \frac{\omega}{\gamma^2+\omega^2} \coth\left(\frac{\beta\omega}{2}\right) \left[1 + e^{-2\gamma t} - 2e^{-\gamma t}\cos(\omega t)\right] \\ & + \langle \boldsymbol{v}^2(0)\rangle e^{-2\gamma t}.\end{aligned} \tag{9.16}$$

As expected, (9.16) is in complete agreement with the result obtained by HA, for the initial condition in which the total density matrix is assumed to be the factorized product of the density matrices for the particle and the heat bath.

The discussion presented above refers to quantum diffusion processes. In the appropriate classical limit, we should be able to recover the results obtained in Chap. 1 for classical diffusion processes. We demonstrate this for the velocity auto-correlation function in the high-temperature ($T \to \infty$) limit, where the coth-function in (9.13)-(9.16) can be replaced by the inverse of its argument. Then, performing the integral in (9.14), we obtain

$$\langle \boldsymbol{v}(t) \cdot \boldsymbol{v}(0) \rangle_0 = \frac{3k_BT}{M} e^{-\gamma t} , \tag{9.17}$$

which agrees with the result in (1.15).

Furthermore, we can use (9.11) to obtain the mean-squared displacement:

$$\begin{aligned} \langle Q^2(t) \rangle = \left\langle \frac{dQ}{dt}\bigg|_0^2 \right\rangle \left(\frac{1-e^{-\gamma t}}{\gamma} \right)^2 \\ + \frac{k_BT}{M} \left[\frac{2}{\gamma^2} \left(\gamma t - 1 + e^{-\gamma t} \right) - \left(\frac{1-e^{-\gamma t}}{\gamma} \right)^2 \right] . \end{aligned} \tag{9.18}$$

If the particle is initially in thermal equilibrium with the bath, the equipartition theorem applies:

$$\left\langle \frac{dQ}{dt}\bigg|_0^2 \right\rangle_0 = \frac{k_BT}{M} . \tag{9.19}$$

We then obtain the familiar result for diffusive motion:

$$\langle Q^2(t) \rangle_0 = \frac{2k_BT}{M\gamma^2} \left(\gamma t - 1 + e^{-\gamma t} \right) . \tag{9.20}$$

9.3 Dynamics of a Charged Particle in a Magnetic Field

An important application of quantum diffusion processes is the dissipative dynamics of a charged particle, e.g., an electron in the presence of an external magnetic field. The quantum dynamics of a charge in a magnetic field is characterized by coherent precession of its velocity vector around the magnetic field. This occurs at a rate given by the cyclotron frequency $\Omega_c = |e|B/(Mc)$, where e is the electric charge, B is the strength of the magnetic field, M is the mass of the particle, and c is the speed of light. Dissipation, which is quantified by the friction coefficient γ, leads to incoherence in this motion. When $\gamma \gg \Omega_c$, we expect classical-like evolution in an otherwise quantum problem. In addition to the parameters γ and Ω_c, there is also the *quantal* frequency $(\beta\hbar)^{-1}$, which was mentioned earlier. This frequency is ubiquitous in a quantum system at a finite temperature, and characterizes the interplay of quantal and thermal fluctuations. Typically, it can be as large as $10^{11}\,\mathrm{s}^{-1}$ even at such low temperatures as $T \simeq 1\,\mathrm{K}$, i.e., quantum coherence effects vanish on time-scales larger than $10^{-11}\,\mathrm{s}$ at $T \simeq 1\,\mathrm{K}$. Thus, we expect to see different behavior in different regimes of the three competing frequency scales Ω_c, γ and $(\beta\hbar)^{-1}$.

In order to study the effects of the magnetic field, it is necessary to generalize the $d = 1$ analysis of Sect. 9.1. We must formulate vector equations of motion, because the dynamics now occurs in the (x, y)-plane if the z-axis is chosen as the field direction. The quantum Langevin equation (9.7) is then rewritten as

$$M\frac{d^2\boldsymbol{Q}}{dt^2} + M\gamma\frac{d\boldsymbol{Q}}{dt} + \frac{\partial U(\boldsymbol{Q})}{\partial \boldsymbol{Q}} - \frac{e}{c}\left(\frac{d\boldsymbol{Q}}{dt} \times \boldsymbol{B}\right) = \boldsymbol{\theta}(t)\,, \tag{9.21}$$

where the additional term on the LHS is the Lorentz force which appears in a manifestly gauge-independent form. It is interesting to note that, although (9.21) can be derived from an underlying Hamiltonian in which the magnetic field occurs through *minimal coupling* in the kinetic energy, the bath properties remain unaffected by the magnetic field. Thus, using indices i, j, etc. to represent the cartesian axes x, y and z, we have

$$\langle\{\theta_i(t), \theta_j(t')\}\rangle = \delta_{ij}\frac{2M\gamma}{\pi}\int_0^\infty d\omega\omega\coth\left(\frac{\beta\omega}{2}\right)\cos[\omega(t-t')]\,, \tag{9.22}$$

$$\langle[\theta_i(t), \theta_j(t')]\rangle = \delta_{ij}\frac{2M\gamma}{i\pi}\int_0^\infty d\omega\omega\sin[\omega(t-t')]\,. \tag{9.23}$$

As commented earlier, the underlying stochastic process is non-Markovian. Therefore, unlike the classical case, there is no appropriate Fokker–Planck description of the dynamics. Hence, for calculational purposes, it is convenient to directly use the Langevin equation (9.21), as we discuss below.

In order to focus on the magnetic-field effects, we ignore the potential energy $U(\boldsymbol{Q})$. As the field points along the z-axis, the motion in the z-direction is that of a free, quantum particle in a dissipative environment. On the other hand, the motion in the (x, y)-plane is described by the equations:

$$\begin{aligned}\frac{d^2x}{dt^2} + \gamma\frac{dx}{dt} - \Omega_c\frac{dy}{dt} &= \frac{\theta_x(t)}{M}\,,\\ \frac{d^2y}{dt^2} + \gamma\frac{dy}{dt} + \Omega_c\frac{dx}{dt} &= \frac{\theta_y(t)}{M}\,.\end{aligned} \tag{9.24}$$

We introduce the variables:

$$\begin{aligned}\xi &= x + iy\,,\\ \theta &= \theta_x + i\theta_y\,,\\ \overline{\gamma} &= \gamma + i\Omega_c\,.\end{aligned} \tag{9.25}$$

Then, (9.24) can be expressed in the compact form

$$\frac{d^2\xi}{dt^2} + \overline{\gamma}\frac{d\xi}{dt} = \frac{\theta(t)}{M}\,, \tag{9.26}$$

with the solution

$$\xi(t) = \bar{\gamma}^{-1}\frac{d\xi}{dt}\bigg|_{t_0}\left[1 - e^{-\bar{\gamma}(t-t_0)}\right] + \int_{t_0}^{t} d\tau e^{-\bar{\gamma}\tau}\int_{t_0}^{\tau} dt' e^{\bar{\gamma}t'}\frac{\theta(t')}{M}, \quad (9.27)$$

assuming $\xi(t_0) = 0$. The time-derivative of (9.27) yields the "velocity" as

$$\frac{d\xi}{dt} = \frac{d\xi}{dt}\bigg|_{t_0} e^{-\bar{\gamma}(t-t_0)} + e^{-\bar{\gamma}t}\int_{t_0}^{t} dt' e^{\bar{\gamma}t'}\frac{\theta(t')}{M}. \quad (9.28)$$

Consider next the correlation function

$$C(t,t') = \langle\dot{\xi}(t)\dot{\xi}^{\dagger}(t')\rangle, \quad t \geq t'. \quad (9.29)$$

where the dot denotes the time-derivative, and the superscript † denotes the adjoint operation. Using the definition of $\xi(t)$ from (9.25), this correlation function can be expressed as

$$C(t,t') = \langle\dot{\xi}(t_0)\dot{\xi}^{\dagger}(t_0)\rangle \exp\left[-\bar{\gamma}(t-t_0) - \bar{\gamma}^*(t'-t_0)\right] + \frac{1}{M^2}e^{-(\bar{\gamma}t+\bar{\gamma}^*t')}\int_{t_0}^{t} d\tau\int_{t_0}^{t'} d\tau' e^{(\bar{\gamma}\tau+\bar{\gamma}^*\tau')}\langle\theta(\tau)\theta^{\dagger}(\tau')\rangle, \quad (9.30)$$

where we have used the facts that the initial velocity $\dot{\xi}(t_0)$ is uncorrelated with $\theta(\tau)$ (from causality), and that

$$\langle\theta(\tau)\rangle = 0. \quad (9.31)$$

We are interested in the limit $t_0 = -\infty$, so we ignore the first term on the RHS of (9.30) for our subsequent analysis. Furthermore, we have (cf. (9.25))

$$\langle\theta(\tau)\theta^{\dagger}(\tau')\rangle = \frac{M\gamma}{\pi}\int_{-\infty}^{\infty} d\omega\omega e^{i\omega(\tau-\tau')}\left[\coth\left(\frac{\beta\omega}{2}\right) - 1\right]. \quad (9.32)$$

Substituting (9.32) in (9.30), integrating over τ and τ' and taking the limit $t_0 = -\infty$, we finally obtain [493]

$$C(t,t') = \frac{1}{M\pi}\int_{-\infty}^{\infty} d\omega\frac{\gamma}{\gamma^2 + (\omega + \Omega_c)^2}\omega\left[\coth\left(\frac{\beta\omega}{2}\right) - 1\right]e^{i\omega(t-t')}. \quad (9.33)$$

Thus, the correlation function is manifestly a function of the time-difference $(t - t')$, implying stationarity. Hence, t' can be set equal to zero.

Now, the correlation function, as defined in (9.29), has an expanded representation

$$C(t,t') = \langle v_x(t)v_x(t')\rangle + \langle v_y(t)v_y(t')\rangle + i\langle[\boldsymbol{v}(t)\times\boldsymbol{v}(t')]_z\rangle. \quad (9.34)$$

Comparing with (9.33), and considering the real and imaginary parts separately, we identify (setting $t' = 0$)

$$\langle v_x(t)v_x(0)\rangle + \langle v_y(t)v_y(0)\rangle = \frac{1}{M\pi}\int_{-\infty}^{\infty} d\omega \frac{\gamma}{\gamma^2 + (\omega + \Omega_c)^2} \times\omega\left[\coth\left(\frac{\beta\omega}{2}\right) - 1\right]\cos(\omega t)\,, \quad (9.35)$$

and

$$\langle [\boldsymbol{v}(t)\times\boldsymbol{v}(0)]_z\rangle = \frac{1}{M\pi}\int_{-\infty}^{\infty} d\omega \frac{\gamma}{\gamma^2 + (\omega + \Omega_c)^2}\,\omega\left[\coth\left(\frac{\beta\omega}{2}\right) - 1\right]\sin(\omega t)\,. \quad (9.36)$$

We next analyze the mean-squared displacement, for which it is useful to consider the following correlation function:

$$\chi(t,t') = \langle \xi(t)\xi^\dagger(t')\rangle\,, \quad t \geq t'\,. \quad (9.37)$$

Equation (9.27) yields the following expression for this quantity:

$$\chi(t,t') = \langle \dot{\xi}(t_0)\dot{\xi}^\dagger(t_0)\rangle \left[1 + e^{(-\overline{\gamma}t - \overline{\gamma}^* t' + 2\gamma t_0)} - e^{-\overline{\gamma}(t-t_0)} - e^{-\overline{\gamma}^*(t'-t_0)}\right] + \frac{1}{M^2}\int_{t_0}^{t} d\tau e^{-\overline{\gamma}\tau}\int_{t_0}^{\tau} dt'' e^{\overline{\gamma}t''}\int_{t_0}^{t'} d\tau' e^{-\overline{\gamma}^*\tau'}\int_{t_0}^{\tau'} dt''' e^{\overline{\gamma}^* t'''}\langle \theta(t'')\theta^\dagger(t''')\rangle\,. \quad (9.38)$$

Using (9.32) and performing the integrals, this expression simplifies as follows [493]:

$$\begin{aligned}\chi(t,t') = {} & \frac{\langle \dot{\xi}(t_0)\dot{\xi}^\dagger(t_0)\rangle}{\gamma^2 + \Omega_c^2}\left[1 + e^{(-\overline{\gamma}t - \overline{\gamma}^* t' + 2\gamma t_0)} - e^{-\overline{\gamma}(t-t_0)} - e^{-\overline{\gamma}^*(t'-t_0)}\right] \\ & + \frac{1}{M\pi}\int_{-\infty}^{\infty} d\omega\omega\left[\coth\left(\frac{\beta\omega}{2}\right) - 1\right]\frac{\gamma}{\gamma^2 + (\omega + \Omega_c)^2} \\ & \times\left\{\frac{1}{\omega^2}\left[1 + e^{i\omega(t-t')} - e^{i\omega(t-t_0)} - e^{i\omega(t'-t_0)}\right]\right. \\ & + \frac{1}{\gamma^2 + \Omega_c^2}\left[1 + e^{(-\overline{\gamma}t - \overline{\gamma}^* t' + 2\gamma t_0)} - e^{-\overline{\gamma}(t-t_0)} - e^{-\overline{\gamma}^*(t'-t_0)}\right] \\ & - \frac{i}{\overline{\gamma}\omega}\left[1 - e^{-i\omega(t'-t_0)}\right]\left[e^{-\overline{\gamma}(t-t_0)} - 1\right] \\ & \left. + \frac{i}{\overline{\gamma}^*\omega}\left[1 - e^{i\omega(t-t_0)}\right]\left[e^{-\overline{\gamma}^*(t'-t_0)} - 1\right]\right\}\,. \quad (9.39)\end{aligned}$$

Taking the $t' = t$ limit and setting $t_0 = 0$, we finally have from (9.39)

$$\langle x^2(t) + y^2(t)\rangle = \frac{\langle \dot{x}(0)^2 + \dot{y}(0)^2\rangle}{\gamma^2 + \Omega_c^2}\left[1 + e^{-2\gamma t} - 2e^{-\gamma t}\cos(\Omega_c t)\right]$$
$$+\frac{1}{M\pi}\int_{-\infty}^{\infty} d\omega \frac{\gamma}{\gamma^2 + (\omega + \Omega_c)^2}\,\omega\left[\coth\left(\frac{\beta\omega}{2}\right) - 1\right]$$
$$\times\left\{\frac{4}{\omega^2}\sin^2\left(\frac{\omega t}{2}\right) + \frac{1}{\gamma^2 + \Omega_c^2}\left[1 + e^{-2\gamma t} - 2e^{-\gamma t}\cos(\Omega_c t)\right]\right.$$
$$+\frac{2}{\omega(\gamma^2 + \Omega_c^2)}\left[2\Omega_c \sin^2\left(\frac{\omega t}{2}\right) - \gamma\sin(\omega t) + \right.$$
$$\left.\left. e^{-\gamma t}\left[\Omega_c\cos(\omega + \Omega_c)t - \Omega_c\cos(\Omega_c t) + \gamma\sin(\omega + \Omega_c)t - \gamma\sin(\Omega_c t)\right]\right]\right\}. \tag{9.40}$$

Again, the physical implications of the above expressions can be clarified by considering the high-temperature limit, viz., the special case of dissipative dynamics of a classical charged particle in a *magnetic field*. This problem is particularly relevant in the context of plasma confinement in a tokomak [494]. Furthermore, the results obtained in Sect. 9.2 for a free particle can be recovered by setting $\Omega_c = 0$ in (9.35) and (9.40).

9.4 Landau Diamagnetism in a Dissipative System

As indicated in Sect. 9.3, the quantum-mechanical problem of a charged particle in the presence of a magnetic field serves as a paradigm for various exactly solvable models [495]. For example, in the context of condensed matter physics, the well-known Landau theory of diamagnetism results from the solution of this problem [496,497]. The physics of Landau levels is of great interest in diverse physical systems, e.g., the quantum Hall effect, high-temperature superconductivity, etc. [498,499]. In this section, we study Landau diamagnetism in the context of dissipative quantum mechanics [500].

Diamagnetism is a material property that arises from the response of an ensemble of charged particles (e.g., electrons) to an applied magnetic field. This field causes cyclotron motion of each particle, thereby creating an orbital magnetic moment, governed by the Faraday-Lenz law [497]. Thus, the system exhibits a negative magnetic susceptibility, which is characteristic of diamagnetism. However, in the context of classical statistical physics, the celebrated Bohr–van Leeuwen (BVL) theorem demonstrates that the diamagnetic susceptibility is identically zero [501,502]. The boundary of the system plays a crucial role in this result – the bulk contribution of the diamagnetic moment exactly cancels the contribution arising from the so-called *skipping orbits* of electrons which hit the boundary and get multiply reflected to constitute an *edge current* [497].

In the quantum-mechanical context, Landau showed that the discreteness of energy levels (and the consequent degeneracy of each level) results in a non-zero diamagnetic susceptibility [496]. Landau's calculation also suggests that the bulk and surface contributions are different in quantum mechanics. Therefore, the cancellation of the two terms is incomplete, unlike in the classical case. This result is unusual in that the boundary of the system, which is generally unimportant for thermodynamic properties, plays a crucial role in determining diamagnetism [503]. Landau's result also demonstrated that the calculation of diamagnetic susceptibility requires an explicitly quantum treatment.

Our study in this section is motivated by the following question: Since dissipation is known to lead to classical-like motion of a quantum system, should one expect the BVL theorem to be restored in the limit of large damping? We show that a proper analysis of this question requires a careful treatment of the boundary, bringing out the subtle roles of the *thermodynamic limit* and *approach to equilibrium* [500]. The confined boundaries are also of great interest in their own right, in view of the current focus on mesoscopic systems. Towards the end of this section, we present results for diamagnetism when the electrons are constrained to move in a $d = 2$ parabolic potential in the plane normal to the applied field.

With this background, let us calculate the diamagnetic moment, for which we need the time-dependent magnetic moment due to the circular electron motion:

$$\begin{aligned} m_z(t) &= \frac{|e|}{2c} \langle x\dot{y} - y\dot{x} \rangle \\ &= \frac{|e|}{4c} \mathrm{Im} \left\langle \dot{\xi}\xi^\dagger + \xi^\dagger\dot{\xi} \right\rangle . \end{aligned} \tag{9.41}$$

The expression for $m_z(t)$ can be computed using the solution for $\xi(t)$ given in (9.27). That result corresponds to the value of $m_z(t)$ in a non-equilibrium situation, appropriate for describing the magneto-transport behavior of the system. However, we focus our attention on equilibrium properties, which can be obtained by taking the limits $t_0 = 0$ and $t = \infty$. The Langevin equation builds in, at the outset, the fluctuation-dissipation theorem, as a consequence of the detailed balance of transitions. Therefore, it is guaranteed that the subsystem of interest does equilibrate with the surrounding heat bath.

Using the results of Sect. 9.3, and after some algebra, we obtain [500]

$$\lim_{t\to\infty} m_z(t) = m_z = -\frac{|e|}{2\pi Mc} \int_{-\infty}^{\infty} d\omega \frac{\gamma}{\gamma^2 + (\omega - \Omega_c)^2} \coth\left(\frac{\beta\omega}{2}\right) . \tag{9.42}$$

The dissipation-free limit ($\gamma \to 0$) corresponds to the original Landau problem. In that case, we should recover the expected result, which can be independently derived from the partition function of equilibrium statistical mechanics. Note that

$$\lim_{\gamma \to 0} \frac{\gamma}{\gamma^2 + (\omega - \Omega_c)^2} = \pi \delta(\omega - \Omega_c) \,. \tag{9.43}$$

Therefore, (9.42) yields for $\gamma = 0$:

$$m_z^0 = -\frac{|e|}{2Mc} \coth\left(\frac{\beta \Omega_c}{2}\right) . \tag{9.44}$$

Interestingly, this is only part of the Landau result [496,497], and arises in the equilibrium calculation if one neglects the role of *boundary electrons*. As has been stressed earlier, the skipping orbits of the boundary electrons make an essential contribution to diamagnetism.

Therefore, in order to obtain the complete result for m_z, we must consider the dissipative dynamics of the charged particle in a finite volume. As the latter is difficult to implement mathematically, we employ a method due to Darwin [504]. Darwin was interested in calculating m_z^0 from the partition function, involving a trace over the density matrix. In order to obtain the correct Landau expression, he introduced a *fictitious* constraining potential $V(x, y) = k(x^2 + y^2)/2$ (where k is a parameter) in the Hamiltonian, completed the calculation of the trace, and then set $k = 0$. We adopt the same approach here, in which case (9.26) is modified to

$$\frac{d^2\xi}{dt^2} + \overline{\gamma}\frac{d\xi}{dt} + \frac{k}{M}\xi = \frac{\theta(t)}{M} \,. \tag{9.45}$$

We should remark that, if we switch off the magnetic field, $\Omega_c = 0$ and $\overline{\gamma} = \gamma$. Hence, (9.45) includes the Langevin equation of a damped quantum harmonic oscillator. Thus, the solution of (9.45) (presented below) also contains the result for the *quantum Brownian motion* of a harmonic oscillator [505]. The solution of (9.45) (with $\xi(t_0 = 0) = 0$) now becomes

$$\xi(t) = \frac{1}{\omega_+ - \omega_-} \left\{ \dot{\xi}(0) \left(e^{\omega_+ t} - e^{\omega_- t} \right) + \int_0^t d\tau \left[e^{\omega_+ (t-\tau)} - e^{\omega_- (t-\tau)} \right] \frac{\theta(\tau)}{M} \right\} , \tag{9.46}$$

where

$$\omega_\pm = -\frac{\overline{\gamma}}{2} \pm \frac{1}{2}\sqrt{\overline{\gamma}^2 - \omega_0^2} \, , \quad \omega_0^2 = \frac{4k}{M} \, . \tag{9.47}$$

Substituting this result in (9.41), and letting $t \to \infty$, we obtain [500]

$$\begin{aligned} m_z = &\frac{|e|\gamma}{2\pi Mc} \frac{1}{|\omega_+ - \omega_-|^2} \mathrm{Im} \int_{-\infty}^{\infty} d\omega \omega \coth\left(\frac{\beta\omega}{2}\right) \left(\frac{1}{\omega + i\omega_+^*} - \frac{1}{\omega + i\omega_-^*} \right) \\ &\times \left(\frac{\omega_+}{\omega - i\omega_+} - \frac{\omega_-}{\omega - i\omega_-} \right) . \end{aligned} \tag{9.48}$$

Before taking the Darwin limit (i.e., $k \to 0$, or $\omega_0 \to 0$) in (9.48), it is important to compute the integral over ω. This is done by closing the contour in the upper half-plane and noting that the coth-function has poles at $\beta\omega/2 = in\pi$, n being an integer. The overall result is

$$m_z = -\frac{|e|\eta}{Mc}\mathrm{Im}\left\{\sum_{n=1}^{\infty}\frac{n^2\pi^2}{(n\pi+\nu_+^*)(n\pi+\nu_-^*)(n\pi-\nu_+)(n\pi-\nu_-)}\right.$$
$$\left.+\frac{1}{(\nu_+^*-\nu_-^*)}\left[\frac{(\nu_+^*)^2\cot(\nu_+^*)}{(\nu_++\nu_+^*)(\nu_+^*+\nu_-)}-\frac{(\nu_-^*)^2\cot(\nu_-^*)}{(\nu_-+\nu_-^*)(\nu_-^*+\nu_+)}\right]\right\}, \tag{9.49}$$

where we have introduced the dimensionless quantities:

$$\eta = \frac{\beta\gamma}{2}, \quad \nu_\pm = \frac{\beta\omega_\pm}{2}. \tag{9.50}$$

Equation (9.49) provides an expression for the orbital (diamagnetic) magnetization of a charged particle in a quantum dissipative environment, and moving in a confining parabolic potential. For $\omega_0 \to 0$, we have

$$\nu_+ \simeq \frac{\nu_0^2}{4}\frac{(\eta - i\nu_c)}{\eta^2+\nu_c^2}, \quad \nu_- \simeq -(\eta + i\nu_c),$$
$$\nu_c = \frac{\beta\Omega_c}{2}, \quad \nu_0 = \frac{\beta\omega_0}{2}. \tag{9.51}$$

For $\omega_0 = 0$, the result for the magnetization is

$$m_z = \frac{|e|}{2Mc}\left[\sum_{n=1}^{\infty}\frac{4n\pi\eta\nu_c}{(\nu_c^2+\eta^2-n^2\pi^2)^2+4n^2\pi^2\nu_c^2}+\frac{\nu_c}{\eta^2+\nu_c^2}\right.$$
$$\left.-\frac{1}{2}\frac{\sinh(2\nu_c)}{\sinh^2\nu_c+\sin^2\eta}\right]. \tag{9.52}$$

Equation (9.52) now has the correct zero-dissipation limit because, when $\eta = 0$, we retrieve the Landau result

$$m_z^0 = \frac{|e|}{2Mc}\left[\frac{1}{\nu_c} - \coth(\nu_c)\right]. \tag{9.53}$$

At this stage, a few comments on (9.52) are in order. First, note that the magnetic-field dependence of m_z can be clarified by simplifying (9.48) as

$$m_z(B) = -\frac{|e|\gamma}{2\pi Mc}\int_{-\infty}^{\infty} d\omega\omega^2\coth\left(\frac{\beta\omega}{2}\right)\frac{1}{(\omega^2-\omega_0^2-\omega\Omega_c)^2+\omega^2\gamma^2}, \tag{9.54}$$

where the cyclotron frequency Ω_c has a linear dependence on B. The transformation $B \to -B$ is equivalent to $\Omega_c \to -\Omega_c$. We then change the integration variable $\omega \to -\omega$ to obtain

$$m_z(-B) = -m_z(B)\,, \tag{9.55}$$

in conformity with time-reversal invariance.

Next, we may rewrite (9.52) as

$$m_z = \frac{|e|}{2Mc}\left\{\sum_{n=1}^{\infty}\frac{4n\pi\eta\nu_c}{(\nu_c^2+\eta^2-n^2\pi^2)^2+4n^2\pi^2\nu_c^2}\right.$$
$$\left. +\mathrm{Re}\left[\frac{1}{\nu_c - i\eta} - \coth(\nu_c - i\eta)\right]\right\}\,. \tag{9.56}$$

Equation (9.56) is in a suggestive form as the last two terms constitute the Landau answer (cf. (9.53)), with the Landau levels being *lifetime-broadened* due to damping, yielding a *Matsubara frequency* [495]. The first term on the RHS of (9.56) (involving the summation over n) provides a nontrivial correction to the modified Landau terms. Using the following summation formula:

$$\coth z = \sum_{n=-\infty}^{\infty}\frac{1}{z+in\pi}$$
$$= \frac{1}{z} + 2\sum_{n=1}^{\infty}\frac{z}{z^2+n^2\pi^2}\,, \tag{9.57}$$

we can express (9.56) in the compact form

$$m_z = -2k_BTB\left(\frac{e}{Mc}\right)^2\sum_{n=1}^{\infty}\frac{1}{(\nu_n+\gamma)^2+\Omega_c^2}\,, \tag{9.58}$$

where

$$\nu_n = 2k_BTn\pi\,. \tag{9.59}$$

This form can be immediately generalized to the *non-Ohmic* case, characterized by a frequency-dependent friction coefficient, in which case (9.58) becomes [506]

$$m_z = -2k_BTB\left(\frac{e}{Mc}\right)^2\sum_{n=1}^{\infty}\frac{1}{[\nu_n+\gamma(i\nu_n)]^2+\Omega_c^2}\,. \tag{9.60}$$

Thus, m_z always remains negative, irrespective of whether the dissipation is Ohmic or non-Ohmic. Furthermore, the magnitude of m_z is reduced from the Landau result (which arises for $\gamma = 0$).

Finally, we should point out that the result given in (9.49), containing the contribution of the confining parabolic potential, is interesting in itself as it can be realized in mesoscopic quantum structures [507,508]. We consider

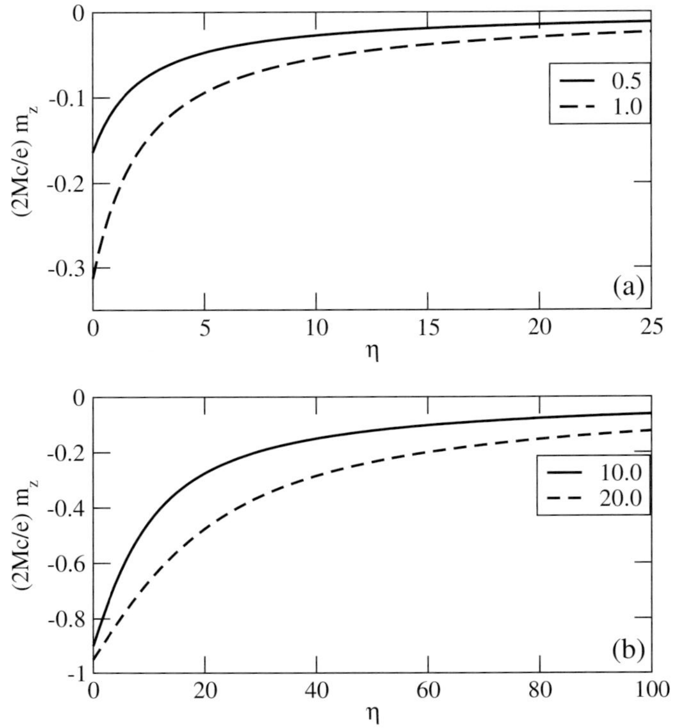

Fig. 9.1. Plot of m_z vs. η for different values of ν_c. The data for m_z is in units of $|e|/(2Mc)$, the Bohr magneton. Results are shown for (**a**) $\nu_c = 0.5, 1.0$, and (**b**) $\nu_c = 10.0, 20.0$

the limiting case of (9.49) for zero damping in order to study the role of the boundary. The corresponding result is

$$m_z = \frac{|e|}{4Mc}\frac{1}{\sqrt{\nu_0^2+\nu_c^2}}\left\{\left(\sqrt{\nu_0^2+\nu_c^2}-\nu_c\right)\coth\left[\frac{1}{2}\left(\sqrt{\nu_0^2+\nu_c^2}-\nu_c\right)\right]\right.$$
$$\left.-\left(\sqrt{\nu_0^2+\nu_c^2}+\nu_c\right)\coth\left[\frac{1}{2}\left(\sqrt{\nu_0^2+\nu_c^2}+\nu_c\right)\right]\right\} . \qquad (9.61)$$

The analysis presented in this section helps clarify the following issues:

1. The approach to equilibrium of a quantum dissipative system, the analysis of which elucidates the subtle role of boundary electrons.
2. The effect of dissipation on Landau diamagnetism, which is an equilibrium property (cf. (9.56)).
3. The combined effect of dissipation and confinement on orbital magnetism.

Before concluding this section, we plot (in Fig. 9.1) m_z vs. η for different values of the dimensionless cyclotron frequency ν_c. The plots are obtained from a numerical evaluation of (9.52). It is seen that $m_z \to 0$ monotonically for large values of the dimensionless damping parameter η, although the approach is slower for larger values of ν_c. This observation underscores the point made earlier concerning the competition between coherence and dissipation, and we will return to this issue in Chap. 10. The other relevant point, which has also been made earlier, is that for $\eta = 0$, (9.52) reduces to the Landau result in (9.53). The latter, for large values of the cyclotron frequency ν_c, yields a saturation value of the magnetization at $-|e|/(2Mc)$. This is evident from Fig. 9.1(b) for $\nu_c = 20$, and corresponds to a physical situation where only the lowest Landau level is preferentially occupied.

10 Coherence and Decoherence

10.1 Introduction

Decoherence can be introduced in the evolution of a quantum system by the environment. This is an important issue in the topical areas of mesoscopic phenomena [509], quantum measurement problems [510–512], wave-packet dynamics in atomic physics [513], electron transfer in chemical and biological reactions [514,515], and quantum information processes [516]. In this chapter, we will discuss decoherence in the context of the problems of condensed matter physics which were discussed in Chaps. 8 and 9. At the outset, it is necessary to clarify the meaning of the terms *coherence* and *decoherence.* Following Imry [517], we consider a beam of particles with momentum $\boldsymbol{k}_i$, incident from the left on two rigid, elastic, point scatterers located at $\boldsymbol{x}_1$ and $\boldsymbol{x}_2$ (see Fig. 10.1). The beam is scattered with a wave-vector $\boldsymbol{k}_f$, resulting in a momentum transfer $\boldsymbol{k} = \boldsymbol{k}_f - \boldsymbol{k}_i$. Assuming that the two amplitudes of scattering are the same, the scattered intensity can be written as

$$S(\boldsymbol{k}) = \left|A(\boldsymbol{k})\left(e^{i\boldsymbol{k}\cdot\boldsymbol{x}_1} + e^{i\boldsymbol{k}\cdot\boldsymbol{x}_2}\right)\right|^2 . \tag{10.1}$$

This quantity consists of two parts:

$$S(\boldsymbol{k}) = S_1(\boldsymbol{k}) + S_2(\boldsymbol{k}) , \tag{10.2}$$

where

$$S_1(\boldsymbol{k}) = 2|A(\boldsymbol{k})|^2 , \tag{10.3}$$

and

$$S_2(\boldsymbol{k}) = 2|A(\boldsymbol{k})|^2 \cos(\boldsymbol{k}\cdot\boldsymbol{d}) , \tag{10.4}$$

with $\boldsymbol{d} = \boldsymbol{x}_2 - \boldsymbol{x}_1$.

Notice that the first contribution to the intensity, $S_1(\boldsymbol{k})$, corresponds to adding the two probabilities (each being $|A(\boldsymbol{k})|^2$). Therefore, this contribution may be regarded as being *classical.* The second contribution, $S_2(\boldsymbol{k})$, depends upon the wave nature of the incident and scattered beams, and is a function of the *phase shift* $\boldsymbol{k}\cdot\boldsymbol{d}$. It is this contribution which is responsible for interference in wave optics and quantum mechanics, and has its origin in coherence.

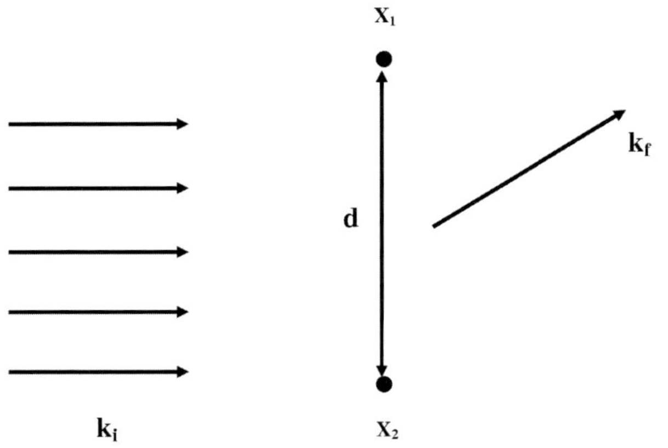

Fig. 10.1. Elastic scattering from two rigid, point scatterers, located at $\boldsymbol{x}_1$ and $\boldsymbol{x}_2 = \boldsymbol{x}_1 + \boldsymbol{d}$. The incident and scattered beams have wave-vectors $\boldsymbol{k}_i$ and $\boldsymbol{k}_f$, respectively

Decoherence (or dephasing) is the process of diminishing of the interference of the two waves. This occurs due to interactions of the system with the degrees of freedom of the environment, e.g., lattice vibrations, electromagnetic fields, other entities not measured in the experiment, etc.

One of the simplest systems for studying coherence and decoherence in quantum mechanics is a particle of mass M in a $d = 1$ symmetric double-well potential, introduced earlier in Chap. 8 (see Fig. 8.1). This potential has the following functional form:

$$V(x) = \frac{1}{2}k(|x| - a)^2 , \tag{10.5}$$

where the minima are located at $x = \pm a$. The two lowest eigenfunctions, and their energy eigenvalues (as a function of a), are shown in Figs. 10.2 and 10.3, respectively [22]. The symmetric (i.e., $\psi_0 + \psi_1$) and antisymmetric (i.e., $\psi_0 - \psi_1$) combinations of the two lowest eigenfunctions are localized about the right minimum and the left minimum of the double-well potential, respectively. These are the *physical* states corresponding to wave packets localized in the right and left wells. Furthermore, these are the only relevant states if the temperature is low enough that the populations of levels higher than the first two energy levels can be ignored. (Subsequently, we will work in units where $\hbar = 1$.)

Thus, as discussed in Sect. 8.1.2, the Hilbert space of the system is spanned by two states. These can be represented as the pseudospin eigenstates of the spin operator σ_z, with $|+\rangle$ and $|-\rangle$ corresponding to the wave packet being localized in the right and left wells, respectively. The system tunnels between the two states with a frequency which is the difference between

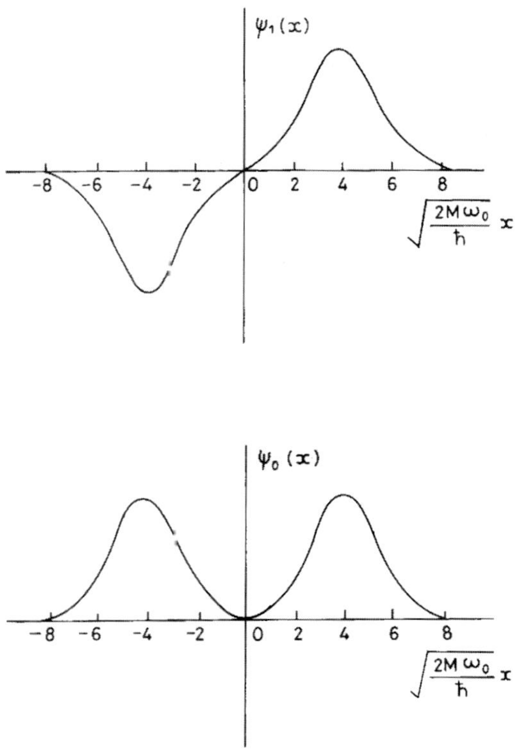

Fig. 10.2. Wavefunctions of the ground state (ψ_0) and the first excited state (ψ_1), as a function of the dimensionless distance $\overline{x} = \sqrt{2M\omega_0/\hbar}\, x$ (where $\omega_0 = \sqrt{k/M}$). The parameter a is fixed as $\sqrt{2M\omega_0/\hbar}\, a = 4$

the lowest energy eigenvalues, and is obtained using the WKB approximation as

$$\Delta \simeq \sqrt{\frac{8\omega_0 V_0}{\pi}} \exp\left(-\frac{2V_0}{\omega_0}\right) , \tag{10.6}$$

where $V_0 = ka^2/2$. We have earlier discussed the concept of coherence in a static picture, using (10.4). We can also consider a time-dependent description of coherence in which the two point scatterers in Fig. 10.1 are not rigid, but rather denote the two positions of (say) an H-atom in the minima of the double-well (see Chap. 8). If the system is coherent, the mean position of the wave packet is an oscillatory function of time (cf. (8.6)):

$$\langle x(t)\rangle = a\cos(\Delta t) . \tag{10.7}$$

On the other hand, when decoherence sets in, the mean position is expected to be exponentially damped:

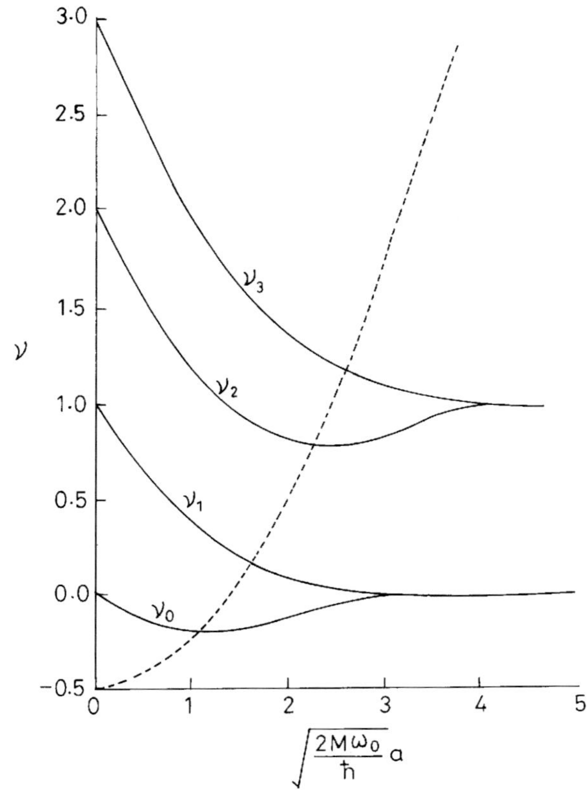

Fig. 10.3. The quantum number $\nu = E/(\hbar\omega_0) - 1/2$, E being the energy eigenvalue, vs. $\overline{a} = \sqrt{2M\omega_0/\hbar}\ a$. The case $\overline{a} = 0$ corresponds to a single harmonic oscillator, whereas the case $\overline{a} \to \infty$ corresponds to two disjoint harmonic oscillators. Thus, for large $\overline{a}$, the eigenvalues collapse pair-wise and become doubly degenerate

$$\langle x(t) \rangle = a e^{-t/\tau} , \tag{10.8}$$

where the time-constant τ depends on the tunneling frequency Δ, the temperature T, and the strength of interaction with other degrees of freedom K, as analyzed for various physical situations in Chap. 8.

In Sect. 8.8, we have discussed a physical realization of the two-state model in terms of the tunneling of an interstitial hydrogen in niobium in the presence of other trapping impurities, e.g., oxygen. Decoherence in this system results from the coupling of the tunneling hydrogen (or proton) with metallic electrons. The transition from coherence to decoherence, as the temperature is increased, was demonstrated for the structure factor of neutron scattering from the tunneling centers. As noted earlier, the bosonic heat bath which induces quantum dissipation (and thereby decoherence) is generic, and models similar decoherence phenomena in diverse systems.

This chapter is organized as follows. In Sect. 10.2, we revisit the quantum dissipative theory of diamagnetism, discussed earlier in Sect. 9.4. We will argue here that the transition from the Landau to the Bohr–van Leeuwen (BVL) regime can be viewed as a transition from coherence to decoherence. Next, recall that the problem discussed for the tunneling center in Sect. 8.8 relates to a subsystem with a finite Hilbert space. On the other hand, the problem discussed in Sect. 9.4 has a subsystem with an infinite-dimensional Hilbert space. A combination of these two paradigms, in which the subsystem has both finite- and infinite-dimensional aspects, is introduced in Sect. 10.3, which addresses the issue of c-axis transport in layered superconductors [518,519]. Again, the observed suppression of c-axis tunneling is interpreted as a transition from coherence to decoherence. In Sect. 10.4, we investigate whether there can be decoherence in the absence of dissipation. Section 10.5 discusses some general issues in this context. Finally, Sect. 10.6 concludes this chapter.

10.2 Landau Diamagnetism in a Dissipative System Revisited

We return to the problem of diamagnetism in solid-state physics, discussed earlier in Sect. 9.4. The problem of interest is the orbital diamagnetism of an ideal gas of electrons in a box, under the influence of an external magnetic field. In this context, recall the BVL theorem which demonstrates that the orbital diamagnetic moment is identically zero in classical statistical mechanics [501,502]. This occurs because the contribution from the circulating electrons in the bulk is exactly canceled by the contribution from the *skipping orbits* due to electrons hitting the walls of the enclosure and being reflected. Subsequently, Landau [496,497] demonstrated that diamagnetism could be explained only in the quantum context. As the position and momentum of the electron do not commute in quantum physics, the bulk contribution is different from the surface contribution, yielding a non-zero diamagnetic moment.

As mentioned earlier, diamagnetism originates from coherent circular motion of the electron in a plane normal to the magnetic field. Then, the relevant question is: what happens to Landau diamagnetism if the coherent precessional motion of the electron is disturbed due to interactions with other degrees of freedom, e.g., defects, other electrons, phonons, etc? This question can be addressed in the SBA formalism, and the total Hamiltonian for the system (electron plus bath) can be written as

$$\mathcal{H} = \mathcal{H}_s + \sum_j \left[\frac{\boldsymbol{p}_j^2}{2m_j} + \frac{1}{2} m_j \omega_j^2 (\boldsymbol{q}_j - \boldsymbol{r})^2 \right] , \tag{10.9}$$

where the subsystem Hamiltonian $\mathcal{H}_s$ is given by

$$\mathcal{H}_s = \frac{1}{2M}\left(\boldsymbol{p} - \frac{e\boldsymbol{A}}{c}\right)^2 . \tag{10.10}$$

In (10.9)-(10.10), $\boldsymbol{p}$ and $\boldsymbol{r}$ are the momentum and position of the electron, which has mass M and charge e. The quantity $\boldsymbol{A}$ is the vector potential associated with the magnetic field $\boldsymbol{B}$. Since the motion of the electron is restricted in the plane normal to the $\boldsymbol{B}$-field, the bath variables (i.e., momenta $\{\boldsymbol{p}_j\}$ and coordinates $\{\boldsymbol{q}_j\}$) are $d = 2$ vectors. In second-quantized notation, we would obtain a bath Hamiltonian analogous to that for the spin-boson model in Sect. 8.1.

From the corresponding quantum Langevin equation, the time-dependent magnetic moment can be computed as discussed in Sect. 9.4. We recall the asymptotic ($t \to \infty$) result from (9.58):

$$m_z = -\frac{2k_BT}{B}\Omega_c^2 \sum_{n=1}^{\infty} \frac{1}{(\nu_n + \gamma)^2 + \Omega_c^2} , \tag{10.11}$$

where

$$\nu_n = 2k_BTn\pi . \tag{10.12}$$

The result in (10.11) is remarkable in that an equilibrium property like the diamagnetic moment has an explicit dependence on a rate constant such as the friction coefficient γ. In most cases, the interaction with the heat bath is treated in the weak-coupling approximation or the so-called Born–Markov (BM) limit [33]. However, the present treatment accounts exactly for the many-body degrees of freedom and it is not meaningful to separately consider the subsystem and the heat bath, as both are linked. Equation (10.11) is reminiscent of another familiar expression in solid-state physics, viz., the Drude formula for the electrical conductivity which also involves the damping parameter γ [520]. Indeed, this observation motivates us to express (10.11) as a *Landau-Drude* formula [521]:

$$-\frac{Mc}{e}m_z = \frac{1}{\nu_c}\sum_{n=1}^{\infty} \frac{1}{1 + (\overline{\mu}_n + r)^2} , \tag{10.13}$$

where we have introduced the dimensionless quantities:

$$\nu_c = \frac{\beta\Omega_c}{2} , \quad \overline{\mu}_n = \frac{n\pi}{\nu_c} . \tag{10.14}$$

The quantity r is a scaled resistance, which is the ratio of the Drude resistivity R_D to the Hall resistivity R_H:

$$\begin{aligned} r &= \frac{R_D}{R_H} , \\ R_D &= \frac{M\gamma}{n_e e^2} , \\ R_H &= \frac{B}{n_e ec} , \end{aligned} \tag{10.15}$$

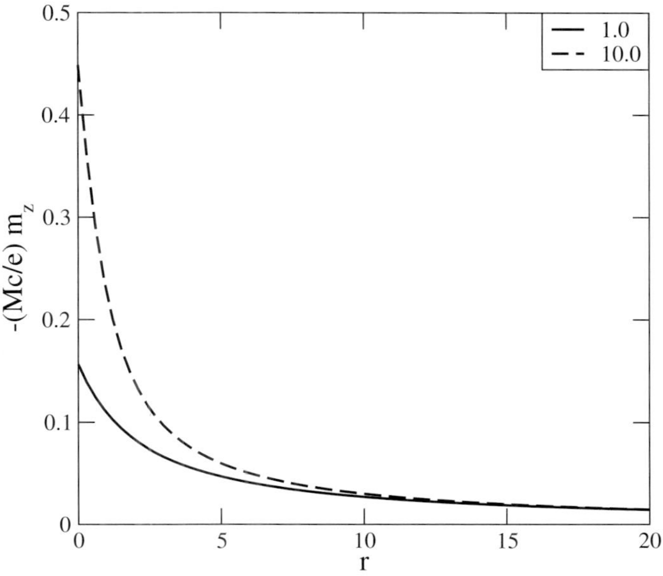

Fig. 10.4. The dimensionless magnetization $-(Mc/e)m_z$ vs. scaled resistance r for two different values of the dimensionless cyclotron frequency $\nu_c = 1.0, 10.0$. The case $r = 0$ corresponds to the Landau limit, and $r = \infty$ corresponds to the BVL limit

where n_e is the number density of electrons. Notice that $r = \eta/\nu_c$, where η is the dimensionless dissipation introduced in (9.50). In Fig. 10.4, we plot the LHS of (10.13) vs. r for two different values of the dimensionless cyclotron frequency ν_c. As resistance is the basic manifestation of dissipation, it is evident that larger values of r will result in stronger decoherence. Furthermore, the persistence of coherence is more pronounced for larger cyclotron frequencies ν_c. Figure 10.4 illustrates how the system makes the transition from the coherent *Landau regime* to the decoherent *BVL regime*, at least asymptotically in time.

Before concluding this section, we should stress that the calculations summarized here and in Sect. 9.4 are based on the simplest treatment of dissipative diamagnetism, where the only quantum aspect is the non-commutativity of position and momentum. A more complete analysis should incorporate the Fermi-Dirac statistics of electron assemblies. The present approach is valid when we have a dilute gas of electrons for which the de Broglie wavelength is much smaller than the average inter-electron distance. It would also be relevant to investigate the opposite limit of a dense electron gas, and analyze the effects of decoherence on quantum statistical correlations.

10.3 Zeno Blocking of c-axis Transport in YBCO

The high-T_c superconductors like $YBa_2Cu_3O_{7-x}$ (or YBCO) are characterized by strong anisotropy as they consist of CuO planes, which are stacked up as layers. One of the manifestations of this anisotropy is the measured value of the resistivity $\rho(T)$ as a function of temperature. Tozer et al. [522] found that both the in-plane resistivity (ρ_{ab}) and the c-axis resistivity (ρ_c, c being normal to the ab-plane of CuO) exhibit the following temperature dependence in the *normal* phase over a large range of temperatures:

$$\rho(T) = \frac{C}{T} + DT\,. \tag{10.16}$$

The value of D, which characterizes the linear dependence, is strongly anisotropic with

$$D_{ab} \simeq 1.4 \times 10^{-6} \quad \Omega \text{ cm K}^{-1}\,, \quad D_c \simeq 3 \times 10^{-5} \quad \Omega \text{ cm K}^{-1}\,. \tag{10.17}$$

Thus, transport along the c-axis is an order of magnitude *slower* than that in the ab-plane. However, the mechanism of transport, as evidenced by the identical temperature dependence, must be the same for both the intra-layer and the inter-layer electrons.

Motivated by the above considerations, we introduce a model in which the inter-layer transport of normal-state electrons occurs via a quantum tunneling process. However, this tunneling is impeded by intra-layer inelastic scattering events. Now, tunneling is a coherent process which is *decohered* when *observed* by the other degrees of freedom involved in inelastic scattering. This mechanism of suppression of tunneling is analogous to the *watched-pot effect* [439] or the *Zeno effect* of quantum measurement theory [438,523].

Following the discussion in Chap. 8, the two planes involved in the tunneling process can be mapped into the two states of a pseudospin Pauli matrix σ_z. Furthermore, tunneling between the two states is mediated by the operator σ_x, which is completely off-diagonal in the representation of σ_z. The normal-state electron is free, but undergoes quantum Brownian motion due to its interaction with the bosonic degrees of freedom of the bath. The full Hamiltonian may then be written as [448]

$$\begin{aligned}\mathcal{H} = \frac{\boldsymbol{p}^2}{2M} - \Delta\sigma_x + \boldsymbol{x}\left[\frac{1+\sigma_z}{2}\sum_{\boldsymbol{k}} G_{\boldsymbol{k}}(a_{\boldsymbol{k}} + a_{\boldsymbol{k}}^+) + \frac{1-\sigma_z}{2}\sum_{\boldsymbol{k}} g_{\boldsymbol{k}}(b_{\boldsymbol{k}} + b_{\boldsymbol{k}}^+)\right] \\ + \sum_{\boldsymbol{k}} \left(\Omega_{\boldsymbol{k}} a_{\boldsymbol{k}}^+ a_{\boldsymbol{k}} + \omega_{\boldsymbol{k}} b_{\boldsymbol{k}}^+ b_{\boldsymbol{k}}\right)\,,\end{aligned} \tag{10.18}$$

where we have introduced different sets of bath variables $\{a_{\boldsymbol{k}}, a_{\boldsymbol{k}}^+\}$ and $\{b_{\boldsymbol{k}}, b_{\boldsymbol{k}}^+\}$ corresponding to the two layers. In order to have a clearer understanding of the structure of the Hamiltonian $\mathcal{H}$, it is useful to project (10.18) to one or the other of the two planes between which tunneling occurs. Thus

$$\langle +|\mathcal{H}|+\rangle = \frac{\boldsymbol{p}^2}{2M} + \boldsymbol{x}\sum_{\boldsymbol{k}} G_{\boldsymbol{k}}(a_{\boldsymbol{k}} + a_{\boldsymbol{k}}^+) + \sum_{\boldsymbol{k}} \Omega_{\boldsymbol{k}} a_{\boldsymbol{k}}^+ a_{\boldsymbol{k}} \,. \tag{10.19}$$

Similarly,

$$\langle -|\mathcal{H}|-\rangle = \frac{\boldsymbol{p}^2}{2M} + \boldsymbol{x}\sum_{\boldsymbol{k}} g_{\boldsymbol{k}}(b_{\boldsymbol{k}} + b_{\boldsymbol{k}}^+) + \sum_{\boldsymbol{k}} \omega_{\boldsymbol{k}} b_{\boldsymbol{k}}^+ b_{\boldsymbol{k}} \,. \tag{10.20}$$

Thus, either of these Hamiltonians separately describes the *quantum Brownian motion* of a free electron, as in Sect. 9.2. Alternatively, as the operator $\boldsymbol{x}$ causes transitions amongst the free-particle momentum states, (10.19) and (10.20) account for inelastic scattering processes within the two planes. On the other hand, the off-diagonal elements of $\mathcal{H}$ are given by

$$\langle +|\mathcal{H}|-\rangle = \langle -|\mathcal{H}|+\rangle = -\Delta \,, \tag{10.21}$$

where Δ is simply the tunneling frequency for coherent propagation along the c-axis. Therefore, $\mathcal{H}$ acts in the product Hilbert space of (i) the discrete two-level system of adjacent CuO layers, (ii) the continuous phase space of a free quantum particle, and (iii) the *independent* sets of bosonic excitations belonging to the two layers.

One other noteworthy point is that, if the intra-plane dynamics is absent and the kinetic energy of the electron goes to zero, the position operator $\boldsymbol{x}$ can be replaced by a constant. Because the bath operators $a_{\boldsymbol{k}}$ ($a_{\boldsymbol{k}}^+$) and $b_{\boldsymbol{k}}$ ($b_{\boldsymbol{k}}^+$) belonging to two distinct layers are taken to be independent, the Hamiltonian in (10.18) then reduces to the usual spin-boson Hamiltonian (cf. (8.7)):

$$\mathcal{H}_{SB} = -\Delta\sigma_x + \sigma_z \sum_{\boldsymbol{k}} g_{\boldsymbol{k}}(b_{\boldsymbol{k}} + b_{\boldsymbol{k}}^+) + \sum_{\boldsymbol{k}} \omega_{\boldsymbol{k}} b_{\boldsymbol{k}}^+ b_{\boldsymbol{k}} \,. \tag{10.22}$$

As has been discussed in Chap. 8, when the damping due to the heat-bath coupling exceeds a certain critical value, the system undergoes a spontaneous symmetry-breaking transition at $T = 0$. Then, tunneling is suppressed and the particle is localized in one of the two wells. While this phenomenon can also be viewed as a quantum Zeno effect, and is subsumed by the general Hamiltonian in (10.18), the effect discussed here concerning the suppression of inter-plane tunneling because of a succession of intra-plane inelastic scattering processes is a more subtle one, with richer consequences.

Thus, the subsystem Hamiltonian has ingredients of both the models studied in Chaps. 8 and 9. The subsystem Hilbert space is a direct product of the discrete space of the two-state tunneling system and the continuum space of the momentum ($\boldsymbol{p}$) and position ($\boldsymbol{x}$) operators of the electron in either of the two planes.

The Hamiltonian in (10.18) has been studied in the framework of the DBGA and extensions thereof [448]. The analysis is similar to that in Sects. 8.2 and 8.3, and we do not replicate it here. Rather, in Fig. 10.5,

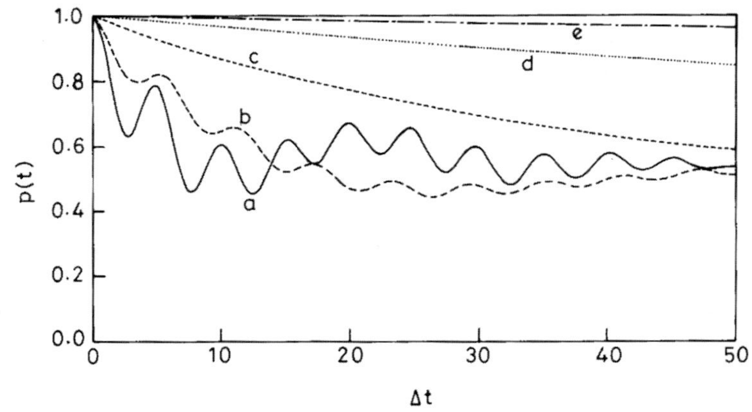

Fig. 10.5. The survival probability $P(t)$ vs. scaled time Δt from the extended-DBGA calculation. The parameter values (see Sects. 8.2 and 8.3) are $\beta\Delta = 50$, and $\omega_c/\Delta = 1000$. Results are shown for 5 different values of the damping parameter: (a) $K = 0.1$, (b) $K = 0.25$, (c) $K = 0.5$, (d) $K = 0.75$, and (e) $K = 1.0$

we plot the survival probability $P(t)$ (defined in (8.8)) vs. the scaled time Δt [448]. We choose the initial condition $P(0) = 1$, so that $P(t)$ is a measure of the electron leakage from a given plane. The oscillations seen in Fig. 10.5 for low values of the coupling constant K are signatures of quantum coherence. The transition from coherence to decoherence occurs for values of $K > 0.5$.

Let us now examine the implications of this analysis for electron transport in YBCO. Since resistive transport is an incoherent process, we expect a corresponding exponential relaxation of the survival probability $P(t)$:

$$P(t) \simeq e^{-\Delta_r^2 \tau_i t} , \tag{10.23}$$

where Δ_r is a renormalized tunneling frequency, and τ_i is the intra-planar inelastic scattering time. If the inter-planar distance is d_p, (10.23) allows us to define a mobility:

$$\mu = d_p {\Delta_r}^2 \tau_i . \tag{10.24}$$

The inter-plane resistivity ρ_c is then determined from $\rho_c \sim \mu^{-1}$. Furthermore, τ_i also determines the in-plane resistivity through the Drude formula:

$$\rho_{ab} = \frac{M_{ab}^*}{n_e e^2 \tau_i} , \tag{10.25}$$

where M_{ab}^* is the effective mass of the electron, and n_e is the number density of electrons. Thus, the parameter τ_i dictates the temperature-dependence of both ρ_c and ρ_{ab}, as has been suggested experimentally. More significantly, ρ_c is considerably larger than ρ_{ab}, because of frequent interruptions of tunneling events.

10.4 Decoherence Without Dissipation

In the examples discussed above, the interaction Hamiltonian was chosen to be off-diagonal in the representation in which the subsystem and the heat-bath Hamiltonians are simultaneously diagonal. Therefore, the interaction Hamiltonian causes transitions between the energy levels of both the subsystem and the heat bath. As a result, there is continuous energy transfer between the subsystem and the heat bath though the latter, being a very large system, has its thermal equilibrium undisturbed. In the terminology of scattering theory, we consider only inelastic scattering processes mediated by the coupling between the subsystem and the heat bath. As a variant to this approach, we next study a physical situation where the interaction Hamiltonian continues to be off-diagonal in the heat-bath representation, but is chosen to be diagonal in the representation of the subsystem. A concrete example can be constructed where the interaction Hamiltonian is taken to be proportional to the subsystem Hamiltonian $\mathcal{H}_s$ itself. Thus, we consider [524,525]

$$\mathcal{H} = \mathcal{H}_s + \mathcal{H}_s \sum_{\boldsymbol{k}} g_{\boldsymbol{k}} \left(b_{\boldsymbol{k}} + b_{\boldsymbol{k}}^{+} \right) + \sum_{\boldsymbol{k}} \omega_{\boldsymbol{k}} b_{\boldsymbol{k}}^{+} b_{\boldsymbol{k}} \,. \tag{10.26}$$

Before we proceed, it is useful to discuss a physical realization of the Hamiltonian in (10.26). Imagine that $\mathcal{H}_s$ represents the Zeeman coupling between a nuclear spin and an external magnetic field, in the context of an NMR experiment. The energy levels will then be given by the eigenvalues of $\mathcal{H}_s$ but, because of the interaction term, the level separation will have a fluctuating component, driven by the heat bath. Though there is no direct energy relaxation in the subsystem, the NMR line-shape is expected to show line-broadening and narrowing, depending on the rate of fluctuations [376].

As usual, we define the *reduced* density operator ρ_s for the subsystem from the total density operator ρ as

$$\rho_s = \mathrm{Tr}_B(\rho) \,, \tag{10.27}$$

where Tr_B denotes a trace over the bath degrees of freedom. The master equation satisfied by ρ_s is as follows [525]:

$$\frac{\partial}{\partial t}\rho_s(t) = -i\left[\mathcal{H}_s, \rho_s\right] + i\dot{\eta}(t)\left[\mathcal{H}_s^2, \rho_s\right] - \dot{\gamma}(t)\left(\mathcal{H}_s^2\rho_s - 2\mathcal{H}_s\rho_s\mathcal{H}_s + \rho_s\mathcal{H}_s^2\right) \,, \tag{10.28}$$

where

$$\dot{\eta}(t) = -\sum_{\boldsymbol{k}} \frac{g_{\boldsymbol{k}}^2}{\omega_{\boldsymbol{k}}} \cos(\omega_{\boldsymbol{k}} t) \,, \tag{10.29}$$

and

$$\dot{\gamma}(t) = \sum_{\boldsymbol{k}} \frac{g_{\boldsymbol{k}}^2}{\omega_{\boldsymbol{k}}} \sin(\omega_{\boldsymbol{k}} t) \coth\left(\frac{\beta\omega_{\boldsymbol{k}}}{2}\right) . \tag{10.30}$$

The form of the master equation in (10.28) will be justified subsequently. The occurrence of the first and third terms on the RHS of (10.28) is as expected. The first term corresponds to unitary evolution, while the third term yields relaxational dynamics due to the heat bath and is of the generic Lindblad form [526]. However, the second term is unusual and arises only because of the quantum nature of the heat bath.

As the subsystem Hamiltonian $\mathcal{H}_s$ commutes with the total Hamiltonian $\mathcal{H}$, there is no energy dissipation in the subsystem. But is there any decoherence? Clearly, this question is meaningful only if the initial state of the subsystem is not an eigenstate of the subsystem Hamiltonian. In this restrictive sense, the transition from coherence to decoherence may be analyzed for some examples of (10.26) [525]. The discussion also corroborates an observation made earlier, i.e., the crossover from coherent to decoherent dynamics in damped quantum systems depends, in general, on the specific dynamical quantity under consideration and the initial preparation.

The process of decoherence discussed here is also known as *dephasing*, especially in the context of mesoscopic physics [509]. There has been some controversy about whether dephasing can occur when the temperature $T \to 0$. This issue can be studied within the present models because of the quantum nature of the heat bath. It is generally believed that purely elastic scattering does not cause decoherence at zero temperature. However, in the model described by (10.26), there are decoherence effects in certain dynamical quantities. This is due to the presence of energy non-conserving terms involving heat-bath variables, even though the subsystem remains dissipationless. The destruction of quantum coherence by environmental influences remains highly topical, and is the subject of ongoing work.

10.4.1 Reduced Density Operator and Master Equation

In order to derive an exact expression for the reduced density operator of the subsystem, it is useful to rewrite the Hamiltonian in (10.26) as

$$\mathcal{H} = \mathcal{H}_s + \mathcal{H}_s \sum_{\boldsymbol{k}} g_{\boldsymbol{k}}(b_{\boldsymbol{k}} + b_{\boldsymbol{k}}^+) + \sum_{\boldsymbol{k}} \omega_{\boldsymbol{k}} b_{\boldsymbol{k}}^+ b_{\boldsymbol{k}} + \mathcal{H}_s^2 \sum_{\boldsymbol{k}} \frac{g_{\boldsymbol{k}}^2}{\omega_{\boldsymbol{k}}} . \tag{10.31}$$

Notice that we have added a *counter term*, i.e., the last term on the RHS of (10.31), to make the system translationally invariant. This can be confirmed by eliminating the coupling term with the aid of the unitary transformation (cf. (8.12)):

$$\mathcal{S} = \exp\left[\mathcal{H}_s \sum_{\boldsymbol{k}} \frac{g_{\boldsymbol{k}}}{\omega_{\boldsymbol{k}}} (b_{\boldsymbol{k}}^+ - b_{\boldsymbol{k}})\right] . \tag{10.32}$$

It is evident that

$$\tilde{\mathcal{H}} = \mathcal{S}\mathcal{H}\mathcal{S}^{-1} = \mathcal{H}_s + \sum_{\boldsymbol{k}} \omega_{\boldsymbol{k}} b_{\boldsymbol{k}}^+ b_{\boldsymbol{k}} \,, \tag{10.33}$$

which is manifestly translationally invariant.

The time-evolution of the density operator can be obtained from the formal solution of the Liouville equation:

$$\rho(t) = e^{-i\mathcal{H}t}\rho(0)e^{i\mathcal{H}t} \,. \tag{10.34}$$

As far as the initial-state density operator $\rho(0)$ is concerned, we make the usual factorization, i.e., the subsystem is assumed to be decoupled from the bath at time $t = 0$. Furthermore, the bath is assumed to be in thermal equilibrium, governed by the Gibbs density matrix. Therefore (cf. (1.A.13)-(1.A.14)),

$$\rho(0) = \rho_B \otimes \rho_s(0) \,, \tag{10.35}$$

where

$$\rho_B = \frac{1}{Z_B} \exp\left(-\beta \mathcal{H}_B\right) \,. \tag{10.36}$$

Using (10.35)-(10.36), and the unitary transformation in (10.33), the density operator can be expressed as

$$\rho(t) = \mathcal{S}^{-1} e^{-i\tilde{\mathcal{H}}t} \mathcal{S}\rho_B \rho_s(0) \mathcal{S}^{-1} e^{i\tilde{\mathcal{H}}t} \mathcal{S} \,. \tag{10.37}$$

Since $\mathcal{H}$ commutes with $\mathcal{H}_s$, we can work in a representation in which $\mathcal{H}_s$ is diagonal. Introducing the eigenvalues and eigenstates of $\mathcal{H}_s$ as

$$\mathcal{H}_s|n\rangle = E_n|n\rangle \,, \tag{10.38}$$

a particular off-diagonal matrix element of $\rho(t)$ can be written as

$$\rho_{nm}(t) = \mathcal{S}_n^{-1} e^{-i\tilde{\mathcal{H}}_n t} \mathcal{S}_n \rho_B \rho_{s,nm}(0) \mathcal{S}_m^{-1} e^{i\tilde{\mathcal{H}}_m t} \mathcal{S}_m \,, \tag{10.39}$$

where

$$\mathcal{S}_n = \exp\left[E_n \sum_{\boldsymbol{k}} \frac{g_{\boldsymbol{k}}}{\omega_{\boldsymbol{k}}} (b_{\boldsymbol{k}}^+ - b_{\boldsymbol{k}})\right] \,, \tag{10.40}$$

and

$$\tilde{\mathcal{H}}_n = E_n + \sum_{\boldsymbol{k}} \omega_{\boldsymbol{k}} b_{\boldsymbol{k}}^+ b_{\boldsymbol{k}} \,. \tag{10.41}$$

The matrix elements for the *reduced* density operator can be obtained by *tracing* (10.39) over the degrees of freedom of the bath. Thus,

$$\rho_{s,nm}(t) = \mathrm{Tr}_B\left[\rho_{nm}(t)\right] \,, \tag{10.42}$$

which, upon employing the cyclic property of the trace, can be rewritten as

$$\rho_{s,nm}(t) = e^{-i(E_n - E_m)t} \mathrm{Tr}_B \left(\rho_B \mathcal{S}_m^{-1} e^{i\mathcal{H}_B t} \mathcal{S}_m \mathcal{S}_n^{-1} e^{-i\mathcal{H}_B t} \mathcal{S}_n \right) \rho_{s,nm}(0) . \tag{10.43}$$

We now use the Heisenberg time-evolution of operators to obtain

$$e^{i\mathcal{H}_B t} \mathcal{S}_m \mathcal{S}_n^{-1} e^{-i\mathcal{H}_B t} = \exp\left\{ (E_m - E_n) \sum_{\boldsymbol{k}} \frac{g_{\boldsymbol{k}}}{\omega_{\boldsymbol{k}}} \left[b_{\boldsymbol{k}}^+(t) - b_{\boldsymbol{k}}(t) \right] \right\} . \tag{10.44}$$

For the system of harmonic oscillators,

$$\begin{aligned} b_{\boldsymbol{k}}^+(t) &= e^{i\omega_{\boldsymbol{k}} t} b_{\boldsymbol{k}}^+ , \\ b_{\boldsymbol{k}}(t) &= e^{-i\omega_{\boldsymbol{k}} t} b_{\boldsymbol{k}} . \end{aligned} \tag{10.45}$$

Therefore, all the $\mathcal{S}$-operators under the trace in (10.43) are structurally similar, viz., they are all exponential operators in which the exponents are linear combinations of $b_{\boldsymbol{k}}$ and $b_{\boldsymbol{k}}^+$. It is convenient to use the standard equality involving exponentials of operators A and B [527]:

$$e^{(A+B)} = e^{-[A,B]/2} e^A e^B , \tag{10.46}$$

which is valid when the commutator $[A, B]$ is a c-number, as is indeed the case for $b_{\boldsymbol{k}}$ and $b_{\boldsymbol{k}}^+$. Using the equality in (10.46), we obtain

$$\begin{aligned} \mathcal{S}_m^{-1} e^{i\mathcal{H}_B t} \mathcal{S}_m \mathcal{S}_n^{-1} e^{-i\mathcal{H}_B t} \mathcal{S}_n &= \exp\left\{ -i(E_m^2 - E_n^2) \sum_{\boldsymbol{k}} \left(\frac{g_{\boldsymbol{k}}}{\omega_{\boldsymbol{k}}} \right)^2 \sin(\omega_{\boldsymbol{k}} t) \right\} \\ \times \exp\left\{ (E_m - E_n) \sum_{\boldsymbol{k}} \frac{g_{\boldsymbol{k}}}{\omega_{\boldsymbol{k}}} \left[\left(e^{i\omega_{\boldsymbol{k}} t} - 1 \right) b_{\boldsymbol{k}}^+ + \left(1 - e^{-i\omega_{\boldsymbol{k}} t} \right) b_{\boldsymbol{k}} \right] \right\} . \end{aligned} \tag{10.47}$$

Furthermore, employing the standard result [29,31]

$$\mathrm{Tr}_B \left[\rho_B e^{(x_{\boldsymbol{k}} b_{\boldsymbol{k}} - x_{\boldsymbol{k}}^* b_{\boldsymbol{k}}^+)} \right] = \exp\left[\frac{|x_{\boldsymbol{k}}|^2}{2} \coth\left(\frac{\beta\omega_{\boldsymbol{k}}}{2} \right) \right] , \tag{10.48}$$

we obtain from (10.43)

$$\begin{aligned} \rho_{s,nm}(t) = \exp\left[-i(E_n - E_m)t \right] &\exp\left[i(E_n^2 - E_m^2)\eta(t) \right] \\ &\times \exp\left[-(E_n - E_m)^2 \gamma(t) \right] \rho_{s,nm}(0) . \end{aligned} \tag{10.49}$$

Here, we have

$$\eta(t) = -\sum_{\boldsymbol{k}} \left(\frac{g_{\boldsymbol{k}}}{\omega_{\boldsymbol{k}}} \right)^2 \sin(\omega_{\boldsymbol{k}} t) , \tag{10.50}$$

$$\gamma(t) = \sum_{\boldsymbol{k}} \left(\frac{g_{\boldsymbol{k}}}{\omega_{\boldsymbol{k}}} \right)^2 \sin^2\left(\frac{\omega_{\boldsymbol{k}} t}{2} \right) \coth\left(\frac{\beta\omega_{\boldsymbol{k}}}{2} \right) . \tag{10.51}$$

Recall the master equation formulated earlier in (10.28). We now recognize that $\rho_s(t)$ obtained in (10.49) can be interpreted as the solution of (10.28). The appropriate expressions for $\dot{\eta}(t)$ and $\dot{\gamma}(t)$ are obtained by taking time-derivatives of (10.50) and (10.51), respectively.

10.4.2 The Harmonic-Oscillator Case

If the subsystem of interest is a $d = 1$ harmonic oscillator, the Hamiltonian $\mathcal{H}_s$ is

$$\mathcal{H}_s = \omega_0 \left(a^+ a + \frac{1}{2} \right) , \tag{10.52}$$

where ω_0 is the oscillator frequency. The master equation (10.28) then becomes

$$\begin{aligned} \frac{\partial}{\partial t}\rho_s(t) = & -i\omega_0[a^+ a, \rho_s] + i\omega_0^2 \dot{\eta}(t)[(a^+ a)^2 + a^+ a, \rho_s] \\ & -\omega_0^2 \dot{\gamma}(t) \left[(a^+ a)^2 \rho_s - 2a^+ a \rho_s a^+ a + \rho_s (a^+ a)^2 \right] . \end{aligned} \tag{10.53}$$

Thus, when the bath is taken to be quantum-mechanical, there is an additional term proportional to $\dot{\eta}(t)$, which is absent in the corresponding master equation when the environment is modeled as a classical stochastic process. Consequently, if the density operator is converted into a P-distribution function [528,529], the latter will obey a non-diffusive equation due to this additional term.

The master equation (10.53) constitutes a Schrödinger-picture description of the quantum dynamics of the system. One could equally well work in the Heisenberg picture, and derive evolution equations for the system operators, in parallel with the discussion in Chap. 7. Thus, substituting the form of $\mathcal{H}_s$ from (10.52) in the Hamiltonian of (10.26), one obtains the following Heisenberg equation of motion for the bath operators:

$$\frac{db_{\boldsymbol{k}}}{dt} = -i\omega_{\boldsymbol{k}} b_{\boldsymbol{k}} - i g_{\boldsymbol{k}} \omega_0 \left(a^+ a + \frac{1}{2} \right) , \tag{10.54}$$

which can be directly integrated (because $a^+ a$ is a constant of motion) to yield

$$b_{\boldsymbol{k}}(t) = e^{-i\omega_{\boldsymbol{k}} t} b_{\boldsymbol{k}}(0) - \frac{g_{\boldsymbol{k}} \omega_0}{\omega_{\boldsymbol{k}}} (1 - e^{-i\omega_{\boldsymbol{k}} t}) \left(a^+ a + \frac{1}{2} \right) . \tag{10.55}$$

Similarly, the Heisenberg equation for the system operator is

$$\frac{da}{dt} = -i\omega_0 a + 2i\dot{\eta}\omega_0^2 \left(a^+ a + \frac{1}{2} \right) a - i\omega_0 a B . \tag{10.56}$$

Here, we have used the definition of $\eta(t)$ in (10.50), and introduced the *noise operator* $B(t)$ as

$$B(t) = \sum_{\boldsymbol{k}} g_{\boldsymbol{k}} \left[e^{-i\omega_{\boldsymbol{k}} t} b_{\boldsymbol{k}}(0) + e^{i\omega_{\boldsymbol{k}} t} b_{\boldsymbol{k}}^+(0) \right] . \tag{10.57}$$

Equation (10.56) is another example of the quantum Langevin equation, introduced in Chap. 9, in which the noise operator $B(t)$ is expressed in terms of the *initial values* of the bath operators. However, in this case, the noise is *multiplicative* as it depends on the system operator a itself [7]. For a thermal bath, the spectral properties of the noise operator are as follows:

$$\langle B(t) \rangle = 0 , \tag{10.58}$$

$$\langle B(t_1)B(t_2) \rangle = \sum_{\boldsymbol{k}} g_{\boldsymbol{k}}^2 \left\{ e^{-i\omega_{\boldsymbol{k}}(t_1-t_2)} \left[1 + \overline{n}(\omega_{\boldsymbol{k}})\right] + e^{i\omega_{\boldsymbol{k}}(t_1-t_2)} \overline{n}(\omega_{\boldsymbol{k}}) \right\} . \tag{10.59}$$

Here, $\overline{n}(\omega_{\boldsymbol{k}})$ is the usual Bose-Einstein factor

$$\overline{n}(\omega_{\boldsymbol{k}}) = \left(e^{\beta\omega_{\boldsymbol{k}}} - 1\right)^{-1} . \tag{10.60}$$

10.4.3 Decoherence Behavior in the Harmonic-Oscillator Case

Recall that, in the present example, the subsystem Hamiltonian commutes with the total Hamiltonian $\mathcal{H}$. Hence, there is no energy dissipation in the subsystem. To study decoherence, it is appropriate to initially prepare the subsystem in a non-eigenstate. For the harmonic-oscillator case, we consider a coherent state:

$$\rho_s(0) = |\alpha\rangle\langle\alpha| , \tag{10.61}$$

and examine the temporal behavior of the survival probability $P(t)$, defined as

$$P(t) = \mathrm{Tr}\left[\rho_s(0)\rho_s(t)\right] . \tag{10.62}$$

Now, the evolution of $\rho_s(t)$ is governed by (10.49), yielding

$$P(t) = \sum_{mn} |\rho_{s,mn}(0)|^2 U_{nm}(t) , \tag{10.63}$$

where

$$U_{nm}(t) = \exp\left[-i\omega_0(n-m)t\right] \exp\left\{i\omega_0^2[(n^2-m^2) + (n-m)]\eta(t)\right\} \times \exp\left[-\omega_0^2(n-m)^2\gamma(t)\right] . \tag{10.64}$$

For numerical evaluation of $P(t)$, we need calculable expressions for $\eta(t)$ and $\gamma(t)$. As usual, we assume that the bath oscillators have a continuous frequency distribution with a spectral density $J(\omega)$ (see (8.21)) such that, for an arbitrary function $f(\omega)$,

$$\sum_{\boldsymbol{k}} g_{\boldsymbol{k}}^2 f(\omega_{\boldsymbol{k}}) \to \int_0^\infty d\omega J(\omega) f(\omega) . \tag{10.65}$$

For the special case of the Ohmic bath discussed earlier, we have (cf. (8.22)-(8.23))

$$J(\omega) = K\omega e^{-\omega/\omega_c} . \tag{10.66}$$

Here, K is the coupling constant, and ω_c is a cut-off frequency. With this form of the spectral density, we can obtain the following expression for $\eta(t)$ from (10.50):

$$\begin{aligned} \eta(t) &= K \tan^{-1}(\omega_c t) , \quad \text{and} \\ \dot{\eta}(t) &= \frac{K\omega_c}{1 + (\omega_c t)^2} . \end{aligned} \tag{10.67}$$

On the other hand, $\gamma(t)$ is not analytically calculable for all times and temperatures, and has to be obtained numerically. However, it is useful to consider the high-temperature limit for which $\coth(\beta\omega/2) \to 2/(\beta\omega)$ in (10.51). In that case,

$$\begin{aligned} \gamma(t) &\simeq \frac{2K}{\beta\omega_c} \sum_{n=0}^{\infty} \frac{(-1)^n}{(2n+2)(2n+1)} (\omega_c t)^{2n+2} , \quad \text{and} \\ \dot{\gamma}(t) &\simeq \frac{2K}{\beta} \tan^{-1}(\omega_c t) . \end{aligned} \tag{10.68}$$

For ready comparison with some well-known cases (to be specified below), it is helpful to consider two limiting situations.

(i) Fast-modulation (FM) limit ($\omega_c t \gg 1$)

In this limit, we have

$$\tan^{-1}(\omega_c t) \simeq \frac{\pi}{2} , \tag{10.69}$$

such that

$$\dot{\gamma}(t) = \frac{K\pi}{\beta} . \tag{10.70}$$

Furthermore, $\dot{\eta}(t)$ vanishes (cf. (10.67)), and the master equation (10.28) reduces to the BM form [530,531]:

$$\frac{\partial}{\partial t}\rho_s(t) = -i[\mathcal{H}_s, \rho_s] - \frac{K\pi}{\beta}(\mathcal{H}_s^2\rho_s - 2\mathcal{H}_s\rho_s\mathcal{H}_s + \rho_s\mathcal{H}_s^2) . \tag{10.71}$$

(ii) Slow-modulation (SM) limit ($\omega_c t \ll 1$)

In this limit, we have

$$\tan^{-1}(\omega_c t) \simeq \omega_c t , \tag{10.72}$$

such that

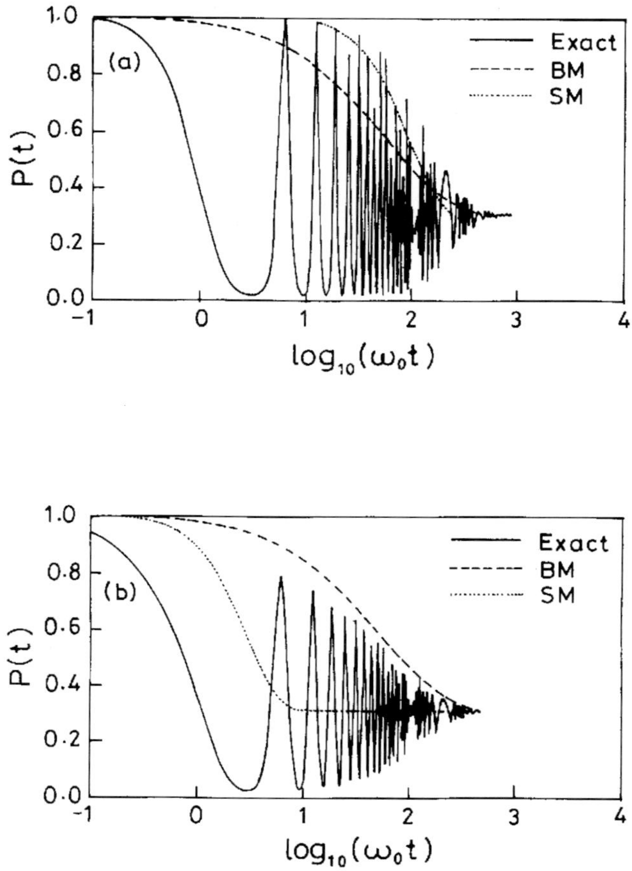

Fig. 10.6. Survival probability $P(t)$ vs. $\log_{10}(\omega_0 t)$ for the exact result with Ohmic dissipation (solid line), the FM or BM limit (dashed line), and the SM limit (dotted line). The parameter values are (**a**) $\omega_c/\omega_0 = 10.0$, and (**b**) $\omega_c/\omega_0 = 0.01$

$$\dot{\gamma}(t) = \frac{2K}{\beta}\omega_c t\,. \tag{10.73}$$

In Fig. 10.6, we present results for the survival probability $P(t)$ in three distinct situations: (a) the exact formulation, as given by (10.63)-(10.64) with Ohmic dissipation (shown as a solid line); (b) the FM or BM limit (shown as a dashed line); and (c) the SM limit (shown as a dotted line). The initial coherent state in (10.61) is determined by $\alpha = 1$. The various parameters are $(\beta\omega_0)^{-1} = 0.1$, and $\pi K \omega_0^2 = 0.1$. In Fig. 10.6(a), $\omega_c/\omega_0 = 10.0$, and the exact dynamics is approximated better by the BM result. On the other hand, in Fig. 10.6(b), $\omega_c/\omega_0 = 0.01$, and the exact dynamics is closer to the SM result. Of course, at the low temperatures considered here, there are more oscillations before the survival probability saturates. At higher temperatures, and for moderate values of ω_c, the curves for the exact, FM and SM dynamics merge because the decoherence rate rapidly increases with time.

10.5 Crossover from Coherence to Decoherence

As mentioned earlier, the destruction of quantum coherence by dissipative interactions with the environment is of interest in diverse problems. Our discussion above has focused on some important examples of the crossover from coherence to decoherence. In each of these examples, the transition was attributed to certain signatures of a physical quantity which can be experimentally measured. In the general context, two questions are relevant:
(a) Can one quantify the criterion for crossover from coherent to incoherent dynamics?
(b) Is this criterion universal, at least for a given system?

The answer to (a) is in the affirmative, at least as far as some model systems are concerned. The appropriate criterion involves the ratio of the damping strength and the characteristic system frequency. However, with regard to question (b), this criterion does not seem to be universal. As a matter of fact, even for a given system, the value of the crossover parameter depends on the particular quantity being studied and its initial preparation [532]. Thus, quantum memory effects seem to play an important role as the system makes a transition from the coherent to the incoherent regimes. To clarify this issue, we again consider the dissipative quantum harmonic oscillator and the spin-boson model, discussed extensively in Chaps. 8-9. Furthermore, we shall focus on the $T = 0$ behavior, where quantum coherence effects are the most prominent.

10.5.1 Dissipative Harmonic Oscillator

First, consider the exactly solvable case of the dissipative quantum harmonic oscillator. Here, we follow the discussion of Egger et al. [532]. Recall that the position operator $Q(t)$ satisfies (9.45) in the limit of zero magnetic field, i.e., $\overline{\gamma} = \gamma$. Thus, the time-dependence of the corresponding classical quantity is governed by the following equation:

$$\left\langle \frac{d^2Q}{dt^2} \right\rangle + \gamma \left\langle \frac{dQ}{dt} \right\rangle + \frac{k}{M}\langle Q(t)\rangle = \frac{\theta(t)}{M}, \tag{10.74}$$

where symbols have their usual meaning.

The angular brackets here describe statistical averages over the ground-state properties ($T = 0$). We know from linear response theory that the response to an external force is characterized by a response function, which is the generalized susceptibility $\chi_{\mathrm{osc}}(t)$ [1]:

$$\langle Q(t)\rangle = \frac{1}{M\omega_0}\int_{-\infty}^{t} dt'\chi_{\rm osc}(t-t')\theta(t')\,. \tag{10.75}$$

From (10.74), we obtain the Fourier transform of $\chi_{\rm osc}(t)$ as

$$\chi_{\rm osc}(\omega) = \frac{\omega_0}{\omega_0^2-\omega^2-i\gamma\omega}\,. \tag{10.76}$$

From the fluctuation-dissipation theorem [1], $\chi_{\rm osc}(\omega)$ is related to the spectral function $S_{\rm osc}(\omega)$, which in turn determines the equilibrium correlation function $C_{\rm osc}(\omega)$. The functional relationship is as follows:

$$\mathrm{Im}\;\chi_{\rm osc}(\omega) = \omega S_{\rm osc}(\omega) = \frac{\omega}{|\omega|}C_{\rm osc}(\omega)\,, \tag{10.77}$$

where

$$C_{\rm osc}(t) = \mathrm{Re}\;\langle Q(t)Q(0)\rangle_0\,. \tag{10.78}$$

Such correlation functions are relevant for many spectroscopy experiments, where the line-shape is determined directly in the frequency domain. Using (10.76)-(10.77), we obtain

$$S_{\rm osc}(\omega) = \frac{\gamma\omega_0}{(\omega_0^2-\omega^2)^2+\gamma^2\omega^2}\,. \tag{10.79}$$

The quantity $S_{\rm osc}(\omega)$ can be used as a signature for the transition from coherence to decoherence. Indeed, as our discussion of neutron scattering of hydrogen tunneling in metals shows (see Sect. 8.8), $S_{\rm osc}(\omega)$ has two *inelastic* peaks at $\omega_m = \pm\omega_0\sqrt{1-\kappa^2/2}$ for *weak* damping, where κ is a dimensionless parameter defined by

$$\kappa = \frac{\gamma}{\omega_0}\,. \tag{10.80}$$

This will be employed as the crossover parameter to quantify the coherence to decoherence transition. When $\kappa\to\kappa_c$, we expect the line-shape to approach a quasi-elastic peak, as was indeed seen in the experimental data on H-tunneling [463]. Now, since the quasi-elastic peak is centered around $\omega=0$, it is useful to make a small-ω expansion of $S_{\rm osc}(\omega)$:

$$S_{\rm osc}(\omega) \simeq \frac{\kappa}{\omega_0^2}\left[1+\left(2-\kappa^2\right)\frac{\omega^2}{\omega_0^2}+O\left(\omega^4\right)\right]\,. \tag{10.81}$$

Thus, the curvature of $S_{\rm osc}(\omega)$ is positive (implying coherence) for $\kappa<\sqrt{2}$, but changes sign at the critical value $\kappa_c=\sqrt{2}$. Clearly, this is also the value of κ at which the peak frequency for inelastic scattering goes to zero:

$$\omega_m(\kappa_c) = 0\,. \tag{10.82}$$

Next, we turn to a different criterion for quantifying the transition from coherence to decoherence. From the discussion in Chap. 7, recall that experiments often yield a time-domain measurement, e.g., in muon-spin depolarization, measurement of remanent magnetization in spin glasses, etc. In such cases, one would be interested in the relaxation of the expectation value $\langle Q(t)\rangle$, starting from a *nonequilibrium* initial state. Applying the force $F(t) = M\omega_0^2 Q_0$ for $t < 0$, the initial condition $\langle Q(0)\rangle = Q_0$ is prepared. Then, the time-dependent quantity of interest is

$$P_{\text{osc}}(t) = \frac{\langle Q(t)\rangle}{Q_0} . \tag{10.83}$$

The solution of (10.74) yields

$$P_{\text{osc}}(t) = \frac{\cos(\Omega t - \phi)}{\cos(\phi)} e^{-\gamma t/2} , \tag{10.84}$$

where

$$\Omega = \sqrt{\omega_0^2 - \frac{\gamma^2}{4}} ,$$
$$\phi = \tan^{-1}\left(\frac{\gamma}{2\Omega}\right) . \tag{10.85}$$

The signature of coherence is now *damped-oscillatory* behavior for weak damping ($\gamma^2 < 4\omega_0^2$ or $\kappa < 2$), which crosses over to *relaxational* (decoherent) behavior for strong damping ($\gamma^2 > 4\omega_0^2$ or $\kappa > 2$). The crossover then occurs at $\kappa_c^* = 2$, where the * is used to distinguish the crossover behavior of $P_{\text{osc}}(t)$ from that of $S_{\text{osc}}(\omega)$. The ratio of the two critical damping strengths is

$$\frac{\kappa_c}{\kappa_c^*} = \frac{1}{\sqrt{2}} , \tag{10.86}$$

thereby demonstrating our earlier statement that the criterion for crossover from coherence to decoherence depends on the particular physical quantity considered.

10.5.2 Spin-Boson Model

Our next example is the symmetric ($\epsilon = 0$) spin-boson model, discussed in Chap. 8. Recall that the physical quantities which are relevant for neutron scattering and muon depolarization are the equilibrium correlation function:

$$C(t) = \langle\sigma_z(t)\sigma_z(0)\rangle_0 , \tag{10.87}$$

and the survival probability:

$$P(t) = \frac{1}{2}\left[1 + \langle\sigma_z(t)\rangle\right] . \tag{10.88}$$

Again, the function $P(t)$ is subject to the initial condition $P(0) = 1$, appropriate to a nonequilibrium preparation. This condition is relevant, e.g., for c-axis transport in YBCO (see Sect. 10.3), where the electron is assumed to reside in one of the CuO planes, prior to tunneling.

As before, $S(\omega) = C(\omega)|\omega|^{-1}$ has the low-frequency expansion:

$$S(\omega) = K\pi\chi_0^2 \left[1 + g(K)\chi_0^2\omega^2 + O\left(\omega^4\right)\right] , \qquad (10.89)$$

where K is the damping parameter (cf. (8.22)); χ_0 is the static susceptibility; and $g(K)$ is a dimensionless parameter. The function $g(K)$ has been computed from renormalization-group techniques [533], as well as form-factor methods [534], and is found to change sign at $K_c = 2/3$.

Finally, we analyze the transition from coherent oscillations to incoherent relaxation for $P(t)$ in the spin-boson context. Within the DBGA, discussed in Sect. 8.2, the relevant crossover parameter is $K_c^* = 1$ [20]. As a matter of fact, the DBGA value $K_c^* = 1$ turns out to be exact [426]. As in the harmonic-oscillator case, we consider the critical ratio:

$$\frac{K_c}{K_c^*} = \frac{2}{3} , \qquad (10.90)$$

which is different from the corresponding quantity for the harmonic-oscillator case (cf. (10.86)). The above analysis emphasizes the fact that the coherence to decoherence transition depends upon the physical quantity being investigated. As in glassy systems, which are characterized by hysteretic behavior, quantum systems also exhibit memory of their initial state of preparation.

10.6 Concluding Remarks

To conclude this chapter, we recall that decoherence (or dephasing) is of great interest in the context of mesoscopic physics. In particular, there has been considerable discussion about whether dephasing occurs when the temperature $T \to 0$. A model of conduction electrons coupled to lattice vibrations (or phonons) was indeed shown to exhibit dephasing at $T = 0$ [535]. However, a contrary perspective was presented immediately thereafter [536,537].

In recent experiments by Mohanty et al. [538,539], the dephasing rate of conduction electrons was measured by magneto-conductance techniques in the weak-localization regime. These experiments show that the dephasing rate of conduction electrons does not vanish as $T \to 0$. The non-zero limiting value is interpreted to result from zero-point electromagnetic fluctuations in the solid. In a metal, these arise from fields caused by other conduction electrons. Hence, the observed decoherence can be attributed to electron-electron interactions. However, using the argument that an electron at the Fermi level cannot lose energy to vacuum fluctuations, the decoherence rate due to electron-electron interactions was earlier calculated to be zero at $T = 0$

[540,541]. Thus, there is a contradiction between the experiments [538,539] and theoretical studies.

Recently, the conditions under which these experiments were carried out have also come under critical scrutiny. The crucial point is that, if the electric currents used for measurement are not sufficiently small, one is outside the linear transport regime [517]. More recent experiments by Ovadyahu [542] use very small currents, and suggest that the decoherence rate may actually vanish as $T \to 0$. Therefore, there appear to be both experimental and theoretical ambiguities regarding the existence of dephasing at $T = 0$ [543,544].

Using models similar to the one discussed in Sect. 10.4, Braun et al. [545] have studied the decoherence of superpositions of macroscopically distinct states. They find that, if two sufficiently distinct states $|S\rangle$ and $|S'\rangle$ are brought to an initial superposition $|\Psi\rangle = C|S\rangle + C'|S'\rangle$, the density operator is initially the projection operator. Then, for a suitable coupling to the environment (as in Sect. 10.4), the density operator $\rho(t)$ quickly decoheres to the mixture $\rho(t) \simeq C^2|S\rangle\langle S| + C'^2|S'\rangle\langle S'|$, on a time-scale τ_{dec}. The subsequent relaxation of this mixture towards an eventual stationary state $\rho(\infty)$ has a much longer characteristic time τ_{diss}. The ratio $\tau_{\mathrm{dec}}/\tau_{\mathrm{diss}}$ becomes smaller for more distinct component states $|S\rangle$ and $|S'\rangle$.

References

1. S. Dattagupta, *Relaxation Phenomena in Condensed Matter Physics* (Academic Press, Orlando 1987)
2. R. Brown, Philos. Mag. **4**, 161 (1828); Philos. Mag. **6**, 161 (1829)
3. A. Einstein, Ann. Physik **17**, 549 (1905); Ann. Physik **19**, 371 (1906)
4. A. Einstein, *Investigations on the Theory of the Brownian Motion* (Dover, New York 1956)
5. M.C. Wang and G.E. Uhlenbeck, Rev. Mod. Phys. **17**, 323 (1945)
6. *Selected Papers on Noise and Stochastic Processes*, ed. by N. Wax (Dover, New York 1954)
7. N.G. Van Kampen, *Stochastic Processes in Physics and Chemistry* (North-Holland, Amsterdam 1981)
8. C.W. Gardiner, *Handbook of Stochastic Methods: For Physics, Chemistry and the Natural Sciences*, Springer Series in Synergetics **13**, ed. by H. Haken (Springer-Verlag, Berlin 1985)
9. R. Kubo, M. Toda and N. Hashitsume, *Statistical Physics Vols. I and II*, Springer Series in Solid-State Sciences **31**, ed. by P. Fulde (Springer-Verlag, Heidelberg 1985)
10. H. Risken, *The Fokker-Planck Equation: Methods of Solution and Applications*, Springer Series in Synergetics **18**, ed. by H. Haken (Springer-Verlag, Berlin 1996)
11. R. Zwanzig, J. Stat. Phys. **9**, 215 (1973)
12. R. Zwanzig, *Nonequilibrium Statistical Mechanics* (Oxford University Press, New York 2001)
13. A.T. Bharucha-Reid, *Elements of the Theory of Markov Processes and Their Applications* (McGraw-Hill, New York 1960)
14. S. Chandrasekhar, Rev. Mod. Phys. **15**, 1 (1943)
15. R. Kubo, J. Phys. Soc. Japan **12**, 570 (1957)
16. G.E. Uhlenbeck and L.S. Ornstein, Phys. Rev. **36**, 823 (1930)
17. *Stochastic Processes: Formalism and Applications*, ed. by G.S. Agarwal and S. Dattagupta, Lecture Notes in Physics **184** (Springer-Verlag, Berlin 1983)
18. H. Goldstein, *Classical Mechanics*, Second Edition (Addison-Wesley, Reading, Massachusetts 1980)
19. A.O. Caldeira and A.J. Leggett, Physica A **121**, 587 (1983)
20. U. Weiss, *Quantum Dissipative Systems* (World Scientific, Singapore 1993)
21. P.A.M. Dirac, *The Principles of Quantum Mechanics* (Clarendon Press, Oxford 1958)
22. E. Merzbacher, *Quantum Mechanics* (John Wiley, New York 1970)
23. P. Debye, *Polar Molecules* (Dover, New York 1929)

24. F. Bloch, Phys. Rev. **105**, 1206 (1957)
25. A.G. Redfield, IBM J. Research Development **1**, 19 (1957)
26. A.G. Redfield, Adv. Magn. Reson. **1**, 1 (1965)
27. P.S. Hubbard, Rev. Mod. Phys. **33**, 249 (1961)
28. S.P. Heims, Phys. Rev. **138**, A587 (1965)
29. R. Kubo, in *Fluctuation, Relaxation and Resonance in Magnetic Systems*, ed. by D. Ter Haar (Oliver and Boyd, Edinburgh 1962), p. 23
30. R. Kubo, J. Phys. Soc. Jpn. **17**, 1100 (1962)
31. R. Kubo, J. Math. Phys. **4**, 174 (1963)
32. N.G. Van Kampen, Phys. Rep. **24**, 171 (1976)
33. G.S. Agarwal, *Quantum Optics*, Springer Tracts in Modern Physics **70**, ed. by G. Hohler (Springer-Verlag, Berlin 1974)
34. R. Zwanzig, J. Chem. Phys. **33**, 1338 (1960)
35. R. Zwanzig, Phys. Rev. **124**, 983 (1961)
36. U. Fano, Phys. Rev. **131**, 259 (1963)
37. H. Mori, Prog. Theor. Phys. **33**, 423 (1965)

Chapter 2

38. H.E. Stanley, *Phase Transitions and Critical Phenomena* (Oxford University Press, Oxford 1971)
39. L.D. Landau and E.M. Lifshitz, *Statistical Physics* (Pergamon Press, Oxford 1980)
40. K. Huang, *Statistical Mechanics, Second Edition* (Wiley, New York 1987)
41. M. Plischke and B. Bergersen, *Equilibrium Statistical Physics, Second Edition* (World Scientific, Singapore 1994)
42. R.J. Glauber, J. Math. Phys. **4**, 294 (1963)
43. M. Suzuki and R. Kubo, J. Phys. Soc. Jpn. **24**, 51 (1968)
44. K. Binder and D.W. Heermann, *Monte Carlo Simulation in Statistical Physics: An Introduction, Fourth Edition* (Springer-Verlag, Berlin 2002)
45. K. Kawasaki, Phys. Rev. **145**, 224 (1966)
46. K. Kawasaki, in *Phase Transitions and Critical Phenomena Vol. 2*, ed. by C. Domb and M.S. Green (Academic Press, London 1972), p. 443
47. J.S. Smart, *Effective Field Theories of Magnetism* (Saunders, Philadelphia 1966)
48. K. Binder, Z. Phys. **267**, 313 (1974)
49. K. Binder, Phys. Rev. B **15**, 4425 (1977)
50. J.-F. Gouyet, Europhys. Lett. **21**, 335 (1993); Phys. Rev. E **51**, 1695 (1995)
51. M. Plapp and J.-F. Gouyet, Phys. Rev. E **55**, 45 (1997); Phys. Rev. E **55**, 5321 (1997)
52. S. Puri, Phys. Rev. E **55**, 1752 (1997)
53. S. Puri and R. Sharma, Phys. Rev. E **57**, 1873 (1998)
54. J.R. Manning, *Diffusion Kinetics for Atoms in Crystals* (D. Van Nostrand, Princeton 1968)
55. C.P. Flynn, *Point Defects and Diffusion* (Clarendon, Oxford 1972)
56. K. Yaldram and K. Binder, Acta Metall. **39**, 707 (1991); J. Stat. Phys. **62**, 161 (1991)
57. K. Tafa, S. Puri and D. Kumar, Phys. Rev. E **63**, 046115 (2001); Phys. Rev. E **64**, 056139 (2001)

58. A.J. Bray, Phys. Rev. Lett. **66**, 2048 (1991)
59. J.F. Annett and J.R. Banavar, Phys. Rev. Lett. **68**, 2941 (1992)
60. M.K. Phani, J.L. Lebowitz, M.H. Kalos and O. Penrose, Phys. Rev. Lett. **45**, 366 (1980)
61. S. Dattagupta, V. Heine, S. Marais and E. Salje, J. Phys. Condens. Matter **3**, 2963 (1991); J. Phys. Condens. Matter **3**, 2975 (1991)
62. J.V. Smith, *Feldspar Minerals Vols. I and II* (Springer-Verlag, Berlin 1974)
63. E. Salje, *Phase Transitions in Ferroelastic and Co-elastic Crystals* (Cambridge University Press, Cambridge 1990)
64. S. Dattagupta, Physica A **194**, 137 (1993)
65. I.S. Gradshteyn and I.M. Ryzhik, *Table of Integrals, Series, and Products*, ed. by A. Jeffrey (Academic Press, London 1994)
66. P.C. Hohenberg and B.I. Halperin, Rev. Mod. Phys. **49**, 435 (1977)
67. J.D. Gunton, M. San Miguel and P.S. Sahni, in *Phase Transitions and Critical Phenomena Vol. 8*, ed. by C. Domb and J.L. Lebowitz (Academic Press, New York 1983), p. 267
68. V.L. Ginzburg and L.D. Landau, Zh. Eksp. Teor. Fiz. **20**, 1064 (1950)
69. S. Puri and Y. Oono, J. Phys. A **21**, L755 (1988)
70. A.J. Bray, Phys. Rev. Lett. **62**, 2841 (1989); Phys. Rev. B **41**, 6724 (1990)
71. J.W. Cahn and J.E. Hilliard, J. Chem. Phys. **28**, 258 (1958); J. Chem. Phys. **31**, 688 (1959)
72. J.W. Cahn, Acta Metall. **9**, 795 (1961); Acta Metall. **10**, 907 (1962)
73. J.S. Langer, M. Bar-on and H.D. Miller, Phys. Rev. A **11**, 1417 (1975)
74. K. Kitahara and M. Imada, Prog. Theor. Phys. Suppl. **64**, 65 (1978)
75. S. Puri, A.J. Bray and J.L. Lebowitz, Phys. Rev. E **56**, 758 (1997)
76. H.E. Cook, Acta Metall. **18**, 297 (1970)
77. M. Blume, V.J. Emery and R.B. Griffiths, Phys. Rev. A **4**, 1071 (1971)
78. L.P. Kadanoff and J. Swift, Phys. Rev. **165**, 310 (1968)
79. B.I. Halperin, P.C. Hohenberg and S.-K. Ma, Phys. Rev. B **10**, 139 (1974)

Chapter 3

80. K. Binder, in *Materials Science and Technology Vol. 5: Phase Transformations of Materials*, ed. by R.W. Cahn, P. Haasen and E.J. Kramer (VCH, Weinheim 1991), p. 405
81. A.J. Bray, Adv. Phys. **43**, 357 (1994)
82. A. Onuki, *Phase Transition Dynamics* (Cambridge University Press, Cambridge 2002)
83. I.M. Lifshitz, Sov. Phys. JETP **15**, 939 (1962); S.M. Allen and J.W. Cahn, Acta Metall. **27**, 1085 (1979)
84. K. Binder and D. Stauffer, Phys. Rev. Lett. **33**, 1006 (1974); Z. Phys. B **24**, 406 (1976)
85. G. Porod, in *Small-Angle X-Ray Scattering*, ed. by O. Glatter and O. Kratky (Academic Press, New York 1982)
86. A.J. Bray and K. Humayun, Phys. Rev. E **47**, R9 (1993)
87. H. Tomita, Prog. Theor. Phys. **72**, 656 (1984); Prog. Theor. Phys. **75**, 482 (1986)
88. T. Ohta, D. Jasnow and K. Kawasaki, Phys. Rev. Lett. **49**, 1223 (1982)
89. Y. Oono and S. Puri, Mod. Phys. Lett. B **2**, 861 (1988)

90. A.J. Bray and S. Puri, Phys. Rev. Lett. **67**, 2670 (1991)
91. H. Toyoki, Phys. Rev. B **45**, 1965 (1992)
92. A.N. Pargellis, P. Finn, J.W. Goodby, P. Panizza, B. Yurke and P.E. Cladis, Phys. Rev. A **46**, 7765 (1992)
93. T.J. Newman, A.J. Bray and M.A. Moore, Phys. Rev. B **42**, 4514 (1990)
94. M. Zapotocky and W. Zakrzewski, Phys. Rev. E **51**, R5189 (1995)
95. R.E. Blundell, A.J. Bray and S. Sattler, Phys. Rev. E **48**, 2476 (1993)
96. Y. Oono and S. Puri, Phys. Rev. Lett. **58**, 836 (1987); Phys. Rev. A **38**, 434 (1988); S. Puri and Y. Oono, Phys. Rev. A **38**, 1542 (1988)
97. G.F. Mazenko, Phys. Rev. Lett. **63**, 1605 (1989); Phys. Rev. B **42**, 4487 (1990); Phys. Rev. B **43**, 5747 (1991)
98. G.F. Mazenko, Phys. Rev. E **58**, 1543 (1998); Phys. Rev. E **61**, 1088 (2000)
99. H. Nishimori and A. Onuki, Phys. Rev. B **42**, 980 (1990); A. Onuki and H. Nishimori, Phys. Rev. B **43**, 13649 (1991)
100. F. Langmayr, P. Fratzl, G. Vogl and W. Miekeley, Phys. Rev. B **49**, 11759 (1994)
101. P. Fratzl, O. Penrose and J.L. Lebowitz, J. Stat. Phys. **95**, 1429 (1999)
102. I.M. Lifshitz and V.V. Slyozov, J. Phys. Chem. Solids **19**, 35 (1961)
103. D.A. Huse, Phys. Rev. B **34**, 7845 (1986)
104. C. Yeung, Phys. Rev. Lett. **61**, 1135 (1988)
105. H. Furukawa, Prog. Theor. Phys. **99**, 358 (1989)
106. A. Shinozaki and Y. Oono, Phys. Rev. Lett. **66**, 173 (1991); Phys. Rev. E **48**, 2622 (1993)
107. T.M. Rogers, K.R. Elder and R.C. Desai, Phys. Rev. B **37**, 9638 (1988)
108. J. Amar, F. Sullivan and R.D. Mountain, Phys. Rev. B **37**, 196 (1988)
109. J.F. Marko and G.T. Barkema, Phys. Rev. E **52**, 2522 (1995)
110. T. Ohta and H. Nozaki, in *Space-Time Organization in Macromolecular Fluids*, ed. by F. Tanaka, M. Doi and T. Ohta (Springer-Verlag, Berlin 1989), p. 51
111. H. Tomita, Prog. Theor. Phys. **90**, 521 (1993)
112. G.F. Mazenko, Phys. Rev. E **50**, 3485 (1994); G.F. Mazenko and R. Wickham, Phys. Rev. E **51**, 2886 (1995); G.F. Mazenko, Phys. Rev. E **62**, 5967 (2000)
113. C. Yeung, Y. Oono and A. Shinozaki, Phys. Rev. E **49**, 2693 (1994)
114. S. Puri, A.J. Bray and F. Rojas, Phys. Rev. E **52**, 4699 (1995); F. Rojas, S. Puri and A.J. Bray, J. Phys. A **34**, 3985 (2001)
115. A. Coniglio and M. Zannetti, Europhys. Lett. **10**, 575 (1989)
116. A.J. Bray and K. Humayun, Phys. Rev. Lett. **68**, 1559 (1992)
117. E.D. Siggia, Phys. Rev. A **20**, 595 (1979)
118. H. Furukawa, Phys. Rev. A **31**, 1103 (1985); Adv. Phys. **34**, 703 (1985)
119. M. Grant and K.R. Elder, Phys. Rev. Lett. **82**, 14 (1999)
120. V.M. Kendon, J-C. Desplat, P. Bladon and M.E. Cates, Phys. Rev. Lett. **83**, 576 (1999)
121. F.J. Higuera, S. Succi and R. Benzi, Europhys. Lett. **9**, 345 (1989)
122. M.R. Swift, E. Orlandini, W.R. Osborn and J. Yeomans, Phys. Rev. E **54**, 5041 (1996)
123. T. Koga and K. Kawasaki, Phys. Rev. A **44**, R817 (1991)
124. S. Puri and B. Dünweg, Phys. Rev. A **45**, R6977 (1992)
125. M. Laradji, S. Toxvaerd and O.G. Mouritsen, Phys. Rev. Lett. **77**, 2253 (1996)
126. S. Bastea and J.L. Lebowitz, Phys. Rev. Lett. **78**, 3499 (1997)

127. D.A. Huse and C.L. Henley, Phys. Rev. Lett. **54**, 2708 (1985)
128. D.S. Fisher and D.A. Huse, Phys. Rev. Lett. **56**, 1601 (1986); Phys. Rev. B **38**, 373 (1988); Phys. Rev. B **38**, 386 (1988)
129. S. Puri, D. Chowdhury and N. Parekh, J. Phys. A **24**, L1087 (1991)
130. S. Puri and N. Parekh, J Phys. A **25**, 4127 (1992)
131. M.F. Gyure, S.T. Harrington, R. Strilka and H.E. Stanley, Phys. Rev. E **52**, 4632 (1995)
132. D. Chowdhury, J. Phys. Condens. Matter **6**, 2435 (1994)
133. T. Kawakatsu, K. Kawasaki, M. Furusaka, O. Okabayashi and T. Kanaya, J. Phys. Condens. Matter **6** 6385 (1994)
134. P. Fratzl and O. Penrose Phys. Rev. B **50**, 3477 (1994); Phys. Rev. B **53**, 2890 (1996)
135. S. Puri, K. Binder and S. Dattagupta, Phys. Rev. B **46**, 98 (1992); S. Puri, N. Parekh and S. Dattagupta, J. Stat. Phys. **77**, 839 (1994)
136. C. Yeung, T.M. Rogers, A. Hernandez-Machado and D. Jasnow, J. Stat. Phys. **66**, 1071 (1992)
137. F.J. Alexander, C.A. Laberge, J.L. Lebowitz and R.K.P. Zia, J. Stat. Phys. **82**, 1133 (1996)
138. A.D. Rutenberg and C. Yeung, Phys. Rev. E **60**, 2710 (1999)
139. B. Schmittmann and R.K.P. Zia, in *Phase Transitions and Critical Phenomena Vol. 17*, ed. by C. Domb and J.L. Lebowitz (Academic Press, London 1995)
140. G. Krausch, Mat. Sc. and Engineering Rep. **R14**, 1 (1995)
141. S. Puri and H.L. Frisch, J. Phys. Condens. Matter **9**, 2109 (1997)
142. J. Lal and R. Bansil, Macromolecules **24**, 290 (1991); R. Bansil, J. Lal and B. Carvalho, Polymer **33**, 2961 (1992)
143. H. Tanaka, Phys. Rev. Lett. **71**, 3158 (1993); Phys. Rev. Lett. **76**, 787 (1996)
144. K. Sekimoto, N. Suematsu and K. Kawasaki, Phys. Rev. A **39**, 4912 (1989)
145. A. Onuki and S. Puri, Phys. Rev. E **59**, R1331 (1999)
146. J. Sharma and S. Puri, Phys. Rev. E **64**, 021513 (2001)
147. T. Hashimoto, T. Takebe and S. Suehiro, J. Chem. Phys. **88**, 5874 (1988); T. Takebe, R. Sawaoka and T. Hashimoto, J. Chem. Phys. **91**, 4369 (1989)
148. X.-L. Wu, D.J. Pine and P K. Dixon, Phys. Rev. Lett. **66**, 2408 (1991)
149. D. Beysens and F. Perrot, J. de Physique **45**, L31 (1994)
150. A. Onuki, J. Phys. Condens. Matter **9**, 6119 (1997)
151. G.S. Grest and P.S. Sahni, Phys. Rev. B **30**, 226 (1984)
152. C. Jeppesen and O.G. Mouritsen, Phys. Rev. B **47**, 14724 (1993)
153. S.K. Das and S. Puri, Phys. Rev. E **65**, 026141 (2002)

Chapter 4

154. A.B. Harris, J. Phys. C **7**, 1671 (1974)
155. A.B. Harris and T.C. Lubensky, Phys. Rev. Lett. **33**, 1540 (1974)
156. G. Grinstein and A. Luther Phys. Rev. B **13**, 1329 (1976)
157. R.B. Stinchcombe, in *Phase Transitions and Critical Phenomena Vol. 7*, ed. by C. Domb and J.L. Lebowitz (Academic Press, London 1983)
158. T. Nattermann and J. Villain, Phase Transitions **11**, 5 (1988)
159. T. Nattermann and P. Rujan, Int. J. Mod. Phys. B **3**, 1597 (1989)
160. T. Nattermann, in *Spin Glasses and Random Fields*, ed. by A.P. Young (World Scientific, Singapore 1997)

161. S.-K. Ma, *Modern Theory of Critical Phenomena* (Addison-Wesley, Reading, Massachusetts 1981)
162. Z.W. Lai, G.F. Mazenko and O.T. Valls, Phys. Rev. B **37**, 9481 (1988)
163. W. Feller, *Introduction to Probability Theory and its Applications* (John Wiley, New York 1968)
164. D. Stauffer, Phys. Rep. **54**, 1 (1979)
165. D. Stauffer, *Introduction to Percolation Theory* (Taylor and Francis, London 1985)
166. A. Coniglio, Phys. Rev. Lett. **46**, 250 (1981)
167. D. Forster, D.R. Nelson and M.J. Stephen, Phys. Rev. A **16**, 732 (1977)
168. D.A. Huse, C.L. Henley and D.S. Fisher, Phys. Rev. Lett. **55**, 2924 (1985)
169. C.L. Henley, Phys. Rev. Lett. **54**, 2030 (1985)
170. R. Rammal and A. Benoit, Phys. Rev. Lett. **55**, 649 (1985)
171. D. Chowdhury and B. Biswal, Physica A **180**, 253 (1992)
172. K. Binder and A.P. Young, Rev. Mod. Phys. **58**, 801 (1986)
173. M. Mezard, G. Parisi and M. Virasoro, *Spin Glass Theory and Beyond* (World Scientific, Singapore 1987)
174. K.H. Fischer and J.A. Hertz, *Spin Glasses* (Cambridge University Press, Cambridge 1991)
175. G. Parisi, Phys. Rev. Lett. **43**, 1754 (1979); J. Phys. A **13**, L115 (1980)
176. D. Sherrington and S. Kirkpatrick, Phys. Rev. Lett. **35**, 1792 (1975)
177. H. Rieger, B. Steckemetz and M. Schreckenberg, Europhys. Lett. **27**, 485 (1994)
178. J. Kisker, L. Santen, M. Schreckenberg and H. Rieger, Phys. Rev. B **53**, 6418 (1996)
179. J.-P. Bouchaud, L.F. Cugliandolo, J. Kurchan and M. Mezard, in *Spin Glasses and Random Fields*, ed. by A.P. Young (World Scientific, Singapore 1997), p. 161
180. A. Crisanti and F. Ritort, J. Phys. A **36**, R181 (2003)
181. Y. Imry and S.-K. Ma, Phys. Rev. Lett. **35**, 1399 (1975)
182. J.Z. Imbrie, Phys. Rev. Lett. **53**, 1747 (1984); Commun. Math. Phys. **98**, 145 (1985)
183. J. Bricmont and A. Kupiainen, Phys. Rev. Lett. **59**, 1829 (1987); Commun. Math. Phys. **116**, 539 (1988)
184. J. Villain, Phys. Rev. Lett. **52**, 1543 (1984)
185. G. Grinstein and J.F. Fernandez, Phys. Rev. B **29**, 6389 (1984)
186. R. Bruinsma and G. Aeppli, Phys. Rev. Lett. **52**, 1547 (1984)
187. D. Andelman and J.-F. Joanny, Phys. Rev. B **32**, 4818 (1985)
188. M. Grant and J.D. Gunton, Phys. Rev. B **29**, 6266 (1984); Phys. Rev. B **35**, 4922 (1987)
189. J.D. Shore and J.P. Sethna, Phys. Rev. B **43**, 3782 (1991)
190. J.D. Shore, M. Holzer and J.P. Sethna, Phys. Rev. B **46**, 11376 (1992)
191. T. Nagai and K. Kawasaki, Physica A **134**, 483 (1986)
192. H. Ikeda, Y. Endoh and S. Itoh, Phys. Rev. Lett. **64**, 1266 (1990)
193. H. Ikeda, J. Phys. C **16**, L1033 (1983)
194. S. Itoh, H. Ikeda, T. Suzuki and Y. Endoh, J. Phys. Soc. Jpn. **60**, 3238 (1991)
195. A.G. Schins, A.F.M. Arts and H.W. de Wijn, Phys. Rev. Lett. **70**, 2340 (1993)
196. D.K. Shenoy, J.V. Selinger, K.A. Grüneberg, J. Naciri and R. Shashidhar, Phys. Rev. Lett. **82**, 1716 (1999)

197. V. Likodimos, M. Labardi and M. Allegrini, Phys. Rev. B **61**, 14440 (2000)
198. V. Likodimos, M. Labardi, X.K. Orlik, L. Pardi, M. Allegrini, S. Emonin and O. Marti, Phys. Rev. B **63**, 064104 (2001)
199. K. Hamano, J. Zhang, K. Abe, T. Mitsui, H. Sakata and K. Ema, J. Phys. Soc. Jpn. **65**, 142 (1996)
200. J. Zhang, K. Hamano, K. Abe, T. Mitsui, H. Sakata and K. Ema, J. Phys. Soc. Jpn. **65**, 149 (1996)
201. S. Fishman and A. Aharony, J. Phys. C **12**, L729 (1979)
202. J.L. Cardy, Phys. Rev. B **29**, 505 (1984)
203. J. Villain, J. Phys. (Paris) Lett. **43**, L551 (1982)
204. P.-Z. Wong and J.W. Cable, Phys. Rev. B **28**, 5361 (1983)
205. R.A. Cowley and W.J.L. Buyers, J. Phys. C **15**, L1209 (1982)
206. R.J. Birgenau, R.A. Cowley, G. Shirane and H. Yoshizawa, J. Stat. Phys. **34**, 817 (1984)
207. Y. Shapira, N.F. Oliveira and S. Foner, Phys. Rev. B **30**, 6639 (1984)
208. M. Lederman, J.V. Selinger, R. Bruinsma, R. Orbach and J. Hammann, Phys. Rev. B **48**, 3810 (1993)
209. Q. Feng, R.J. Birgenau and J.P. Hill, Phys. Rev. B **51**, 15188 (1995)
210. J.-K. Zuo, G.-C. Wang and T.-M. Lu, Phys. Rev. Lett. **60**, 1053 (1988)
211. P. Schleger, R.A. Hadfield, H. Casalta, N.H. Andersen, H.F. Poulsen, M. von Zimmermann, J.R. Schneider, R. Liang, P. Dosanjh and W.N. Hardy, Phys. Rev. Lett. **74**, 1446 (1995)
212. F. Brochard and P.G. de Gennes, J. Phys. (Paris) Lett. **44**, 785 (1983)
213. P.G. de Gennes, J. Phys. Chem. **88**, 6469 (1984)
214. A.J. Liu, D.J. Durian, E. Herbolzheimer and S.A. Safran, Phys. Rev. Lett. **65**, 1897 (1990)
215. L. Monette, A.J. Liu and G.S. Grest, Phys. Rev. A **46**, 7664 (1992)
216. A. Chakrabarti, Phys. Rev. Lett. **69**, 1548 (1992)
217. B.J. Frisken, F. Ferri and D.S. Cannell, Phys. Rev. Lett. **66**, 2754 (1991)
218. B.J. Frisken and D.S. Cannell, Phys. Rev. Lett. **69**, 632 (1992)
219. Z. Zhuang, A.G. Casielles and D.S. Cannell, Phys. Rev. Lett. **77**, 2969 (1996)
220. B.J. Frisken, D.S. Cannell, M.Y. Lin and S.K. Sinha, Phys. Rev. E **51**, 5866 (1995)
221. A.E. Bailey, B.J. Frisken and D.S. Cannell, Phys. Rev. E **56**, 3112 (1997)
222. *The Monte Carlo Method in Condensed Matter Physics*, ed. by K. Binder, Springer Topics in Applied Physics **71** (Springer-Verlag, Berlin 1992)
223. S. Puri, in *Models and Techniques of Statistical Physics*, ed. by S.M. Bhattacharjee (Narosa, New Delhi 1997), p. 126
224. G.S. Grest and D.J. Srolovitz, Phys. Rev. B **32**, 3014 (1985)
225. D.J. Srolovitz and G.S. Grest, Phys. Rev. B **32**, 3021 (1985)
226. D. Chowdhury, M. Grant and J.D. Gunton, Phys. Rev. B **35**, 6792 (1987)
227. D. Chowdhury and S. Kumar, J. Stat. Phys. **49**, 855 (1987)
228. J.H. Oh and D.-I. Choi, Phys. Rev. B **33**, 3448 (1986)
229. A.J. Bray and K. Humayun, J. Phys. A **24**, L1185 (1991)
230. U.M.B. Marconi, Phys. Rev. E **57**, 1290 (1998)
231. S. Jain, Phys. Rev. E **59**, R2493 (1999); Phys. Rev. E **60**, R2445 (1999)
232. B. Derrida, A.J. Bray and C. Godreche, J. Phys. A **27**, L357 (1994)
233. A.J. Bray, B. Derrida and C. Godreche, Europhys. Lett. **27**, 177 (1994)
234. B. Biswal, S. Puri and D. Chowdhury, Physica A **229**, 72 (1996)

235. H. Hayakawa, J. Phys. Soc. Jpn. **60**, 2492 (1991)
236. T. Iwai and H. Hayakawa, J. Phys. Soc. Jpn. **62**, 1583 (1993); J. Phys. Soc. Jpn. **62**, 2971 (1993)
237. E.T. Gawlinski, K. Kaski, M. Grant and J.D. Gunton, Phys. Rev. Lett. **53**, 2266 (1984)
238. E.T. Gawlinski, S. Kumar, M. Grant, J.D. Gunton and K. Kaski, Phys. Rev. B **32**, 1575 (1985)
239. E. Pytte and J.F. Fernandez, Phys. Rev. B **31**, 616 (1985)
240. S.R. Anderson, Phys. Rev. B **36**, 8435 (1987)
241. K. Binder, Z. Phys. B **50**, 343 (1983)
242. G. Grinstein and S.-K. Ma, Phys. Rev. Lett. **49**, 685 (1982)
243. E. Oguz, A. Chakrabarti, R. Toral and J.D. Gunton, Phys. Rev. B **42**, 704 (1990)
244. S. Puri and N. Parekh, J. Phys. A **26**, 2777 (1993)
245. M. Rao and A. Chakrabarti, Phys. Rev. Lett. **71**, 3501 (1993)
246. E. Oguz, J. Phys. A **27**, 2985 (1994)
247. M. Rao and A. Chakrabarti, Phys. Rev. E **48**, R25 (1993)

Chapter 5

248. T. Young, Philos. Trans. R. Soc. London, Ser. A **95**, 69 (1805)
249. J.W. Cahn, J. Chem. Phys. **66**, 3667 (1977)
250. H. Nakanishi and M.E. Fisher, Phys. Rev. Lett. **49**, 1565 (1982)
251. K. Binder, in *Phase Transitions and Critical Phenomena Vol. 8*, ed. by C. Domb and J.L. Lebowitz (Academic Press, London 1983), p. 1
252. P.G. de Gennes, Rev. Mod. Phys. **57**, 827 (1985)
253. H.W. Diehl, in *Phase Transitions and Critical Phenomena Vol. 10*, ed. by C. Domb and J.L. Lebowitz (Academic Press, London 1986), p. 75
254. S. Dietrich, in *Phase Transitions and Critical Phenomena Vol. 12*, ed. by C. Domb and J.L. Lebowitz (Academic Press, London 1988), p. 1
255. R.A.L. Jones, L.J. Norton, E.J. Kramer, F.S. Bates and P. Wiltzius, Phys. Rev. Lett. **66**, 1326 (1991)
256. K. Binder, J. Non-Equilib. Thermodyn. **23**, 1 (1998)
257. G. Krausch, C.-A. Dai, E.J. Kramer and F.S. Bates, Phys. Rev. Lett. **71**, 3669 (1993)
258. F. Bruder and R. Brenn, Phys. Rev. Lett. **69**, 624 (1992)
259. W. Straub, F. Bruder, R. Brenn, G. Krausch, H. Bielefeldt, A. Kirsch, O. Marti, J.L. Mlynek and J.F. Marko, Europhys. Lett. **29**, 353 (1995)
260. G. Krausch, J. Mlynek, W. Straub, R. Brenn and J.F. Marko, Europhys. Lett. **28**, 323 (1994)
261. M. Geoghegan, R.A.L. Jones and A.S. Clough, J. Chem. Phys. **103**, 2719 (1995)
262. M. Geoghegan, H. Ermer, G. Jüngst, G. Krausch and R. Brenn, Phys. Rev. E **62**, 940 (2000)
263. A. Karim, J.F. Douglas, B.P. Lee, S.C. Glotzer, J.A. Rogers, R.J. Jackman, E.J. Amis and G.M. Whitesides, Phys. Rev. E **57**, R6273 (1998)
264. B.D. Ermi, G. Nisato, J.F. Douglas, J.A. Rogers and A. Karim, Phys. Rev. Lett. **81**, 3900 (1998)
265. M. Böltau, S. Walheim, J. Mlynek, G. Krausch and U. Steiner, Nature **391**, 877 (1998)

266. B. Aichmayer, P. Fratzl, S. Puri and G. Saller, Phys. Rev. Lett. **91**, 015701 (2003)
267. P. Wiltzius and A. Cumming, Phys. Rev. Lett. **66**, 3000 (1991)
268. A. Cumming, P. Wiltzius, F.S. Bates and J.H. Rosedale, Phys. Rev. A **45**, 885 (1992)
269. B.Q. Shi, C. Harrison and A. Cumming, Phys. Rev. Lett. **70**, 206 (1993)
270. C. Harrison, W. Rippard and A. Cumming, Phys. Rev. E **52**, 723 (1995)
271. W.-J. Ma, P. Keblinski, A. Maritan, J. Koplik and J.R. Banavar, Phys. Rev. E **48**, R2362 (1993)
272. P. Keblinski, W.-J. Ma, A. Maritan, J. Koplik and J.R. Banavar, Phys. Rev. Lett. **72**, 3738 (1994)
273. S.M. Troian, Phys. Rev. Lett. **71**, 1399 (1993); Phys. Rev. Lett. **72**, 3739 (1994)
274. P. Guenoun, D. Beysens and M. Robert, Phys. Rev. Lett. **65**, 2406 (1990); Physica A **172**, 137 (1991)
275. H. Tanaka, Phys. Rev. Lett. **70**, 53 (1993); Phys. Rev. Lett. **70**, 2770 (1993); Europhys. Lett. **24**, 665 (1993)
276. H. Tanaka, J. Phys. Condens. Matter **13**, 4637 (2001)
277. G.M. Xiong and C.D. Gong, Phys. Rev. B **39**, 9384 (1989)
278. Z. Jiang and C. Ebner, Phys. Rev. B **39**, 2501 (1989)
279. R.C. Ball and R.L.H. Essery, J. Phys. Condens. Matter **2**, 10303 (1990)
280. J.D. Weeks, in *Ordering in Strongly Fluctuating Condensed Matter Systems*, ed. by T. Riste (Plenum, New York 1980), p. 293
281. H. Van Beijeren and I. Nolden, in *Structure and Dynamics of Surfaces Vol. II*, ed. by W. Schommers and P. Blanckenhagen (Springer, Heidelberg 1987), p. 259
282. K. Binder and P.C. Hohenberg, Phys. Rev. B **6**, 3461 (1972)
283. M.E. Fisher, J. Stat. Phys. **34**, 667 (1984)
284. K. Binder and H.L. Frisch, Z. Phys. B **84**, 403 (1991)
285. S. Puri and K. Binder, Phys. Rev. A **46**, R4487 (1992); Phys. Rev. E **49**, 5359 (1994)
286. I. Schmidt and K. Binder, Z. Phys. B **67**, 369 (1987)
287. S. Puri and K. Binder, Z. Phys. B **86**, 263 (1992)
288. M.E. Fisher and H. Nakanishi, J. Chem. Phys. **75**, 5857 (1981)
289. H. Nakanishi and M.E. Fisher, J. Chem. Phys. **78**, 3279 (1983)
290. A.O. Parry and R. Evans, Phys. Rev. Lett. **64**, 439 (1990); Phys. Rev. Lett. **66**, 2175 (1991)
291. M.R. Swift, A.L. Owczarek and J.O. Indekeu, Europhys. Lett. **14**, 475 (1991)
292. M. Müller, K. Binder and E.V. Albano, Int. J. Mod. Phys. B **15**, 1867 (2001)
293. K. Binder, D. Landau and M. Müller, J. Stat. Phys. **110**, 1411 (2003)
294. S. Puri and K. Binder, J. Stat. Phys. **77**, 145 (1994)
295. A. Bhattacharya, M. Rao and A. Chakrabarti, Phys. Rev. E **49**, 524 (1994)
296. M.A. Vaksman and W.E. McMullen, Phys. Rev. E **49**, 4724 (1994)
297. H.W. Diehl and H.K. Janssen, Phys. Rev. A **45**, 7145 (1992)
298. R.A.L. Jones, E.J. Kramer, M.H. Rafailovich, J. Sokolov and S.A. Schwarz, Phys. Rev. Lett. **62**, 280 (1989)
299. R.A.L. Jones and E.J. Kramer, Philos. Mag. B **62**, 129 (1990)
300. H.L. Frisch, P. Nielaba and K. Binder, Phys. Rev. E **52**, 2848 (1995)
301. H.P. Fischer, P. Maass and W. Dieterich, Phys. Rev. Lett. **79**, 893 (1997)
302. S. Puri and K. Binder, Phys. Rev. Lett. **86**, 1797 (2001); Phys. Rev. E **66**, 061602 (2002)

303. S. Puri, K. Binder and H.L. Frisch, Phys. Rev. E **56**, 6991 (1997)
304. I.E. Dzyaloshinskii, E.M. Lifshitz and L.P. Pitaevskii, Adv. Phys. **10**, 165 (1961)
305. R. Lipowsky, J. Phys. A **18**, L585 (1985)
306. R. Lipowsky and D.A. Huse, Phys. Rev. Lett. **57**, 353 (1986)
307. G. Brown and A. Chakrabarti, Phys. Rev. A **46**, 4829 (1992)
308. G. Brown, A. Chakrabarti and J.F. Marko, Phys. Rev. E **50**, 1674 (1994)
309. G. Brown and A. Chakrabarti, J. Chem. Phys. **101**, 3310 (1994); J. Chem. Phys. **102**, 1440 (1995)
310. Y. Oono and Y. Shiwa, Mod. Phys. Lett. B **1**, 49 (1987)
311. Y. Oono and M. Bahiana, Phys. Rev. Lett. **61**, 1109 (1988)
312. J.F. Marko, Phys. Rev. E **48**, 2861 (1993)
313. C. Sagui, A.M. Somoza, C. Roland and R.C. Desai, J. Phys. A **26**, L1163 (1993)
314. B.P. Lee, J.F. Douglas and S.C. Glotzer, Phys. Rev. E **60**, 5812 (1999)
315. Z. Zhang and A. Chakrabarti, Phys. Rev. E **50**, R4290 (1994)
316. H. Chen and A. Chakrabarti, Phys. Rev. E **55**, 5680 (1997)
317. E. Velasco and S. Toxvaerd, Phys. Rev. Lett. **71**, 388 (1993); J. Phys. Condens. Matter **6**, A205 (1994)
318. H. Tanaka and T. Araki, Europhys. Lett. **51**, 154 (2000)
319. S. Toxvaerd, Phys. Rev. Lett. **83**, 5318 (1999)
320. S. Bastea, S. Puri and J.L. Lebowitz, Phys. Rev. E **63**, 041513 (2001)
321. O. Kuksenok, J.M. Yeomans and A.C. Balazs, Phys. Rev. E **65**, 031502 (2002)
322. O. Kuksenok and A.C. Balazs, Phys. Rev. E **68**, 011502 (2003)
323. R. Toral and A. Chakrabarti, Phys. Rev. B **43**, 3438 (1991)
324. H.L. Frisch, S. Puri and P. Nielaba, J. Chem. Phys. **110**, 10514 (1999)
325. S. Puri and H.L. Frisch, J. Chem. Phys. **79**, 5560 (1993)
326. E.L. Ince, *Ordinary Differential Equations* (Dover, New York 1956)

Chapter 6

327. Y. Kuramoto, *Chemical Oscillations, Waves and Turbulence* (Springer-Verlag, Berlin 1984)
328. M.C. Cross and P.C. Hohenberg, Rev. Mod. Phys. **65**, 851 (1993)
329. W. van Saarloos, in *Spatiotemporal Patterns in Nonequilibrium Systems*, ed. by P.E. Cladis and P. Palffy-Muhoray (Addison-Wesley, Reading, MA 1994), p. 19
330. L.M. Pismen, *Vortices in Nonlinear Fields* (Oxford University Press, Oxford 1999)
331. I.S. Aranson and L. Kramer, Rev. Mod. Phys. **74**, 99 (2002)
332. S.C. Müller, T. Plesser and B. Hess, Physica D **24**, 71 (1987)
333. G.S. Skinner and H.L. Swinney, Physica D **48**, 1 (1991)
334. R.W. Walden, P. Kolodner, A. Passner and C.M. Surko, Phys. Rev. Lett. **55**, 496 (1985)
335. E. Moses and V. Steinberg, Phys. Rev. A **34**, 693 (1986)
336. F.T. Arecchi, G. Giacomelli, P.L. Ramazza and S. Residori, Phys. Rev. Lett. **65**, 2531 (1990); Phys. Rev. Lett. **67**, 3749 (1991)
337. P.S. Hagan, SIAM J. Appl. Math **42**, 762 (1982)
338. I.S. Aranson, L. Aranson, L. Kramer and A. Weber, Phys. Rev. A **46**, R2992 (1992); A. Weber, L. Kramer, I.S. Aranson and L. Aranson, Physica D **61**, 279 (1992)

339. H. Chate, Nonlinearity **7** 185 (1994)
340. H. Chate, in *Spatiotemporal Patterns in Nonequilibrium Systems*, ed. by P.E. Cladis and P. Palffy-Muhoray (Addison-Wesley, Reading, MA 1994), p. 33
341. H. Chate and P. Manneville, Physica A **224**, 348 (1996); P. Manneville and H. Chate, Physica D **96**, 30 (1996)
342. S.K. Das, S. Puri and M.C. Cross, Phys. Rev. E **64**, 046206 (2001)
343. S. Puri, S.K. Das and M.C. Cross, Phys. Rev. E **64**, 056140 (2001)
344. S.K. Das and S. Puri, Phys. Rev. E **65**, 046123 (2002)
345. S. Puri, Phys. Lett. A **164**, 211 (1992)
346. M. Suzuki, Prog. Theor. Phys. **56**, 477 (1976)
347. K. Kawasaki, M.C. Yalabik and J.D. Gunton, Phys. Rev. A **17**, 455 (1978)
348. S. Puri and C. Roland, Phys. Lett. A **151**, 500 (1990)
349. M.J. Ablowitz and H. Segur, *Solitons and the Inverse Scattering Transform*, SIAM Studies in Applied Mathematics (SIAM Press, Philadelphia 1981)
350. S. Puri, Int. J. Mod. Phys B **4**, 1483 (1990)
351. E.P. Gross, Nuovo Cimento **20**, 454 (1961); L. Pitaevskii, Sov. Phys. JETP **13**, 451 (1961)
352. *Bose-Einstein Condensation*, ed. by A. Griffin, D.W. Snoke and S. Stringari (Cambridge University Press, Cambridge 1995)
353. I.S. Aranson, L. Kramer and A. Weber, Phys. Rev. E **47**, 3231 (1993); Phys. Rev. E **48**, R9 (1993)
354. G. Huber, P. Alstrom and T. Bohr, Phys. Rev. Lett. **69**, 2380 (1992)
355. V.N. Biktashev, in *Nonlinear Waves II*, ed. by A.V. Gaponov-Grekhov and M.I. Rabinovich (Springer-Verlag, Heidelberg 1989), p. 87
356. L.M. Pismen and A.A. Nepomnyashchy, Phys. Rev. A **44**, R2243 (1991)
357. T. Bohr, G. Huber and E. Ott, Europhys. Lett. **33**, 589 (1996); Physica D **106**, 95 (1997)
358. P.G. Kevrekidis, A.R. Bishop and K.O. Rasmussen, Phys. Rev. E **65**, 016122 (2001)
359. B.A. Malomed, Phys. Rev. E **50**, R3310 (1994)
360. H. Sakaguchi and B.A. Malomed, Physica D **118**, 250 (1998)
361. G.F. Mazenko, Phys. Rev. E **64**, 016110 (2001)
362. G.F. Mazenko, Phys. Rev. Lett. **78**, 401 (1997)
363. C. Brito, I.S. Aranson and H. Chate, Phys. Rev. Lett. **90**, 068301 (2003)
364. A.T. Winfree, S. Caudle, G. Chen, P. McGuire and Z. Szilagyi, Chaos **6**, 617 (1996)
365. F. Siegert and C.J. Weijer, Physica D **49**, 224 (1991)
366. R.A. Gray and J. Jalife, Int. J. of Bifurcation and Chaos **6**, 415 (1996)
367. L.M. Pismen and J. Rubinstein, Physica D **47**, 353 (1991)
368. M. Gabbay, E. Ott and P.N. Guzdar, Phys. Rev. Lett. **78**, 2012 (1997); Physica D **118**, 371 (1998); Phys. Rev. E **58**, 2576 (1998)
369. A.T. Winfree, Physica D **84**, 126 (1995)
370. F. Fenton and A. Karma, Phys. Rev. Lett. **81**, 481 (1998); Chaos **8**, 20 (1998)
371. J. Koplik and H. Levine, Phys. Rev. Lett. **71**, 1375 (1993); Phys. Rev. Lett. **76**, 4745 (1996)
372. S. Puri, unpublished

Chapter 7

373. R. Blinc and B. Zeks, Adv. Phys. **21**, 693 (1972)
374. B.K. Chakrabarti, A. Dutta and P. Sen, *Quantum Ising Phases and Transitions in Transverse Ising Models*, Lecture Notes in Physics **41** (Springer-Verlag, Heidelberg 1996)
375. D. Bitko, T.F. Rosenbaum and G. Aeppli, Phys. Rev. Lett. **77**, 940 (1996)
376. A. Abraham, *The Theory of Nuclear Magnetism* (Oxford University Press, London 1961)
377. J.M. Luttinger and L. Tisza, Phys. Rev. **70**, 954 (1946)
378. D. Bitko, Ph.D. thesis, University of Chicago, Illinois (1992, unpublished)
379. S. Sachdev, *Quantum Phase Transitions* (Cambridge University Press, Cambridge 1999)
380. R.N. Bhatt, in *Spin Glasses and Random Fields*, ed. by A.P. Young (World Scientific, Singapore 1997)
381. S.L. Sondhi, S.M. Girvin, J.P. Carini and D. Shahar, Rev. Mod. Phys. **69**, 315 (1997)
382. S.A. Carter, T.F. Rosenbaum, J.M. Honig and J. Spalek, Phys. Rev. Lett. **67**, 3440 (1991)
383. S.A. Carter, T.F. Rosenbaum, P. Metcalf, J.M. Honig and J. Spalek, Phys. Rev. B **48**, 16841 (1993)
384. H.v. Löhneysen, T. Pietrus, G. Portisch, H.G. Schlager, A. Schröder, M. Sieck and T. Trappmann, Phys. Rev. Lett. **72**, 3262 (1994)
385. B. Bogenberger and H.v. Löhneysen, Phys. Rev. Lett. **74**, 1016 (1995)
386. M.B. Maple, J. Low Temp. Phys. **99**, 223 (1995)
387. A.V. Chubukov and S. Sachdev, Phys. Rev. Lett. **71**, 169 (1993)
388. A. Sokol and D. Pines, Phys. Rev. Lett. **71**, 2813 (1993)
389. M. Suzuki, Prog. Theo. Phys. **56**, 1454 (1976)
390. M. Barma and B.S. Shastry, Phys. Rev. B **18**, 3351 (1978)
391. A. Aharony and B.I. Halperin, Phys. Rev. Lett. **35**, 1308 (1975)
392. W. Wu, B. Ellman, T.F. Rosenbaum, G. Aeppli and D.H. Reich, Phys. Rev. Lett. **67**, 2076 (1991)
393. W. Wu, D. Bitko, T.F. Rosenbaum and G. Aeppli, Phys. Rev. Lett. **71**, 1919 (1993)
394. V. Banerjee and S. Dattagupta, J. Phys. Condens. Matter **10**, 8351 (1998); in *Nonlinearities in Complex Systems*, ed. by S. Puri and S. Dattagupta (Narosa, New Delhi 1997)
395. *Non-Debye Relaxation in Condensed Matter*, ed. by T.V. Ramakrishnan and M. Rajlakshmi (World Scientific, Singapore 1987)
396. J.A. Mydosh, in *Amorphous Magnetism*, ed. by R.A. Levy and R. Hasegawa (Plenum, New York 1977)
397. W. Götze, in *Liquids, Freezing and the Glass Transition*, ed. by J.P. Hansen, D. Levesque and J. Zinn-Justin (North-Holland, Amsterdam 1991)
398. V. Narayanamurti and R.O. Pohl, Rev. Mod. Phys. **42**, 201 (1970)
399. R.C. Zeller and R.O. Pohl, Phys. Rev. B **4**, 2029 (1971)
400. P.W. Anderson, B.I. Halperin and C.M. Varma, Philos. Mag. **25**, 1 (1972)
401. W.A. Phillips, J. Low Temp. Phys. **7**, 351 (1972)
402. B.W. Southern, J. Phys. C **9**, 4011 (1976)
403. S. Kirkpatrick and D. Sherrington, Phys. Rev. B **17**, 4384 (1978)

404. H. Sompolinsky, Phys. Rev. Lett. **47**, 935 (1981)
405. H. Sompolinsky and A. Zippelius, Phys. Rev. B **25**, 6860 (1982)
406. H.-J. Sommers, Phys. Rev. Lett. **58**, 1268 (1987)
407. J.R. de Almeida and D.J. Thouless, J. Phys. A **11**, 983 (1978)
408. K.D. Usadel, Solid State Commun. **58**, 629 (1986)
409. T. Yamamoto and H. Ishii, J. Phys. C **20**, 6053 (1987)
410. Y.Y. Goldschmidt and P.-Y. Lai, Phys. Rev. Lett. **64**, 2467 (1990)
411. V. Dobrosavljevic and D. Thirumalai, J. Phys. A **23**, L767 (1990)
412. T. Giamarchi and P. Le Doussal, Phys. Rev. B **53**, 15206 (1996)
413. T.K. Kopec, Phys. Rev. E **54**, 3367 (1996)
414. H. Umezawa, H. Matsumoto and M. Tachiki, *Thermofield Dynamics and Condensed States* (North-Holland, Amsterdam 1982)
415. T.K. Kopec, J. Phys. C **21**, 297 (1988)
416. A. Aharony, Phys. Rev. B **13**, 2092 (1976)
417. M.J. Stephen and A. Aharony, J. Phys. C **14**, 1665 (1981)
418. S. Dattagupta, Ind. J. of Physics **43**, 141 (1999)
419. E. Courtens, Phys. Rev. Lett. **52**, 69 (1984)
420. H.J. Brükner, E. Courtens and H.-G. Unruh, Z. Phys. B **73**, 337 (1988)
421. T.K. Kopec, B. Tadic, R. Pirc and R. Blinc, Z. Phys. B **78**, 493 (1990)
422. V. Banerjee and S. Dattagupta, Phase Transitions **62**, 233 (1997)
423. I.A. Akhiezer and A.I. Spolnik, Sov. Phys. Solid State **25**, 81 (1983)
424. V. Dobrosavljevic and R.M. Stratt, Phys. Rev. B **36**, 8484 (1987)
425. S. Dattagupta, B. Tadic, R. Pirc and R. Blinc, Phys. Rev. B **44**, 4387 (1991); Phys. Rev. B **47**, 8801 (1993)

Chapter 8

426. A.J. Leggett, S. Chakravarty, A.T. Dorsey, M.P.A. Fisher, A. Garg and W. Zwerger, Rev. Mod. Phys. **59**, 1 (1987)
427. *Hydrogen in Metals Vol. I*, Topics in Applied Physics **28**, ed. by G. Alefeld and J. Völkl (Springer-Verlag, Berlin 1978)
428. W. Press, *Single-Particle Rotations in Molecular Crystals*, Springer Tracts in Modern Physics **92** (Springer-Verlag, Heidelberg 1981)
429. R.P. Feynman, *Lecture Notes in Physics* (Addison-Wesley, Reading, MA 1964)
430. T. Holstein, Ann. Phys. (N.Y.) **8**, 325 (1959); Ann. Phys. (N.Y.) **8**, 343 (1959)
431. J. Kondo, in *Fermi Surface Effects*, Springer Series in Solid State Sciences **77**, ed. by A. Yoshimori and J. Kondo (Springer-Verlag, Heidelberg 1988)
432. L.-D. Chang and S. Chakravarty, Phys. Rev. B **31**, 154 (1985)
433. F. Guinea, Phys. Rev. Lett. **53**, 1268 (1984)
434. F. Guinea, V. Hakim and A. Muramatsu, Phys. Rev. Lett. **54**, 263 (1985); Phys. Rev. B **32**, 4410 (1985)
435. T. Qureshi and S. Dattagupta, Phys. Rev. B **49**, 12848 (1994)
436. S. Chakravarty, Phys. Rev. Lett. **49**, 681 (1982)
437. A.J. Bray and M.A. Moore, Phys. Rev. Lett. **49**, 1545 (1982)
438. B. Misra and E.C.G. Sudarshan, J. Math. Phys. **18**, 756 (1977)
439. M. Simonius, Phys. Rev. Lett. **40**, 980 (1978)
440. R.P. Feynman and A.R. Hibbs, *Quantum Mechanics and Path Integrals* (McGraw-Hill, New York 1965)

441. R.P. Feynman and F.L. Vernon, Ann. Phys. (N.Y.) **24**, 118 (1963)
442. S. Chakravarty and A.J. Leggett, Phys. Rev. Lett. **52**, 5 (1984)
443. G.D. Mahan, *Many-Particle Physics, Second Edition* (Plenum, New York 1990)
444. C. Aslangul, N. Pottier and D. Saint-James, J. Phys. (Paris) **47**, 1657 (1986)
445. S. Dattagupta, H. Grabert and R. Jung, J. Phys. Condens. Matter **1**, 1405 (1989)
446. U. Weiss and M. Wollensak, Phys. Rev. Lett. **62**, 1663 (1989)
447. T. Qureshi, Phys. Rev. B **53**, 3183 (1996)
448. M. Sanjay Kumar, S. Dattagupta and N. Kumar, Phys. Rev. B **65**, 134501 (2002)
449. S. Dattagupta and T. Qureshi, Physica B **174**, 262 (1991)
450. H. Grabert, S. Linkwitz, S. Dattagupta and U. Weiss, Europhys. Lett. **2**, 631 (1986)
451. T. Qureshi and S. Dattagupta, Pramana **35**, 579 (1990)
452. *Handbook of Mathematical Functions*, ed. by M. Abramowitz and I.A. Stegun (Dover, New York 1971)
453. A.C. Hearn, *Reduce User's Manual* (Rand Publications, Santa Monica 1985)
454. T. Qureshi and S. Dattagupta, J. Phys. Condens. Matter **3**, 1079 (1991)
455. T. Qureshi and S. Dattagupta, Phys. Rev. B **47**, 1092 (1993)
456. J. Kondo, Physica B **125**, 279 (1984); Physica B **126**, 377 (1984)
457. R.E. Lechner, D. Richter and C. Riekel, *Neutron Scattering and Muon Spin Rotation*, Springer Tracts in Modern Physics **101** (Springer-Verlag, Heidelberg 1983)
458. T. Springer and D. Richter, *Methods in Experimental Physics* (Academic Press, New York 1987)
459. C. Morkel, H. Wipf and K. Neumaier, Phys. Rev. Lett. **40**, 947 (1978)
460. A. Magerl, J.J. Rush, J.M. Rowe, D. Richter and H. Wipf, Phys. Rev. B **27**, 927 (1983)
461. H. Wipf and K. Neumaier, Phys. Rev. Lett. **52**, 1308 (1984)
462. H. Wipf, D. Steinbinder, K. Neumaier, P. Gutsmiedl, A. Magerl and J. Dianoux, Europhys. Lett. **4**, 1379 (1987)
463. D. Steinbinder, H. Wipf, A. Magerl, D. Richter, A.J. Dianoux and K. Neumaier, Europhys. Lett. **6**, 535 (1988)
464. S. Dattagupta, Bull. Mater. Sci. **19**, 15 (1996)
465. H. Grabert, Phys. Rev. B **46**, 12753 (1992)
466. H. Grabert and H.R. Schober, in *Hydrogen in Metals Vol. II*, ed. by H. Wipf (Springer-Verlag, Heidelberg 1994)
467. C.P. Flynn and A.M. Stoneham, Phys. Rev. B **1**, 3966 (1970)
468. Y. Kagan and M.I. Klinger, J. Phys. C **7**, 2791 (1974)
469. Z. Qi, J. Völkl, R. Lässer and H. Wenzl, J. Phys. F **13**, 2053 (1983)
470. S. Dattagupta and H.R. Schober, Phys. Rev. B **57**, 7606 (1998)
471. J. Kondo, Prog. Theor. Phys. **32**, 37 (1964)
472. P.W. Anderson and G. Yuval, Phys. Rev. Lett. **23**, 89 (1969)
473. P.W. Anderson, G. Yuval and D.R. Hamann, Phys. Rev. B **1**, 4464 (1970)
474. M. Blume, V.J. Emery and A. Luther, Phys. Rev. Lett. **25**, 450 (1970)
475. F. Gonzalez-Jimenez, P. Imbert and F. Hartmann-Boutron, Phys. Rev. B **9**, 95 (1974)
476. F. Hartmann-Boutron, Phys. Rev. B **10**, 2113 (1974)

477. F. Gonzalez-Jimenez, F. Hartmann-Boutron and P. Imbert, Phys. Rev. B **10**, 2122 (1974)
478. H. Alloul, Physica B **86-88**, 449 (1977)
479. P. Bonville, P. Imbert, G. Jehanno, F. Gonzalez-Jimenez and F. Hartmann-Boutron, Phys. Rev. B **30**, 3672 (1984)
480. W.A. Phillips, Rep. Prog. Phys. **50**, 1657 (1987)
481. J.L. Black, in *Glassy Metals Vol. I*, Topics in Applied Physics **46**, ed. by H.J. Güntherodt and H. Beck (Springer-Verlag, Heidelberg 1981), p. 167

Chapter 9

482. I.R. Senitzky, Phys. Rev. **119**, 670 (1960)
483. F. Schwabl and W. Thirring, Ergeb. Exakten. Naturwiss. **36**, 219 (1964)
484. G.W. Ford, M. Kac and P. Mazur, J. Math. Phys. **6**, 504 (1965)
485. P. Ullersma, Physica **32**, 27 (1966)
486. A.O. Caldeira and A.J. Leggett, Phys. Rev. A **31**, 1059 (1985)
487. F. Haake and R. Reibold, Phys. Rev. A **32**, 2462 (1985)
488. F. Haake and M. Zukowski, Phys. Rev. A **47**, 2506 (1993)
489. G.W. Ford, J.T. Lewis and R.F. O'Connell, Phys. Rev. A **37**, 4419 (1988)
490. V. Hakim and V. Ambegaokar, Phys. Rev. A **32**, 423 (1985)
491. B.L. Hu, J.P. Paz and Y. Zhang, Phys. Rev. D **45**, 2843 (1992)
492. G.-L. Ingold, in *Quantum Transport and Dissipation* (Wiley-VCH, Weinheim 1998)
493. S. Dattagupta and J. Singh, Pramana **47**, 211 (1996)
494. S. Ichimaru, *Basic Principles of Plasma Physics: A Statistical Approach* (Benjamin-Cummings, Reading, MA 1973)
495. J.J. Sakurai, *Modern Quantum Mechanics, Second Edition* (Addison-Wesley, Reading, MA 1994)
496. L. Landau, Z. Phys. **64**, 629 (1930)
497. J.H. Van Vleck, *Theory of Electric and Magnetic Susceptibilities* (Oxford University Press, London 1932)
498. *Mesoscopic Phenomena in Solids*, ed. by B.L. Altshuler, P.A. Lee and R.A. Webb (North-Holland, Amsterdam 1991)
499. T.M. Hong and J.M. Wheatley, Phys. Rev. B **42**, 6492 (1990); Phys. Rev. B **43**, 5702 (1991)
500. S. Dattagupta and J. Singh, Phys. Rev. Lett. **79**, 961 (1997)
501. N. Bohr, Ph.D. Thesis 1911, in *Collected Works Vol. I* (North-Holland, Amsterdam 1972)
502. H.J. van Leeuwen, J. Phys. (Paris) **2**, 361 (1921)
503. R. Peierls, *Surprises in Theoretical Physics* (Princeton University Press, Princeton 1979)
504. C.G. Darwin, Proc. Cambridge Philos. Soc. **27**, 86 (1930)
505. H. Grabert, P. Schramm and G.-L. Ingold, Phys. Rep. **168**, 115 (1988)
506. X.L. Li, G.W. Ford and R.F. O'Connell, Phys. Rev. A **41**, 5287 (1990); Phys. Rev. E **53**, 3359 (1996)
507. Ch. Sikorski and U. Merkt, Phys. Rev. Lett. **62**, 2164 (1989)
508. U. Merkt, Physica B **189**, 165 (1993)

Chapter 10

509. Y. Imry, *Introduction to Mesoscopic Physics* (Oxford University Press, London 1997)
510. W.H. Zurek, Physics Today **44**, 36 (1991)
511. W.H. Zurek, S. Habib and J.P. Paz, Phys. Rev. Lett. **70**, 1187 (1993)
512. J.P. Paz and W.H. Zurek, Phys. Rev. Lett. **82**, 5181 (1999)
513. B.M. Garraway and K.-A. Suominen, Rep. Prog. Phys. **58**, 365 (1995)
514. D. Chandler, in *Liquids, Freezing and the Glass Transition*, ed. by J.P. Hansen, D. Levesque and J. Zinn-Justin (North-Holland, Amsterdam 1991)
515. A. Garg, J.N. Onuchic and V. Ambegaokar, J. Chem. Phys. **83**, 4491 (1985)
516. A. Garg, Phys. Rev. Lett. **77**, 964 (1996)
517. Y. Imry, cond-mat 0202044 (2002)
518. N. Kumar and A.M. Jayannavar, Phys. Rev. B **45**, 5001 (1992)
519. N. Kumar, T.P. Pareek and A.M. Jayannavar, Mod. Phys. Lett. B **11**, 347 (1997); Phys. Rev. B **57**, 13399 (1998)
520. N.W. Ashcroft and N.D. Mermin, *Solid State Physics* (Holt, Rinehart and Winston, New York 1976)
521. S. Dattagupta, A.M. Jayannavar and N. Kumar, Current Science **80**, 861 (2001)
522. S.W. Tozer, A.W. Kleinsasser, T. Penney, D. Kaiser and F. Holtzberg, Phys. Rev. Lett. **59**, 1768 (1987)
523. M.V. Berry, in *Fundamental Problems in Quantum Theory*, ed. by D.M. Greenberger and A. Zeilinger (New York Academy of Sciences, New York 1995)
524. J. Shao, M.-L. Ge and H. Cheng, Phys. Rev. E. **53**, 1243 (1996)
525. G. Gangopadhyay, M. Sanjay Kumar and S. Dattagupta, J. Phys. A **34**, 5485 (2001)
526. G. Lindblad, Commun. Math. Phys. **48**, 119 (1976)
527. A. Messiah, *Quantum Mechanics* (North-Holland, Amsterdam 1961)
528. R.J. Glauber, Phys. Rev. **130**, 2529 (1963)
529. E.C.G. Sudarshan, Phys. Rev. Lett. **10**, 277 (1963)
530. A. Tameshtit and J.E. Sipe, Phys. Rev. A **47**, 1697 (1993); Phys. Rev. A **49**, 89 (1994)
531. G. Gangopadhyay and S.H. Lin, Pramana **49**, 399 (1997)
532. R. Egger, H. Grabert and U. Weiss, Phys. Rev. E **55**, R3809 (1997)
533. T.A. Costi and C. Kieffer, Phys. Rev. Lett. **76**, 1683 (1996)
534. F. Lesage, H. Saleur and S. Skorik, Phys. Rev. Lett. **76**, 3388 (1996); Nucl. Phys. B **474**, 602 (1996)
535. N. Kumar, D.V. Baxter, R. Richter and J.O. Strom-Olsen, Phys. Rev. Lett. **59**, 1853 (1987)
536. J. Rammer, A.L. Shelankov and A. Schmid, Phys. Rev. Lett. **60**, 1985 (1988)
537. G. Bergmann, Phys. Rev. Lett. **60**, 1986 (1988)
538. P. Mohanty, E.M.Q. Jariwala and R.A. Webb, Phys. Rev. Lett. **78**, 3366 (1997)
539. P. Mohanty and R.A. Webb, Phys. Rev. B **55**, R13452 (1997)
540. G. Bergmann, Phys. Rep. **107**, 1 (1984)
541. B.L. Altshuler, A.G. Aronov and D.E. Khmelnitsky, J. Phys. C **15**, 7367 (1982)

542. Z. Ovadyahu, Phys. Rev. B **63**, 235403 (2001)
543. D.S. Golubev and A.D. Zaikin, Phys. Rev. Lett. **81**, 1074 (1998); Phys. Rev. B **59**, 9195 (1999); Phys. Rev. Lett. **82**, 3191 (1999)
544. P. Cedraschi, V.V. Ponomarenko and M. Büttiker, Phys. Rev. Lett. **84**, 346 (2000)
545. D. Braun, F. Haake and W.T. Strunz, Phys. Rev. Lett. **86**, 2913 (2001)

Index

Springer Series in
MATERIALS SCIENCE

Editors: R. Hull R. M. Osgood, Jr. J. Parisi H. Warlimont

10 **Computer Simulation of Ion-Solid Interactions**
By W. Eckstein

11 **Mechanisms of High Temperature Superconductivity**
Editors: H. Kamimura and A. Oshiyama

12 **Dislocation Dynamics and Plasticity**
By T. Suzuki, S. Takeuchi, and H. Yoshinaga

13 **Semiconductor Silicon**
Materials Science and Technology
Editors: G. Harbeke and M. J. Schulz

14 **Graphite Intercalation Compounds I**
Structure and Dynamics
Editors: H. Zabel and S. A. Solin

15 **Crystal Chemistry of High-T_c Superconducting Copper Oxides**
By B. Raveau, C. Michel, M. Hervieu, and D. Groult

16 **Hydrogen in Semiconductors**
By S. J. Pearton, M. Stavola, and J. W. Corbett

17 **Ordering at Surfaces and Interfaces**
Editors: A. Yoshimori, T. Shinjo, and H. Watanabe

18 **Graphite Intercalation Compounds II**
Editors: S. A. Solin and H. Zabel

19 **Laser-Assisted Microtechnology**
By S. M. Metev and V. P. Veiko
2nd Edition

20 **Microcluster Physics**
By S. Sugano and H. Koizumi
2nd Edition

21 **The Metal-Hydrogen System**
By Y. Fukai

22 **Ion Implantation in Diamond, Graphite and Related Materials**
By M. S. Dresselhaus and R. Kalish

23 **The Real Structure of High-T_c Superconductors**
Editor: V. Sh. Shekhtman

24 **Metal Impurities in Silicon-Device Fabrication**
By K. Graff 2nd Edition

25 **Optical Properties of Metal Clusters**
By U. Kreibig and M. Vollmer

26 **Gas Source Molecular Beam Epitaxy**
Growth and Properties of Phosphorus Containing III–V Heterostructures
By M. B. Panish and H. Temkin

27 **Physics of New Materials**
Editor: F. E. Fujita 2nd Edition

28 **Laser Ablation**
Principles and Applications
Editor: J. C. Miller

29 **Elements of Rapid Solidification**
Fundamentals and Applications
Editor: M. A. Otooni

30 **Process Technology for Semiconductor Lasers**
Crystal Growth and Microprocesses
By K. Iga and S. Kinoshita

31 **Nanostructures and Quantum Effects**
By H. Sakaki and H. Noge

32 **Nitride Semiconductors and Devices**
By H. Morkoç

33 **Supercarbon**
Synthesis, Properties and Applications
Editors: S. Yoshimura and R. P. H. Chang

34 **Computational Materials Design**
Editor: T. Saito

35 **Macromolecular Science and Engineering**
New Aspects
Editor: Y. Tanabe

36 **Ceramics**
Mechanical Properties, Failure Behaviour, Materials Selection
By D. Munz and T. Fett

37 **Technology and Applications of Amorphous Silicon**
Editor: R. A. Street

38 **Fullerene Polymers and Fullerene Polymer Composites**
Editors: P. C. Eklund and A. M. Rao

Springer Series in
MATERIALS SCIENCE

Editors: R. Hull R. M. Osgood, Jr. J. Parisi H. Warlimont

39 **Semiconducting Silicides**
Editor: V. E. Borisenko

40 **Reference Materials in Analytical Chemistry**
A Guide for Selection and Use
Editor: A. Zschunke

41 **Organic Electronic Materials**
Conjugated Polymers and Low Molecular Weight Organic Solids
Editors: R. Farchioni and G. Grosso

42 **Raman Scattering in Materials Science**
Editors: W. H. Weber and R. Merlin

43 **The Atomistic Nature of Crystal Growth**
By B. Mutaftschiev

44 **Thermodynamic Basis of Crystal Growth**
P–T–X Phase Equilibrium and Non-Stoichiometry
By J. Greenberg

45 **Thermoelectrics**
Basic Principles and New Materials Developments
By G. S. Nolas, J. Sharp, and H. J. Goldsmid

46 **Fundamental Aspects of Silicon Oxidation**
Editor: Y. J. Chabal

47 **Disorder and Order in Strongly Nonstoichiometric Compounds**
Transition Metal Carbides, Nitrides and Oxides
By A. I. Gusev, A. A. Rempel, and A. J. Magerl

48 **The Glass Transition**
Relaxation Dynamics in Liquids and Disordered Materials
By E. Donth

49 **Alkali Halides**
A Handbook of Physical Properties
By D. B. Sirdeshmukh, L. Sirdeshmukh, and K. G. Subhadra

50 **High-Resolution Imaging and Spectrometry of Materials**
Editors: F. Ernst and M. Rühle

51 **Point Defects in Semiconductors and Insulators**
Determination of Atomic and Electronic Structure from Paramagnetic Hyperfine Interactions
By J.-M. Spaeth and H. Overhof

52 **Polymer Films with Embedded Metal Nanoparticles**
By A. Heilmann

53 **Nanocrystalline Ceramics**
Synthesis and Structure
By M. Winterer

54 **Electronic Structure and Magnetism of Complex Materials**
Editors: D.J. Singh and D. A. Papaconstantopoulos

55 **Quasicrystals**
An Introduction to Structure, Physical Properties and Applications
Editors: J.-B. Suck, M. Schreiber, and P. Häussler

56 **SiO_2 in Si Microdevices**
By M. Itsumi

57 **Radiation Effects in Advanced Semiconductor Materials and Devices**
By C. Claeys and E. Simoen

58 **Functional Thin Films and Functional Materials**
New Concepts and Technologies
Editor: D. Shi

59 **Dielectric Properties of Porous Media**
By S.O. Gladkov

60 **Organic Photovoltaics**
Concepts and Realization
Editors: C. Brabec, V. Dyakonov, J. Parisi and N. Sariciftci

Printing: Saladruck, Berlin
Binding: Stein+Lehmann, Berlin